Smart Innovation, Systems and Technologies

Volume 288

The Smart Innovation, Systems and Technologies book series encompasses the topics of knowledge, intelligence, innovation and sustainability. The aim of the series is to make available a platform for the publication of books on all aspects of single and multi-disciplinary research on these themes in order to make the latest results available in a readily-accessible form. Volumes on interdisciplinary research combining two or more of these areas is particularly sought.

The series covers systems and paradigms that employ knowledge and intelligence in a broad sense. Its scope is systems having embedded knowledge and intelligence, which may be applied to the solution of world problems in industry, the environment and the community. It also focusses on the knowledge-transfer methodologies and innovation strategies employed to make this happen effectively. The combination of intelligent systems tools and a broad range of applications introduces a need for a synergy of disciplines from science, technology, business and the humanities. The series will include conference proceedings, edited collections, monographs, handbooks, reference books, and other relevant types of book in areas of science and technology where smart systems and technologies can offer innovative solutions.

High quality content is an essential feature for all book proposals accepted for the series. It is expected that editors of all accepted volumes will ensure that contributions are subjected to an appropriate level of reviewing process and adhere to KES quality principles.

Indexed by SCOPUS, EI Compendex, INSPEC, WTI Frankfurt eG, zbMATH, Japanese Science and Technology Agency (JST), SCImago, DBLP.

All books published in the series are submitted for consideration in Web of Science.

More information about this series at https://link.springer.com/bookseries/8767

Agnessa O. Inshakova · Elena I. Inshakova
Editors

New Technology for Inclusive and Sustainable Growth

Technological Support, Standards and Commercial Turnover

Springer

Editors
Agnessa O. Inshakova
Institute of Law
Volgograd State University
Volgograd, Russia

Elena I. Inshakova
Institute of Economy and Management
Volgograd State University
Volgograd, Russia

ISSN 2190-3018 ISSN 2190-3026 (electronic)
Smart Innovation, Systems and Technologies
ISBN 978-981-16-9807-1 ISBN 978-981-16-9808-8 (eBook)
https://doi.org/10.1007/978-981-16-9808-8

This Springer imprint is published by the registered company Springer Nature Singapore Pte Ltd.
The registered company address is: 152 Beach Road, #21-01/04 Gateway East, Singapore 189721, Singapore

This book is dedicated to Oleg V. Inshakov (1952–2018) with love and gratitude.

Preface

The present collective research is aimed at solving a complex interdisciplinary scientific problem, which consists in the lack of balanced mechanisms of digital technologies impact on effective state regulation of production and non-production spheres of ecological entrepreneurship, taking into account ensuring inclusive growth and sustainable development of Russia. The problem state in the country of ecological entrepreneurship requires a logical search for solutions, including the use of digital economy and digital law. To solve this problem, the book uses the methodological arsenal of various branches of science, primarily economic and legal. Within the framework of the stated topic, a range of tasks is solved, the setting of which determines its assignment to the key directions of the Strategy for Scientific and Technological Development of the Russian Federation. The monographic project is aimed at solving specific problems in the context of the chosen direction of research, including the state scale: the fundamental bases of the state, law, and public authorities in the fourth industrial revolution in the context of inclusive growth and sustainable development of the country are analyzed; the relationship of ecological entrepreneurship with the tasks of building digital economy in Russia is established; the role of digital technologies in the development of ecological entrepreneurship is specified. The chosen methodological approach allows to achieve a synergistic effect and create a common innovation space within the framework of the environmentally oriented direction of the digital economy; to determine the ways and degree of impact of the digital environment on the state regulation of environmental entrepreneurship; to develop measures to improve the efficiency and condition of productive and non-productive areas of environmental entrepreneurship through digital technologies, to secure the state of the environment with the help of investment attraction (crowdfunding and tokenization), blockchain technology, and smart contracts; to argue for the necessity of comprehensive application in Russia of end-to-end technologies, including all types of neo-industrial technologies, allowing for accelerated development of the ecological entrepreneurship sector; and to suggest the most effective digital technologies to ensure the accelerated development of the ecological goods and services market in the Russian Federation.

The proposed collective research is closely connected with the formation of scientific and technological advances, providing economic growth and social development of the Russian Federation. The solution of specific tasks set in the Strategy for Scientific and Technological Development of the Russian Federation involves the collection of empirical data on the current situation in the interaction between society and nature, interdisciplinary research on the impact of digital technologies on the environment, life, and health of citizens, the state of public and state institutions called to participate in environmental protection. Understanding the overall picture will allow analyzing the current situation with the state of ecological entrepreneurship, to formulate a number of specific proposals to the public authorities, ensuring the consistent expansion of the "green" sectors of the economy through the intensive involvement of the potential of Russian entrepreneurship. The collected empirical materials, doctrinal developments of the authors, studied foreign experience, proposals for the modernization of Russian legislation, and existing economic regulators will contribute to the transition of Russia to advanced digital and intelligent production technologies.

The Part I of the book "Legislation on the guard of national security, protection of rights and interests of participants of the modern socio-economic structure" analyzes the constitutional reform in Russia in the context of the transformation of the provisions of the Russian Constitution aimed at protecting and strengthening the sovereignty and independence of the state as the fundamental elements of legal provision of inclusive growth and sustainable development. Domestic and foreign constitutional practice is compared. Effective legal mechanisms for preserving national constitutional identity and independence of the state, taking into account the best legal practices and accumulated political and legal experience, are proposed.

In the chapter of the monograph "Development of a model of sustainable development in organizations through the digital integration of management system standards", the authors substantiate that the modern market environment of the functioning of organizations is a continuously changing system. This variability determines the relevance of the search for the most effective tools for its sustainable development. The process of managing these economic systems has recently undergone significant changes. This process has become multidimensional; it requires a comprehensive analysis of all factors that affect its external and internal environment. Among these factors, the digitalization of procedures for the implementation of production and logistics processes through standardization can be mentioned.

The author's team pointedly posed the question of the extent to which the fourth industrial revolution will affect the foundations of the state and law. The fundamental foundations of the structure of the state, law, and public power under the conditions of the fourth industrial revolution have been studied. It is concluded that under the conditions of the fourth industrial revolution, the classical and fundamental importance of the state as an institution for resolving public affairs is far from disappearing, but it also becomes particularly relevant to the country's sustainable development. New approaches to management and flexible technologies do not change the foundations of the state and law, but effectively complement them.

The transformation of law and legislation in the context of digitalization in the aspect of intellectual property protection is investigated in the context of the lack of a unified conceptual framework related to the digitalization of law. It is argued that digitalization will not fundamentally change civil law until the person himself, who stands at the center of the civil law environment, changes. It is justified that many of the provisions defining the conceptual foundations of this process are reflected in economic terms, which guarantees their sustainability, since freedom of choice of behavior is assumed, which cannot be said of the legal field.

The authors address the problem of legal regulation of technologies with elements of artificial intelligence. The status of artificial intelligence in terms of law is defined, as well as the range of persons who can and must bear responsibility as a result of its improper and illegal use.

Criminal and procedural problems of counteraction to economic extremism among young people are analyzed taking into account the possibilities of criminal and procedural activity as the factor of influence on the stability of development and security of society and economy through revealing, suppression, and investigation of economic crimes, including extremist ones. The problems of the normative legal organization of criminal procedure activity and the quality of its implementation, influencing the efficiency of combating economic extremism among young people, were identified and the directions and ways of their solution were determined.

The consumers' right to information in distance selling of goods is investigated as a factor of inclusive growth. It is substantiated that distance selling allows consumers to reduce the cost of purchasing socially important goods, which increases the availability of such goods to all groups of the population, regardless of the location of the consumer, which in turn is interpreted by the authors as a factor of inclusive growth. It is proposed to consider the right of consumers to receive reliable and complete information about a product as a subjective right that provides equal opportunities to meet household needs and improve well-being. The key role of consumers' right to information in the system of measures to ensure the satisfaction of their legitimate interests in the distance sale of goods is emphasized, and specific proposals for improving Russian legislation are formulated.

The problem of protecting the rights of indigenous peoples as an essential element of inclusive economic growth is raised. It is argued that traditional approaches to economics have long focused on the role of market forces as a driver of economic growth, rather than on ensuring equitable development outcomes driven by market forces. Policies of total non-governmental intervention in the economy result in rising income inequality and persistent poverty, which weaken social cohesion. It is argued that economic growth must be inclusive, include growth in well-being, and protection of the rights of all vulnerable groups, with an emphasis on protecting the rights of populations such as indigenous peoples.

Research attention is also paid to the creative industries in the context of the global digitalization of copyright. Particular emphasis is placed on solving the problem of intellectual property protection that accompanies the development of digital creative industries while preserving historical heritage.

One of the chapters is devoted to the issues of proof in insurance disputes, taking into account the ongoing transformation of the law in the context of digitalization. The peculiarities of the methods of proof are revealed. Recommendations are given on the implementation of a new model of insurance contracts and exchange of documents in the process of their execution, taking into account the development of online technologies, digitalization processes, and pandemic—the factors of modern reality affecting the insurance industry. The study aims to create optimal legal structures of electronic insurance contracts.

The legal personality of credit organizations in the civil legislation of the Republic of Tajikistan is studied as a participant of civil legal relations. Inadmissibility of application of the category of legal capacity to credit organizations is stated. The author's approach to solving the issue of the acquisition of legal personality by credit organizations is presented.

The Part II of the book "Legal and technological support of sustainable development of ecological entrepreneurship and environmental protection measures in business" is aimed at the study of modern problems of digitalization of environmental protection in the Russian Federation, first of all, digitalization of environmental law. Such unprecedented environmental problems as climate change and increased pollution of air, water, and soil, which pose a threat to human life and health, are addressed. The advantages of using digital environmental technologies for national environmental monitoring, increasing the safety of industrial enterprises, in agriculture, and for improving the comfort of life in an urban environment are substantiated. It is argued that digitalization programs for environmental protection can also have the greatest effect in logistics, transportation, natural resource use, production, and consumption waste management (especially when implementing circular economy standards in practice), as well as in a number of other sectors. It is recommended to add to the environmental legislation of the Russian Federation and its subjects one or more chapters providing for specific areas of application of digital technologies (environmental oversight, monitoring, accounting, etc.).

The role and significance of modern technology in counteracting global climate change are examined which reveals the reasons for the complex nature of climate change, consisting in the growth of average annual temperatures, among which the key importance is the growth of human impact on nature, manifested in an increase in greenhouse gas emissions. Priority directions of development of international cooperation in the field of climate change mitigation are substantiated, and measures on adaptation to the climate change that has already occurred are developed. A systematic approach to the problem of climate protection at the level of federal laws in Russia is proposed. It raises universal problems in the field of combating global climate change, typical for all countries of the world, and requires a solution, including the problem of climate refugees, compensation for the damage caused by climate change, and a number of others.

Modern problems of the digitalization of agriculture in the Russian Federation are considered. It is noted that digital information technologies of the twenty-first century have great potential for economic growth through the automation and optimization of production processes, improved monitoring and control, and allowing to improve the manageability of individual production and the economy as a whole. The economic and legal ways of implementing digital projects in agriculture, allowing to improve the quantity and quality of agricultural production, logistics of its realization, as well as to increase the social security of rural residents, reduce and ease their workload by robotizing production, and significantly increase labor productivity are investigated. We analyze the departmental project of the Ministry of Agriculture of Russia "Digital Agriculture", aimed at the digital transformation of agriculture through the introduction of digital technologies and platform solutions to ensure a technological breakthrough in the agricultural sector and achieve growth productivity in "digital" agricultural enterprises. Amendments and additions to the Federal Law "On the Development of Agriculture" are proposed to clarify and specify incentives and benefits for developers and consumers of digital technologies.

The study devoted to legal and technological support for sustainable development of rural areas in Russia proposes a set of legal measures for environmental protection on agricultural land (including measures to preserve the fertility of agricultural land) and in rural settlements. The author's definition of sustainable development of rural areas is formulated. It is noted that the transition of rural areas to sustainable development is hampered by the lack of an effective system of interagency cooperation and coordination of individual issues related to the development of rural areas. The authors reveal the factors of low profitability of agriculture, limited opportunities for modernization and innovative development of the industry, low level of remuneration of industry workers, and the negative formation of the tax base of local budgets. The problems and prospects of the development of agrarian tourism as one of the most important elements of ensuring the transition of rural areas to sustainable development are investigated. It is argued that the concept of sustainable development of rural areas can be a tool to mitigate the severity of socio-economic and environmental problems in rural areas, balancing the interests of citizens, businesses, and government. A multi-level system of criteria and indicators of sustainable development of rural areas is proposed.

A study on integrated water resources management in West African basin organizations analyzes efforts to codify customary norms that form the body of international water law at the international level. The legal framework of West Africa, which includes various water charters, is examined, updating the existing legal framework on environmental protection, water resources, recognition and preservation of pastoralist rights, livestock production, and water user participation in resource management decision-making. A paradigm shift and new trends in international water law are substantiated. Problems of inter-branch legal regulation of international water flow are solved.

Prospects of Russia's ratification of the Minamata Convention 2013 are weighed. It is noted that this is an important step in the transition to mercury-free alternatives in production processes in order to implement the concept of sustainable development.

Attention is also paid to the extraction of minerals from outside the Earth, in particular, the use of space resources in the context of sustainable economic growth prospects. It is predicted that such use could begin in the very near future and become a profitable new type of economic zone aimed at inclusive and economic growth. Due to the fact that the space sector allows non-governmental sectors (businesses) to benefit from space, it is argued that it should be a question of detailed regulation, including ensuring its long-term sustainability. A comprehensive analysis and comparison of national legislation and policies of foreign states on the use of space resources are conducted.

The legal regulation of "green banking" in the legislation of the Russian Federation is analyzed. Attention is paid to the grounds and procedure for financing environmental measures from state and local budgets (in Russia and other countries), as well as the use of public-private partnerships to finance individual environmental measures.

The Part III of the collective research "New procedural and technological standards of court proceedings in the conditions of digital development in the legislation of Russia and foreign countries" considers the procedural standards of civil court proceedings in Internet courts of China. The research purpose of the prepared chapter is to create new methods of resolving commercial disputes with links to the Internet, namely e-commerce, thanks to which the number of transactions and, consequently, the volume of sales of goods and services have increased. The efforts of the authors are aimed at finding and developing new procedural frameworks, including the norms of the civil procedure code and the law of administrative proceedings, the need for which is caused by the emergence of new judicial instances. Another chapter focuses on the specifics of the legal framework for case management in Indonesia's electronic courts. Electronic case management is explored as the orderly use of an electronic system for electronic case management, receipt of court documents, including requests, copies, duplicates and verdicts, delivery, and storage of documents. To this end, the entire process is analyzed in the context of the use of digital service and information technology management standards established by the Chief Justice.

The applicable law in international commercial arbitration in Panama is also studied in the context of digitalization. Cases, in which the parties explicitly establish the law applicable to the dispute, in particular through digital technologies, are explored, as well as cases in which there is no direct choice of law by the parties and various approaches are possible, such as conflict of laws rules, the closest connection criterion, cumulative method, international conventions, and transnational principles of law.

The consistent introduction of artificial intelligence technologies in Russia and abroad creates necessary prerequisites for the robotization of civil proceedings. Changes in legal and ethical norms of this phenomenon are revealed, as well as the successful experience of China on replacement of judges by computers in human guise. The advantages and disadvantages of the process of developing special laws and legal acts regulating the integration of robots into judicial processes are determined.

New technologies for resolving international energy disputes between companies and states, which constitute one-fifth of international commercial disputes, are also analyzed. A comprehensive analysis of existing forms and methods of resolving international energy disputes is conducted and the optimal mechanism for their consideration, taking into account the balance of interests of both the state and the investor, is determined. The definition of the concept of an international energy dispute is formulated; the advantages and disadvantages of the most common methods of resolving this type of dispute are highlighted. Innovative solutions and automated processes aimed at resolving this type of dispute have been studied.

The impact of inclusive criminal procedure on the stability of the development of society through the detection and suppression of crimes, compensation for damages caused by crime, and protection of rights and freedoms of individuals and legal entities has been analyzed. The topical problems of modern criminal procedural policy and its interrelation with the purpose of criminal proceedings are considered. The author substantiates the proposals aimed at strengthening the inclusiveness of criminal procedure by improving its legal regulation as well as the inclusion of criminal procedure in the system of instruments of state control over crime and protection of national security.

The instrumental significance of the jurisprudence of the International Court of Justice in the context of achieving global sustainable development goals is examined in the example of cases of international boundary delimitation disputes. The law enforcement practice of the International Court of Justice on international border disputes is explored. The tendencies predetermining the dynamics of international legal regulation of relations in the sphere of settlement of international disputes are revealed. The factors determining the transformation of legal means of international dispute resolution in the context of achieving the goals of sustainable development as defined by the UN Charter are noted. The impact of modern technology on the efficiency of corporate disputes in arbitration courts of the Russian Federation is analyzed. It is proved that the new digital means of functioning of state arbitration courts have increased the efficiency of consideration of corporate disputes, which undoubtedly increases the attractiveness of doing business in the form of "corporation" and equalizes the volume of sales of goods and services. Recommendations are formulated to reform the procedural legislation in the field of corporate dispute resolution, which requires improvement.

The impact of new technologies with the help of Legal Design tools on the established approach to the preparation of procedural documents and the legal community as a whole has been studied. The questions are answered: What is Legal Design and what are the reasons for its emergence and development? What is the ideal procedural document? What are the main problems that can be solved with the help of Legal Design? How to work with the tools of Legal Design and when it is appropriate?

The Part IV of the book "Commercial Circulation in the Context of the Fourth Industrial Revolution: Financial and Informational Support" presents the study of private law control as a factor of stability of commercial circulation. The concept of private-legal control, its features, and characteristics are defined.

The concept and types of remote-digital investment transactions in the information space were developed. It was substantiated that the use of computer programs by the parties to these transactions at the preparatory stage, at the stage of video negotiations, and at the stage of subsequent execution of the agreements reached by the parties is obligatory. Programs provide transfer of data and commands by means of working out of algorithms which movement inside is carried out at number 1 and is not carried out at number 0. In this connection, investment transactions in the information space can be called digital. Classifications of investment transactions in information space on several grounds are proposed.

The author analyzes the process of digitalization of banking services and the possibility of international mortgage transactions in a remote format in the context of the transition to Industry 4.0, as well as gaps in the current legislation in the field of identification that do not allow the possibility of remote banking services in full. Proposals are formulated to develop new approaches to identification, including the use of biometric data and the formation of a bank of identifiers in "digital profiles".

The problems of the circulation of utilitarian digital rights, which are considered as an investment mechanism to ensure the inflow of free money of the population into the real sector of the economy in order to protect them from inflation and increase their level of well-being, are raised. The main directions of improvement of legislation in terms of the use of distributed registries (blockchain technology) in Internet commerce are highlighted. According to the authors, technologies of distributed registers are especially in demand in Internet commerce, including cybersecurity tools.

The question of legal regulation and popularization of crypto assets in the modern information space is raised. The investment turnover of tokenized shares represented by special digital algorithms, bonds, other securities, and financial instruments is studied.

The main directions for improving legislation on the application of key technologies for the implementation of the fourth industrial revolution in terms of regulation of counteraction to DDoS attacks carried out on the websites of business entities operating in the online retail sector are highlighted. In the course of the research, data from StormWall's clients operating in various e-commerce segments was used. It was found that during the pandemic, between February and October 2020, the number of DDoS attacks on the websites of business entities operating in the online retail industry increased fourfold compared to the same period in 2019. It has been substantiated that the increase in cyber attacks is primarily due to illegal forms of competition, DDoS attacks are often generated by business entities themselves in the online retail sector. As a type of cyberattacks, hacker attacks are investigated with the aim of rendering an online retailer's website inoperable for the purpose of extortion. Among the varieties of cyberattacks identified, attempts to steal the personal data of online retailers' customers in order to further penetrate their bank accounts are analyzed. Proposals are made to improve the legal regulation of cybersecurity of distance transactions of purchase and sale of goods, carried out in the information and telecommunications network "Internet".

Volgograd, Russia

Agnessa O. Inshakova
Elena I. Inshakova

Contents

Editors and Contributors

About the Editors

Agnessa O. Inshakova Professor, Doctor of Law Agnessa O. Inshakova is Honorary Worker spheres of education of the Russian Federation, head of the master's degree program "Civil law, business law, private international law," head of the scientific school "Law of economic entities in the context of technological and socio-economic transformations," head of the scientific and educational center "Modernization of the legal system of modern Russia" of the Volgograd State University, Institute of Law, and head of the Basic Department of the Southern Scientific Center of the Russian Academy of Sciences. She is Author of more than 400 scientific, educational and methodological publications, including: more than 30 monographs, including in leading foreign publications indexed in the international scientific-metric databases SCOPUS/Web of Science Core Collection in English (as a leading author and editor); 36 textbooks, workshops and training manuals (including more than 10 publications with the stamp recommendations of the Ministry of Science and Higher Education, as well as leading Federal Publishers, including Yurayt, Zertsalo, etc.). She is Author of more than 200 articles in journals in the List Higher Attestation Commission of the Ministry of Education and Science of the Russian Federation; 82 articles in journals indexed in the international database Scopus, including Q1 and Q2; 43 articles in journals indexed in the international database Web of Science Core Collection, including Q1 and Q2. It ranks first in the country in the number of publications in SCOPUS and Web of Science Core Collection in the specialty—Law. She has high scientific-metric indicators: citation index in RSCI-1702, Scopus-260, Web of Science Core Collection-54; h-index in RSCI-17; Scopus-8; Web of Science Core Collection-4.

Elena I. Inshakova Professor, Doctor of Economics, Elena I. Inshakova is Honoured Worker of Higher School of the Russian Federation; Professor of the Department of Economic Theory, Regional Economy and Entrepreneurship at Volgograd State University, Volgograd, Russian Federation; Professor of Basic Department of the

Southern Scientific Center of the Russian Academy of Sciences (RAS) at Volgograd State University; Chief Researcher of the Institute of Social and Economic Development of Region at Volgograd State University. She is the author of more then 200 scientific and educational publications: about 20 monographs, including foreign publications (in English) indexed in the largest international abstract and citation databases Scopus and Web of Science Core Collection as a leading author and editor; 15 textbooks and training manuals for Master's degree, graduate and postgraduate programmes; more than 30 articles in the journals reviewed by the Higher Attestation Commission of the Russian Federation and included in the database Russian Science Citation Index (RSCI) on the Web of Science platform. She is the author of more than 20 articles in publications indexed in the international citation and analytical database Scopus; 10 articles in publications indexed in Web of Science Core Collection. Scientific and metric indicators: citations in RSCI-1094, Scopus-100, Web of Science Core Collection-63; h-index: RSCI-15, Scopus-4, Web of Science Core Collection-2.

Contributors

Aslan Abashidze Peoples' Friendship University of Russia, Moscow, Russia

Naser A. R. Al Ali Law Institute of the Russian University of Transport, Moscow, Russia

Fridon R. Ananidze Peoples' Friendship, University of Russia, Moscow, Russia

Denis A. Appazov Volgograd State University, Volgograd, Russia

Aleksey V. Barkov Financial University Under the Government of the Russian Federation, Moscow, Russia

Boris I. Bashilov Department of Civil Law and Procedure and International Private Law, Peoples' Friendship University of Russia (RUDN University), Moscow, Russia

Vladimir E. Berezko Moscow State Institute of International Relations MGIMO, Moscow, Russia

Alisa M. Berman Department of Civil Law and Procedure and International Private Law, Peoples' Friendship University of Russia (RUDN University), Moscow, Russia

Galina N. Chernichkina Russian State University of Justice, Moscow, Russia

Irina Chernykh RUDN University, Moscow, Russia

Davud A. Davudov Department of Civil and Private International Law, Volgograd State University, Volgograd, Russia

Edgar Josue Young Dominguez RUDN University, Moscow, Russia

Valentin A. Dzedik Volgograd State University, Volgograd Region, Russia

Elena P. Ermakova Department of Civil Law and Procedure and International Private Law, Peoples' Friendship University of Russia (RUDN University), Moscow, Russia

Pavel P. Fantrov Department of Criminal Procedure and Criminalistics, Volgograd State University, Volgograd, Russian Federation

Evgenia E. Frolova Peoples' Friendship University of Russia, (RUDN University), Moscow, Russia;
The Institute of legislation and comparative law under the government of the Russian Federation, Moscow, Russia

Marsel I. Garaev Peoples' Friendship University of Russia, Moscow, Russia

Alexander I. Goncharov Volgograd State University, Volgograd, Russia

Marina V. Goncharova Plekhanov Russian University of Economics, Volgograd, Russia

Oleg V. Grabko Judge of the Twelfth Arbitration Court of Appeal, Saratov, Russia

Yana S. Grishina Russian State University of Justice Moscow, Moscow, Russia

Irina A. Gronic Peoples' Friendship University of Russia (RUDN University), Moscow, Russia

Denis Gugunskiy RUDN University, Moscow, Russia

Agnessa O. Inshakova Head of Department of Civil and International Private Law, ScD in Law, Volgograd State University, Volgograd, Russia;
Department of Law, Volga Branch of Volgograd State University, Volzhsky, Russia

Olesya P. Kazachenok Department of Civil and Private International Law, Volgograd State University, Prospekt Universitetsky, Volgograd, Russia

Svetlana Yu. Kazachenok Department of Civil and International Private Law, Volgograd State University, Volgograd, Russia

Aleksandr S. Kiselev Moscow State Regional University, Moscow, Russia;
Financial University Under the Government of the Russian Federation, Moscow, Russia

Andrey A. Klishas Head of Law and State Theory Department, Peoples' Friendship, University of Russia (RUDN), Scientific Director of the Institute of Law at RUDN, Moscow, Russian Federation

Tatyana V. Kokoreva Volgograd State University, Volgograd, Russia

Nikolay V. Kotelnikov Volgograd Academy of the Ministry of Internal Affairs of Russia, Volgograd, Russia

Mikhail N. Kuznetsov Peoples' Friendship, University of Russia (RUDN University), Moscow, Russia

Valentina A. Lazareva Department of Criminal Procedure and Criminalistics, Samara National Research University Named After Academician S. P. Korolev, Samara, Russia

Denis E. Matytsin Insitute of Law, Volgograd State University, Volgograd, Russia; Department of Civil Law and Procedure, Volga Branch of the International Law Institute, Volzhsky, Russia

Shukhratpur Mukhammadyusuf Law Institute, RUDN, Moscow, Russia

Elvira O. Osadchenko Volgograd State University, Volgograd, Russia

Ernest C. Peguita Peoples' Friendship University of Russia, (RUDN University), Moscow, Russia

Aleksandr N. Pershin Department of Criminalistics, Moscow State Law Academy Named After Kutafina O.E, Moscow, Russia

Anna L. Polina-Stashevskaya Department of Civil Law and Procedure and International Private Law, RUDN University, Moscow, Russia

Viktor F. Pon'ka Head of the Department of Legal Support of the Russian Academy of Sciences, Honorary Worker of Higher Professional Education, Moscow, Russia; Russian Academy of Sciences, Honorary Worker of Higher Professional Education, Moscow, Russia

Alexander A. Pukhart Law Institute, RUDN, Moscow, Russia

Rimma S. Rahmatulina Financial University Under the Government of the Russian Federation, Moscow, Russia

Ekaterina P. Rusakova Peoples' Friendship University of Russia (RUDN University), Moscow, Russia

Anatoliy Ya. Ryzhenkov Volgograd State University, Volgograd, Russia

Anatoly Ya. Ryzhenkov Department of Civil and Private International Law (The Basic Department of the Southern Scientific Center of the Russian Academy of Sciences), Volgograd State University, Volgograd, Russia

Andrey N. Sadkov Volgograd State University, Volgograd, Russia

Maxim V. Sevostyanov Volgograd State University, Volgograd, Russia

Vladimir M. Shinkaruk Department of Criminal Procedure and Criminalistics, Volgograd State University, Volgograd, Russian Federation

Aleksey P. Sokolov Military University of the Ministry of Defense of the Russian Federation, Moscow, Russia

A. M. Solntsev Peoples' Friendship University of Russia, (RUDN University), Moscow, Russia

Alexander M. Solntsev Peoples' Friendship University of Russia, (RUDN University), Moscow, Russia

Natalya A. Solovyeva Department of Criminal Procedure and Criminalistics, Volgograd State University, Universitetsky Prospect, Volgograd, Russia

Irina V. Usacheva Volgograd State University, Volgograd Region, Russia

Louise N. Voznesenskaya Peoples' Friendship University of Russia (RUDN University), Moscow, Russia;
Postgraduate of the Department of Civil Law and Procedure and International Private Law, Moscow, Russia

Mahshid Yazdanimoghadam Department of Civil Law and Procedure and International Private Law, Peoples' Friendship University of Russia (RUDN University), Moscow, Russia

P. N. Yusifova Peoples' Friendship University of Russia, (RUDN University), Moscow, Russia

Part I
Legislation Safeguarding National Security, Defense of the Rights and Interests of Participants in the Modern Socio-economic System

Chapter 1
Constitutional Reform 2020 in Russia: Enhancing the Sovereignty and Independence of the State

Andrey A. Klishas

Abstract The article examines the results of the constitutional reform implemented in Russia in 2020 as part of transforming the Russian Constitution to protect and enhance the sovereignty and independence of the state. The author has analyzed domestic and foreign constitutional practice and concluded that the updated text of the Russian Constitution creates effective legal mechanisms to preserve the national constitutional identity and independence of the state while also reflecting legal best practices and the accumulated political and legal experience.

1.1 Introduction

Sovereignty is one of the most important attributes of the state, legally inherent to its existence. At the same time, the definition of "sovereignty" is still very vague, which can often lead to its misinterpretation [3].[1] This problem is not at all theoretical, but quite a practical one as each state, based on its understanding of sovereignty, sets the limits for the acceptable intervention by various stakeholders in domestic and international relations [4].[2]

[1] In this context, we cannot but agree with V.Ye. Chirkin, who pointed out that "The concept of state sovereignty can only be holistic. It is impossible to be partially sovereign—a country can either have or lack sovereignty.".

[2] Speficially, I.V. Leksin notes that "Today, the idea of state sovereignty is driven by multiple practices: legal duties of citizens toward the state (in a "pre-sovereign" state, the ruler had the power over his subjects), policing and law enforcement by state authorities, state licensing, state monitoring and oversight, etc.".

A. A. Klishas (✉)
Head of Law and State Theory Department, Peoples' Friendship, University of Russia (RUDN), Scientific Director of the Institute of Law at RUDN, 117198 Moscow, Russian Federation
e-mail: klishas_aa@pfur.ru

A. O. Inshakova and E. I. Inshakova (eds.), *New Technology for Inclusive and Sustainable Growth*, Smart Innovation, Systems and Technologies 288,
https://doi.org/10.1007/978-981-16-9808-8_1

We can see how states are making significant concessions as part of international integration, yielding to supranational authorities matters that have been traditionally seen as matters of purely national importance [3].[3]

Russia, as a state under a significant political international pressure, should pay particular attention to defending its interests in the international arena and to protecting its fundamental national values [10].[4] The regulation proposed by the amendments to the Russian Constitution and assuming the primacy of Russian constitutional norms over international treaties are a natural step toward strengthening the Russian national legal system. This approach relies on the doctrine of national constitutional identity, which has been consistently promoted by the Constitutional Court of the Russian Federation and is widely used in developed legal systems. The above doctrine was first formulated in the Decision of the Constitutional Court of the Russian Federation of April 16, 2016 on a case that dealt with the enforceability of the judgment by the European Court of Human Rights in *Anchugov and Gladkov v. Russia* under the Russian Constitution.

1.2 Methodology

The empirical basis of this study is presented by the current regulatory legal regulation as the Russian Federation, as well as a number of European states, in particular, France, Germany, Spain, the Czech Republic, Portugal, and a number of others. The methodological basis of the work consists of such methods as the formal-logical method, the method of system analysis, and the structural–functional method.

[3] In this connection, the statement by V.Ye. Chirkin to the effect that "State sovereignty is inseparably tied to one of the basic principles of international law—non-interference in the internal affairs of states. It means that the state has the supreme power in a given jurisdiction and is legally independent of other states and structures. Only the state establishes the foundations of its regime and the rule of law in society.".

[4] The issues of sovereignty, including those related to setting up legal mechanisms for its protection, are becoming increasingly important for the Russian state and its national policy given the existing international tensions and openly unfriendly policies run by our foreign partners. While discussing particularities of the process for adopting constitutional amendments and its content, Ye.V. Vinogradova rightly emphasizes the particular importance of ensuring the sovereignty and integrity of the state in the presence of a certain growing threat to the sovereignty of the Russian Federation.

1.3 Results

1.3.1 Primacy of Constitutional Norms Over International Law

The amendments to Article 79 do not call into question the place and role of international law in the national legal system. Part 4 of Article 15 of the Russian Constitution remains effective. It states that generally recognized principles and norms of international law and international treaties of the Russian Federation are an integral part of the Russian legal system; if an international treaty of the Russian Federation establishes rules other than those provided for by law, then the rules of the international treaty shall apply. But the same article stipulates that the Russian Constitution has the supreme legal force and direct effect and is applicable throughout the country. Laws and other legal acts adopted in the Russian Federation must not contradict the Russian Constitution. Similarly, international treaties of the Russian Federation may not contradict the Constitution. As the Constitutional Court of the Russian Federation has repeatedly pointed out, the norms of international law do not abolish the primacy of the Russian Constitution for the Russian legal system and are implemented subject to the recognition of its supreme legal force.

The amendments to Article 79 of the Constitution are part of the consistent integration of the principle of primacy of the national Constitution over international treaties in the Russian legal system, including in terms of their interpretation by intergovernmental bodies, accompanied by the creation of a monitoring mechanism to manage legal conflicts that might question the primacy of the Russian Constitution. It should be emphasized that the proposed amendment is applicable to decisions made by intergovernmental bodies under international treaties of the Russian Federation. Therefore, it is not about eliminating the contradictions between the provisions of international law and provisions of the Constitution. A mechanism is introduced to eliminate potential conflicts in the interpretation of international law provisions and the Russian Constitution because contradictions do not arise from the content of specific international legal documents, but from competing interpretations of their provisions in supranational and national legal systems.

Therefore, the primacy of constitutional norms over the norms of international law is in itself a constitutional mechanism that guarantees that the Constitution can be actually implemented, that the social consensus reflected in these provisions and the results achieved through harmonization of different constitutional values remain stable, and that state sovereignty becomes a reality [6, 11].[5]

---

[5] As V.D. Zorkin rightly noted in his article, *The Letter and the Spirit of the Constitution*, at the level of popular consciousness a community united by living in a certain territory defends its distinctive features through "the desire to formulate its religious, national, or regional (for example, European) identity and to preserve and strengthen the traditional values of family, culture, everyday life, etc., while at the level of governmental authorities it is manifested through the desire to prevent the erosion of national and state sovereignty and to assert the constitutional and legal identity of the state".

As noted by the Constitutional Court of the Russian Federation in its Opinion of March 16, 2020, this mechanism is designed not to support a refusal to implement international treaties and decisions of intergovernmental jurisdictional bodies based on such treaties, but to develop a constitutionally acceptable way for the Russian Federation to implement these decisions while steadily reinforcing the supreme legal force of the Russian Constitution and the Russian legal system, which incorporates unilateral and multilateral international treaties of Russia as its integral part, including those that provide for relevant powers of intergovernmental structures.

In describing the relevant constitutional amendments given the nature of current relations between the Constitutional Court of the Russian Federation and the ECHR, a number of authors rightly draw attention to the double standards in certain decisions of the ECHR, coupled with the lack of a pan-European consensus on certain issues, which can prevent from preserving constitutional pluralism by affecting the principle of national constitutional identity. In this connection, P.A. Vinogradova and A.N. Tulayev assume that the Constitutional Court of the Russian Federation is rightfully assigned the relevant authority intended to find a reasonable trade-off between the national and supranational regulation, manifested not only through implementing decisions of a supranational subsidiary judicial body, but also through checking its compliance with the constitutional foundations of the Russian Federation and the principles of legal regulation of civil and political rights and freedoms set forth in the Russian Constitution [5].

Other researchers who analyze the Russian constitutional reform of 2020 from the perspective of the relationship between national and supranational legal systems have found, and not unreasonably, that the constitutionalization of the relevant legal mechanism is due to the practice of the ECHR's interpretation of the European Convention on Human Rights, guided by "abstract political morality of unclear origin", rather than by the desire to understand in good faith the actual intentions of the parties when signing a relevant international treaty [9]. Russia has, therefore, firmly decided to avoid unilateral interventions and double standards, as well as the use of any restrictive measures in international relations [9].

At the same time, we cannot agree with the position of some researchers, who believe that establishing such provisions is unjustified and actually lead, in their opinion, to the refusal of the Russian Federation to comply with its international legal obligations. In particular, E. Teague is critical about this constitutional amendment assuming that it can be used as a legal basis for non-enforcement of ECHR judgments, without taking into account the trends of judicial activism of ECHR judges and the established international practice of national legal systems abandoning unconditional enforcement of decisions made by intergovernmental subsidiary judicial bodies, although he still quite reasonably notes that amending Article 15 of

Yu.L. Shulzhenko, while assessing the amendments made to the Russian Constitution in the context of its legal protection, including the provisions on the precedence of constitutional norms over international law, rightly drew attention to the nature of changes in the constitutional text, which were prepared with a particular focus on the particularities of the Russian state system.

the Russian Constitution is impracticable due to objective reasons while the relevant constitutional amendments are driven by the existing practice of the Russian Constitutional Court [7].

Recognizing the supremacy of international treaties over national legislation is a common constitutional practice in most modern democracies. In this case, prior to assuming international obligations, they are assessed for their compliance with the constitution. A number of foreign jurisdictions impose restrictions on the application of international treaties in case they contradict constitutional provisions.

For example, Article 54 of the French Constitution provides that if the Constitutional Council, at the request of the President of the Republic, the Prime Minister, or the President of one of the Chambers, or 60 Deputies or 60 Senators, declares that an international obligation contains a provision contrary to the French Constitution, then the permission to ratify or approve that international obligation may only be granted after the Constitution is revised accordingly.

In its turn, Article 25 of the German Constitution provides that the generally recognized rules of international law are an integral part of federal law. They take primacy over national laws and directly give rise to rights and obligations for persons residing in the Federation. At the same time, the Constitution of the Federal Republic of Germany does not provide for the primacy of international law over the Constitution itself. A similar legal regulation is established by Article 94 of the Constitution of the Netherlands.

Article 93 of the Spanish Constitution establishes that organic law grants the right to conclude treaties providing for the right to participate in international organizations or institutions whose functions do not contradict the provisions of the Constitution.

At the same time, the practice of constitutional courts in a number of European states shows that in case of changes in the meaning of international obligations undertaken making such obligations contradictory to the provisions of the Constitution, it is permissible to abandon the application of relevant provisions in order to ensure compliance with the Constitution [8].[6]

In France, for example, the Council of State, in its decision of October 30, 1998, clearly stated that "the primacy of international legal rules under Article 55 of the French Constitution does not apply within the national legal system to the constitutional legal rules". This position has been confirmed in subsequent case law, in particular in the Fraisse case examined by another higher French court, the Court of Cassation, in 2000.

In 2004, the Federal Constitutional Court of Germany ruled in the case of Görgül, a Turkish citizen who, after the courts of the Federal Republic of Germany had refused to hand over his extramarital child to him, filed a complaint with the ECHR. The latter regarded the position of the German court as a violation of the relevant provisions of the ECHR on the right to family life. When considering the constitutionality of the

[6] In this context, T.A. Vasilieva is right in pointing out that "whenever serious changes are made to the EU constituent documents, expanding the powers of the Union's institutional structures, the constitutional review bodies of Member States receive appeals that claim the unconstitutionality of relevant provisions based on the concept of state sovereignty".

ECHR decision, the Federal Constitutional Court of Germany found that in Germany the status of ECHR decisions is equal to the status of federal law. The provisions of international treaties may not contradict the provisions of the Federal Constitution, much less affect their validity.

Italy also followed a similar path as its Constitutional Court in a number of its decisions addressing the issue of whether an ECHR decision may have priority in the national legal system (for example, in the case of Scordino v. Italy decided by the ECHR in 2006), the Italian Constitutional Court recognized that any inconsistency between national law and the Convention, as interpreted by the ECHR, must be regarded as a violation of constitutional law and the Constitution.

There is no doubt that all these decisions unambiguously set out the supremacy of the constitutions of these countries over the provisions of international treaties. A similar position is stated in the ECHR's cases Hirst v. United Kingdom, Chester v. Secretary of State for Justice and McGeoch v. The Lord President of the Council and another, Görgülü v Germany, and Maggio and Others v. Italy.

Amendments aimed to establish the primacy of the Russian Constitution over provisions of international treaties will allow Russia to more effectively comply with its international obligations, as they imply the need to take into account national constitutional particularities and the possibility to refuse to implement legal positions that are in conflict with the Russian Constitution.

1.3.2 Constitutionalizing the Principle of Inalienability of Territory

Amid the rapid universalization, expansion of the activity scope, and influence of supranational institutions around the world, their interference in the internal affairs of individual states, as well as the increasing international tensions, a key focus area in expanding the existing mechanisms to preserve and protect state sovereignty as part of the constitutional reform of 2020 consisted in constitutionalizing the principle of inalienability of territory of Russia. The amendment law has supplemented Article 67 of the Russian Constitution with Part 2.1, which provides that the Russian Federation ensures protection of its sovereignty and territorial integrity and also introduces a constitutional ban on actions (except for delimitation, demarcation, and re-demarcation of Russia's state border with neighboring states) that aim to alienate parts of the Russian Federation, as well as on calls for such actions.

The territory is an inalienable attribute of a state, which has the right to dispose of it using its sovereignty. Territory may not be regarded as a state property, but provides the spatial basis for state sovereignty and is linked to all aspects of its organization and activities. It is characterized by the unity of geopolitical, economic, political, and legal space. Hence, one of the most important guarantees to ensure the state's integrity is to ensure the inviolability of its national territory. That is why all matters related to the territory of the Russian Federation are governed by the Russian Federation.

The introduction of provisions on the protection of the nation's territorial integrity in the Constitution contributes to maintaining the constitutional continuity, since the principle of inviolability of borders and territorial integrity of the state was first established by the Soviet Constitution of 1977 (Article 29). Adopting the principle of territorial integrity as a pillar of the constitutional system is also a common practice in other countries. Moreover, while the possibility of the territory being alienated is allowed, the actual procedure is in any case difficult to implement.

Territorial integrity implies the preservation of territorial unity, as well as the unacceptability of unilateral secession of any part of a nation, i.e. secession of its constituent territory. The desire of a state to create legal mechanisms aimed to prevent alienation of its territory cannot and should not be regarded as an iron curtain between the country and the rest of the world, and much less as a sign that the nation would breach its existing international legal obligations.

However, the amendments on territorial integrity are primarily intended to take into account the interests of people living in Russia and to mitigate the risk of their legal statuses being changed as a result of changes in the jurisdiction of the relevant territory. Thereby, the state demonstrates that under no circumstances, including international pressure, tensions, or the need to conclude an international treaty beneficial for the country, the Russian government will not abandon its citizens residing in the Russian territory. Russia must take into account the legitimate expectations of Russian citizens regarding the maintenance of stable ties with the state.

Russia must take all possible measures to confirm that the practices of transfer optation and forced change of citizenship due to a change in the jurisdiction of the territory where the relevant people resided are unacceptable.

In this connection, amendments to the Russian Constitution are designed to introduce the unacceptability not only of the alienation of national territory itself by making such a prohibition constitutional in nature, but also of the very calls for such actions. This structure of a constitutional provision would minimize the potential risks of decisions made by any political forces in power that are so painful for the entire society. It refers to any known means of territorial alienation: forcible seizure (in whole or in part) by a foreign state; voluntary transfer of part of the territory to a foreign state; forcible division of a state and the creation of other independent states in its territory contrary to its national interests, etc. This amendment creates a political and legal basis for the independent development of the nation enabling it to exercise its territorial supremacy.

In its Opinion of March 16, 2020, the Constitutional Court pointed out that the proposed introduction of a ban in Article 67 (Part 21) of the Russian Constitution not only on actions (except for delimitation, demarcation, and re-demarcation of Russia's state border with neighboring states) that aim to alienate parts of the Russian Federation, but also on calls for such actions, while being a restriction of freedom of speech, is nevertheless in line with the constitutional goals of similar restrictions, which need to be evaluated for constitutional acceptability against not only the provisions of Article 29 (Part 2) of the Russian Constitution, but also against its Article 13 (Part 5), which prohibits the creation and operation of public associations whose goals and activities are aimed, *inter alia*, at violating the integrity of the Russian

Federation. Regardless of these normative provisions, it is acceptable to impose such restrictions in line with Article 4 (part 3) of the Russian Constitution, under which the Russian Federation ensures the integrity and inviolability of its territory.

At the same time, the constitutional reform of 2020 was initially associated with the need to significantly transform the existing legal regulation, bringing it in line with the logic of constitutional amendments. One of the first changes in the current legislation in this connection was the clarification of the concept of extremist activity, taking into account the constitutionalization of the principle of inalienability of territory of Russia or a part of the country.

In particular, Federal Law No. 299-FZ dated July 31, 2020, amending Article 1 of the Federal Law *On Combating Extremist Activities* amended the second paragraph of Article 1 of Federal Law No. 114-FZ *On Combating Extremist Activities* dated July 25, 2002, pursuant to which the notion of extremist activities would also include violation of the territorial integrity of Russia (including alienation of part of its territory), except for delimitation, demarcation, or re-demarcation of the state border of the Russian Federation with neighboring countries. Therefore, the terminology of this law was brought in line with the Russian Constitution, which eliminated the uncertainty related to the interpretation of the principle of territorial integrity of Russia.

The provisions on the territorial integrity of the state also exist in foreign constitutional practices. In some jurisdictions, constitutions include provisions that directly derive from the inviolability of state borders, while in others, mechanisms for protecting the territorial integrity of the state are set up by the constitutional review authorities.

Article 2 of the Bulgarian Constitution provides that Bulgaria's territorial integrity is inviolable.

According to Article 273 of the Portuguese Constitution, national defense aims to ensure national independence, territorial integrity, freedom, and security of the population against any external aggression or threat, while respecting the constitutional law, democratic procedures, and international treaties.

In accordance with Article 1 of the Czech Constitution, the primary duty of the state is to ensure the sovereignty and territorial integrity of the Czech Republic, to protect its democratic foundations, and to protect life, health, and property.

In accordance with the provisions of the Preamble to the Latvian Constitution, the Latvian people protect their sovereignty, national independence, territory, territorial integrity, and democracy in Latvia.

Some foreign constitutions formally allow for potential secession of a part of the territory, but the implementation of similar provisions in developed legal systems is significantly complicated. For example, the Constitution of the Federal Republic of Germany does not contain provisions on the admissibility or inadmissibility of secession, but the Federal Constitutional Court of Germany in its decision No. 2 BvR 349/16 dated December 16, 2016, determined that the lands within the Federal Republic of Germany cannot decide to secede from the state. According to the Federal

Constitutional Court of the Federal Republic of Germany, this affects the constitutional procedure set by the German Constitution, which has been established by the people of the entire state, to which the sovereignty belongs.

A similar approach was adopted by the Spanish Constitutional Court, which, in its Ruling No. 42 dated March 25, 2014, despite the results of the referendum in Catalonia, pointed out that secession of the Catalan territory from Spain is inadmissible.

The commitment to the principle of the inadmissibility of alienation of the territory of the state was also expressed by the European Commission for Democracy through Law (Venice Commission) in its report *A General Legal Reference Framework to Facilitate the Settlement of Ethno-Political Conflicts in Europe*. Thus, the approach according to which the principle of territorial integrity is one of the fundamental principles of constitutional law and is associated with the inadmissibility of secession of any state territory was set up as the main pan-European position on this issue [1]. As noted by the Venice Commission, while a number of constitutions guarantee the right to self-determination, this concept excludes the right of secession. Relevant findings were also presented in the report *Self-Determination and Secession in Constitutional Law*, which noted that the right of secession is contrary to the constitutional principle of state integrity [2].

The proposed rule prohibiting alienation of territory of the Russian Federation is fully consistent with the right of peoples to self-determination. The right to self-determination through secession was originally seen as a mechanism of decolonization. Today, developed legal systems proceed from the assumption that every nation can freely exercise its right to self-determination within a state, for example, by establishing an autonomy. The only exceptions are cases where such self-determination is not possible due to discriminatory policies against a particular territorial community and threats to the lives and health of people.

Therefore, developed legal systems and best practices in constitutional law define the principle of territorial integrity as a key foundation of statehood.

1.3.3 Strengthening the Connection Between Public Servants and the State

Meanwhile, another important aspect of ensuring state sovereignty of the Russian Federation is to further strengthen the connection between citizens performing publicly important functions and the state. The activities of those performing the most important public functions for the state, according to the established consensus in Russian society, should be free from any external influence or connection with foreign states and carried out exclusively in the interests of the Russian Federation and its people.

The Constitution of the Russian Federation in this regard has been supplemented with provisions that introduce a ban for persons holding public service positions in the Russian Federation or a number of other positions related to public administration, to be nationals of a foreign state or have a status granting them the right to permanent residence in a foreign state, or open and hold accounts, keep money or other valuables with foreign banks located outside Russia. The introduction of similar additional requirements aims to strengthen the policy, consistently pursued by the Russian Federation, of limiting the foreign influence on the country's public service. Reflecting the relevant norms in the Constitution will make it possible to emphasize the importance of ensuring the preservation of national sovereignty, the inadmissibility of any interference in its internal affairs, as well as the need to implement public functions exclusively in the interests of the Russian Federation and its citizens. Legal connections with a foreign state, including through citizenship of such foreign state or a residence permit or another document confirming the right to permanent residence of a Russian citizen in such foreign state, imply that the person may have commitments to two states at a time. This gives rise to a potential conflict situation if the interests of these states do not coincide and the person that vested with public powers will be forced to take these circumstances into account when making decisions, which is unacceptable, as such a situation is contradictory to the task of protecting the sovereign interests of Russia.

Similar provisions are already present in federal laws, but their introduction in the Constitution will create a solid legal basis for the inadmissibility of revising these restrictions introduced to protect the national interests from the potential influence of foreign states on persons exercising public powers, as well as to exclude potential conflicts of interest for public servants.

One way or another, having the nationality of another state shows the connection of a person with the relevant community and allows identifying the focus of the person's vital interests. That is why the requirement for state and municipal employees not to have citizenship (nationality) of a foreign state serves as an additional guarantee that they will properly perform their public functions. As noted by the Constitutional Court of the Russian Federation, since a Russian citizen having the citizenship of a foreign state has political and legal connections simultaneously with the Russian Federation and with the respective foreign state, to which such citizen also has constitutional and other commitments under the laws of this foreign state, the meaning of Russian citizenship for such citizen as a political and legal expression of the value of his or her connection with the home state can be rightfully considered to be significantly reduced (Ruling No. 797-O–O of the Russian Constitutional Court dated December 4, 2007).

The Constitutional Court, confirming this position in its opinion of March 16, 2020, noted that the same is equally true of the ban on opening and holding accounts (deposits), keeping money and other valuables with foreign banks located outside Russia (the procedure for implementation of which should be established by federal law), since it also implies that the incumbent of the respective position has vital interests outside the Russian Federation, which makes him or her vulnerable to the influence of third parties.

Along with the above restrictions, the constitutional amendments introduce restrictions for certain officials such as the requirement to permanently reside in the Russian Federation, which ensures that they know the real situation and existing problems of the country and are interested in addressing them. In addition, for certain positions are also introduced age restrictions and requirements to have higher professional education, which are associated with the nature of the service and the level of responsibility and imply the need to have a certain life and work experience to make informed decisions.

A similar approach to the exercise of public powers is widespread in the practice of foreign countries in organizing and operating public service functions. As a rule, the requirements for candidates to state and municipal service positions are introduced through laws adopted by the parliament, but in some countries, the restrictions for candidates are set out directly in their constitutions.

Article 9 of the Constitution of Colombia establishes that "the Colombian Nationality shall be forfeited on receipt of a certificate of naturalization in a foreign country and in case of permanent residence abroad."

According to Section 44 of the Australian Constitution, "any person who is under any acknowledgment of allegiance, obedience, or adherence to a foreign power, or is a subject or a citizen or entitled to the rights or privileges of a subject or citizen of a foreign power" becomes incapable of being elected to public service positions.

In the U.S., the federal Constitution does not discuss the status of the highest state official as it falls within the jurisdiction of each particular state. For example, according to Article 5, Section 2 of the California Constitution, a U.S. national who has lived in the state for at least 5 years immediately prior to the election can be elected Governor of California. The Governor of the State of California may not hold any other public office. No governor may serve more than two terms.

It is proposed that the requirements to civil servants, along with the requirement to have no foreign nationality, residence permit, or other document confirming the right of permanent residence of a citizen of the Russian Federation in a foreign state, be expanded by the introduction of a ban on opening and holding accounts (deposits), keeping money and other valuables with foreign banks located outside Russia.

A ban on keeping and holding money abroad is a common practice applicable to a considerable number of persons holding public offices. For instance, Federal Law No. 79-FZ dated May 7, 2013, prohibits opening and holding accounts (deposits), keeping cash and other valuables with foreign banks located outside Russia, holding and/or using foreign financial instruments, in particular, for persons who hold public offices in the Russian Federation, public offices of constituent entities of the Russian Federation as well as a number of other public servants. Introducing such a prohibition in the current legislation is primarily aimed at ensuring Russia's national security, increasing the effectiveness of the public administration system, and combating corruption.

Amid the international tensions, consistent implementation of sanctions against Russia, and attempts to exert pressure from a number of foreign states, it is necessary and reasonable to introduce the relevant requirements to the President of the Russian

Federation, Deputies of the State Duma, Senators of the Russian Federation, senior officials of constituent entities of the Russian Federation, and other officials into the Constitution.

1.4 Conclusion

The constitutional reform of 2020, which is a logical outcome of more than twenty-five years of the development of Russian constitutionalism, has led to a significant upgrade of the existing constitutional and legal reality, taking into account the latest and most advanced legal practices and accumulated political and legal experience, affecting various aspects of the organization and operation of public service. Matters related to enhancing the independence and autonomy of the state have become, if not the key ones, a crucial focus area of constitutional reforms promoted by the amendments to the Constitution.

Russia has been consistently implementing a national policy to ensure the comprehensive protection of its sovereignty and national identity. Improvements to the existing mechanisms for maintaining national independence as provided for under the Constitution will be a key driver behind further progressive development of Russian statehood, as well as the implementation of an independent national policy impervious to attempts by foreign states to influence the adoption of certain decisions.

References

1. CDL-INF 016-e (2000). https://www.venice.coe.int/webforms/documents/CDL-INF(2000)016.aspx. Accessed: 13 Aug 2020
2. CDL-INF 2 (2000). https://www.venice.coe.int/webforms/documents/default.aspx?pdffile=CDL-INF(2000)002-e. Accessed: 13 Aug 2020
3. Chirkin, V.Y.: Integratsionniye mezhgosudarstvenniye protsessy, Konstitutsii i gosudarstvenniy suverenitet [Intergovernmental Integration Processes, Constitutions, and State Sovereignty]. Zhurnal zarubezhnogo zakonodatelstva i sravnitelnogo pravovedeniya **3**(76), 5–19 (2019)
4. Leksin, I.V.: Suverenitet i gosudarstvo: kontseptualniye vzaimosvyazi i raskhozhdeniya [Sovereignty and state: conceptual interrelations and differences. Russian law]. [Rossiyskoye pravo: Obrazovaniye. Praktika. Nauka], No. 3 (105). 17–29 (2018)
5. Vinogradova, P., Tulaev, A.: The legal positions of the Constitutional Court of the Russian Federation on the execution of decisions made by the European Court of Human Rights. Russ. Law J. **9**(1), 138–164 (2021). https://doi.org/10.17589/2309-8678-2021-9-1-138-164
6. Shulzhenko, Y.L.: Pravovaya okhrana Konstitutsii Rossiyskoy Federatsii [Legal protection of the Constitution of the Russian Federation]. Gosudarstvo i pravo [State and Law], **5**. 7–18 (2021). https://doi.org/10.31857/S102694520014853-9
7. Teague, E.: Russia's constitutional reforms of 2020. Russ. Polit. 9. **5**(3), 301–328 (2020). https://doi.org/10.30965/24518921-00503003.
8. Vasilyeva, T.A.: Kontseptsiya suvereniteta v usloviyakh globalizatsii i yevropeuskoy integratsii [The concept of sovereignty in the context of globalization and European integration]. Konstitutsionnoye i munitsipalnoye pravo [Constitutional and municipal law] **2**, 7–9 (2016)

9. Verlaine, M., Shashkova, A., Kudryashova, E.: Amendments of Russian constitution concerning international law and the BRICS. Revista de Investigações Constitucionais **7**(2), 337–353 (2021). https://doi.org/10.5380/rinc.v7i2.73355
10. Vinogradova, Ye.V.: Konstitutsionniye popravki. Rossiyskiy opyt gosudarstvennykh preoprazovaniy. Gosudarstvo i pravo [Constitutional Amendments. The Russian Experience of State Transformations] Gosudarstvo i pravo [State and Law] **5**, 19–25 (2021)
11. Zorkin, V.D.: Bukva i dukh Konstitutsii [The Letter and the Spirit of the Constitution]. Rossiyskaya gazeta [Russian newspaper], **10**(7689), 1, 5 (2018)

Chapter 2
Elaboration of a Model for Sustainable Development in Organizations Through Digital Integration of Management Systems Standards

Valentin A. Dzedik and Irina V. Usacheva

Abstract The modern market environment of the functioning of organizations is a continuously changing system. It is a reason to search for the most effective tools for its sustainable development. The management process of these economic systems has recently changed significantly. It has become multidimensional and requires a comprehensive analysis of all factors influencing its external and internal environment. These include increased focus on general digitalization and globalization of production and logistics processes by standardization. In this paper, the concept of sustainable development in organizations is considered in terms of management based on ESG principles, digitalization, and standardization of business processes of an enterprise. A classification of the main standards containing requirements for management systems has been made regarding the main components of the concept of sustainable development.

2.1 Introduction

The current global economic environment is characterized by a range of interrelated phenomena at the microeconomic level. These include an increased focus on the concept of sustainable development, general digitalization, and globalization of production and logistics processes by standardization [1].

The concept of sustainable development was introduced in the 1970s and implies the managed development of economic actors in terms of environmental, social, and economic components. The desired state is defined as a stable global equilibrium [2].

V. A. Dzedik (✉) · I. V. Usacheva
Volgograd State University, Volgograd Region, Russia

I. V. Usacheva
e-mail: Zeppelin89@volsu.ru

A. O. Inshakova and E. I. Inshakova (eds.), *New Technology for Inclusive and Sustainable Growth*, Smart Innovation, Systems and Technologies 288,
https://doi.org/10.1007/978-981-16-9808-8_2

According to the Brundtland Commission report, sustainable development or process used to achieve sustainability means "meeting our own needs without compromising the ability of future generations to meet their own needs" [3].

This concept is often divided into three dimensions: "economic, environmental and social" or "ecology, economy and equity [4]. The concept of "economic sustainability" therefore often focuses on the use of natural resources providing both renewable and depletable physical resources for economic production.

The concept of "environmental sustainability" focuses on "life support systems" such as the atmosphere or soil that must be preserved for economic production or human life.

The concept of "social sustainability" focuses on the impact of economic systems on a human, and this category includes efforts to eradicate poverty and hunger and to combat inequality [5].

The exploring basis of sustainable development of the company requires the applying of such digital technologies like Big Data, machine learning, artificial intelligence, etc. [6].

The most important advantage of digital technologies is the possibility to quickly identify the causes of the potential instability of the organization [7]. Applying these technologies reduces losses and errors caused by the human factor. It encourages a greater sustainability of companies in an unstable environment [8]. Standardization of the production and logistics processes of the organization is one of the specialized processes of sustainable development. A standardized system for ensuring production and logistics processes allows effective management and achievement of high performance of the organization's activity. Therefore, it is necessary to combine a range of specialized processes, taking into account economic, environmental, and social concepts in the context of modern digitalization to achieve sustainable development in an organization.

2.2 Methodology

At the moment the most significant aspect of the concept of sustainable development are the 17 UN Sustainable Development Goals [9].

In a sense, the ESG model (Environmental, Social, Governance) can be considered related or derived from the concept of sustainable development.

The ESG model is a set of company's activity standards those are used socially conscious investors to test potential investments [10]. Environmental criteria help to identify the way the company protects nature. Social criteria explore the way the company manages relationships with employees, suppliers, customers, and society. Corporate governance deals with the company management, pay levels for governance, audit, and internal control and shareholder rights [11]. Each of the three criteria of the ESG model has several factors which affect the sustainable development of the company. The environmental factor illustrates how the organizations act regarding sustainable development goals without taking into account how they use the energy, exploit waste or emit polluting gases during their activities. The environmental factor covers the following aspects:

- deforestation.
- air and water pollution.
- conscious consumption (energy and water).
- energy efficiency.
- use of nonrenewable natural resources.
- biodiversity conservation policies.
- disaster management measures.

The social factor is important in order to guarantee the rights and personal safety, training, and an equitable treatment of staff. This regards to the practices of more transparent and bilateral relations with the communities involved by consumers and society in the production process.

The following actions of the company can be attributed as part of the social sphere within the ESG model:

- diversity and inclusion.
- employee involvement.
- labor policies.
- respect for human rights.
- employee training.

The company management factor from the ESG point of view refers to the way the company is managed by its partners, senior management, and other leaders, as well as the relationship between senior management and other employees. Effective corporate governance practices include.

- ethics and transparency.
- diversity on the board of directors.
- independence of government
- the structure of tax audit committees
- senior Management Compensation Policy
- the presence of an official complaint channel and the prevention of corruption.

Together, these factors determine the degree of focus of companies on increasing sustainability in terms of social impact and corporate governance.

2.3 Results

Within the concept of sustainable development from a socio-economic perspective, digitalization or informatization of an organization can be interpreted as "a set of processes of integration into social systems and structures of a variety of information phenomena, information tools and solutions for the development of social space, meeting the information needs of social development" [12], or as "… a complex managed technical–technological and socio-economic process of using information and computer technologies, which creates conditions for the further receipt, processing, consumption, dissemination and storage of information" [13]. Some authors emphasize the transformational aspect of informatization, for example, "... policies and processes aimed at building and developing a telecommunications infrastructure that unites geographically distributed information resources. The process of informatization is a consequence of the development of information technologies and the transformation of the technological, product-oriented method of production in the post-industrial period ..." [8].

Standardization, including enterprise management systems, has become a powerful factor of globalization over the past 30 years [14]. This is due to the fact that a product, technology or management approach that has become an industry, national or international standard automatically turns an economic entity, product, technology or approach that proposed and ensured its recognition as a standard into a leader in the competition.

International standardization is deeply integrated into the global processes of sustainable development. For example, each of the standards on the International Organization for Standardization (ISO) website is described by relating it to the UN Sustainable Development Goals. For example, the ISO 9001:2015 standard "Quality management systems. Requirements" is correlated with the goals of "Poverty eradication", "Industrialization, innovation and infrastructure", "Responsible consumption and production" and "Preservation of marine ecosystems" [15].

However, this approach seems to the authors to be partially excessively detailed and, at times, misleading. For example, returning to the ISO 9001:2015 standard, it is not at all obvious on what bases it is assigned to the sustainable development goal "Preservation of marine ecosystems", since there are no direct or indirect indications of this in the text of the standard. In this regard, it seems appropriate to the authors to classify the most popular international standards for management systems (not necessarily related to the jurisdiction of the International Organization for Standardization) regarding the three main components of the concept of sustainable development, which is implemented in Table 2.1.

Some standards are out of the proposed classification, among which may be called as specialized standards on the sustainable development, for example, ISO 37101:2016 "Sustainable development in communities. Management system. General principles and requirements" [22], as well as a new generation of standards related to models of digitalization management systems, for example, the draft ISO 42001 standard "Information Technology–Artificial intelligence–Management system" under development. This allows to formulate a multidimensional model of the concept of sustainable development, which includes not only the first dimension—the measurement of components (Environment, Society, Governance), but also the second dimension—the measurement of specialized processes of sustainable development and the third dimension—the measurement of digitalization, which integrates the other two dimensions to increase their productivity and efficiency (Fig. 2.1).

Table 2.1 Classification of the main standards containing the requirements for management systems in relation to the main components of the concept of sustainable development

No. Ser. No	The title of the standard	Ecology	Society	Economy
1	ISO 9001:2015 "Quality management systems—Requirements" [15]			V
2	IATF 16,949:2016 "Quality management system requirements for automotive production and relevant service parts organizations" [16]			V
3	AS9100C "Quality Management Systems—Requirements for Aviation, Space and Defense Organizations" [17]			V
4	ISO 13485:2016 "Medical devices—Quality management systems—Requirements for regulatory purposes" [18]		V	V
5	ISO 14001:2015 "Environmental management systems—Requirements with guidance for use" [19]	V		
6	ISO 45001:2018 "Occupational health and safety management systems—Requirements with guidance for use" [20]		V	
7	ISO 50001:2018 "Energy management systems—Requirements with guidance for use" [21]	V		V

Source The research results were obtained personally by the author

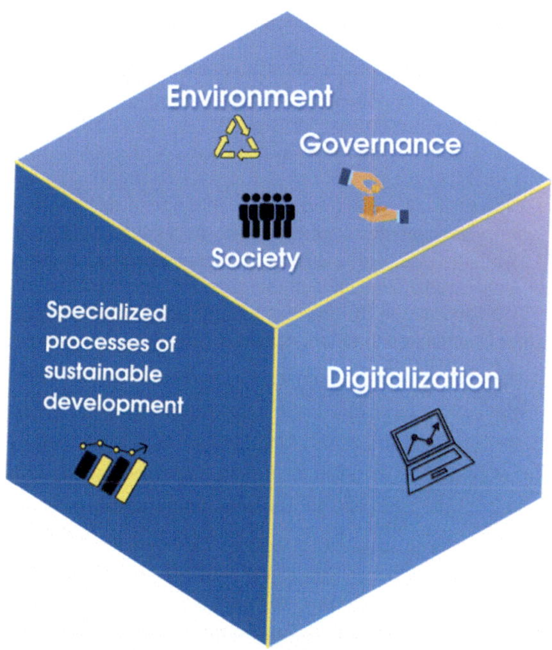

Fig. 2.1 Multidimensional model of the concept of sustainable development. *Source* The research results were obtained personally by the author

2.4 Conclusion

Therefore, it is necessary to combine a range of specialized processes, taking into account economic, environmental, and social concepts in the context of modern digitalization to achieve sustainable development in an organization. The proposed multidimensional model of this concept is necessary for the integration of the main components in order to increase their productivity and efficiency.

References

1. Dzedik, V.: Analysis of long-term prospects for the development of quality management systems. Standards and quality. Appl. 959, 78–80 (2017)
2. Meadows, D., Randers, J., William, W., Behrens, I.: The limits to growth. A report for the Club of Rome's Projects on the Predicament of Mankind, p. 180. Universe Books. New York (1972)
3. Development and International Economic Co-operation: Environment. Report of the UN Commission (1987). https://www.un.org/ru/ga/pdf/brundtland.pdf. Accessed 20 Oct 2021
4. Petrina, O., Savkina, E.: To the question of sustainable development of social and economic systems Vestnik Universiteta. Appl. 2, 37–41 (2017)
5. The Concept of Environmental Sustainability. JSTOR (1995). https://www.jstor.org/stable/2097196?readnow=1&refreqid=excelsior%3Abbd32975fd1ecabf60fbfbd2095ee0f4&seq=3#page_scan_tab_contents. Accessed 20 Oct 2021
6. Usacheva, I., Ponomareva, L., Volkova, A.: "Smart Technologies" for society. State and Economy. ISC 2020.155, 777–785 (2020). https://doi.org/10.1007/978-3-030-59126-7_86
7. Inshakova, A.: Law as the basis of infrastructure support of the digital economy and technology of the internet of things. Legal Concept **18**(3), 6–11 (2019). https://doi.org/10.15688/lc.jvolsu.2019.3.1
8. Ivanov, I.N., Orlova, L.V., Ivanov, S.I.: Digitalization and project management as factors of sustainable development of the organization. Vestnik universiteta, no. 5, pp. 12–18 (2021). https://doi.org/10.26425/1816-4277-2021-5-12-18. https://cyberleninka.ru/article/n/tsifrovizatsiya-i-proektnoe-upravlenie-kak-faktory-ustoychivogo-razvitiya-organizatsii. Accessed 20 Oct 2021
9. Sustainable Development Goals. United Nations. https://www.un.org/sustainabledevelopment/ru/sustainable-development-goals/. Accessed 25 Oct 2021
10. Kalinina, A.E., Inshakova, A.O., Goncharov, A.I.: Polysubject jurisdictional blockchain: electronic registration of facts to reduce economic conflicts. In: Popkova E. (ed.) Ubiquitous Computing and the Internet of Things: Prerequisites for the Development of ICT. Studies in Computational Intelligence, vol. 826. Springer, Cham (2019). https://doi.org/10.1007/978-3-030-13397-9_24
11. Chen, J.: What Are Environmental, Social, and Governance (ESG) Criteria? Investopedia n.d. (2020). https://www.investopedia.com/terms/e/environmental-social-and-governance-esg-criteria.asp. Accessed 26 Oct 2021
12. Lopatina, N.: Informatization management as a priority task of social development. Information resources of Russia. Appl. 1, 23–25 (2005).
13. Sysoeva, S.: The nature of the category "informatization" in the post-industrial economy. Herald of Omsk University. Appl. 3, 330–335 (2011)
14. Anisimov, A.P., Matytsin, D.E.: The concept and types of environmental entrepreneurship. In: Inshakova, A.O., Frolova, E.E. (eds.) Smart Technologies for the Digitisation of Industry: Entrepreneurial Environment. Smart Innovation, Systems and Technologies, vol. 254. Springer, Singapore (2022). https://doi.org/10.1007/978-981-16-4621-8_25

15. ISO 9001:2015: Quality management systems—requirements. International Organization for Standardization (2019). https://www.iso.org/standard/62085.html. Accessed 24 Oct 2021
16. IATF 16949:2016: Quality management system requirements for automotive production and relevant service parts organizations. International Organization for Standardization (2016). https://www.iso.org/standard/52844.html. Accessed 26 Oct 2021
17. AS9100C: Quality Management Systems - Requirements for Aviation, Space and Defense Organizations. International Organization for Standardization (2013). https://www.iso.org/files/live/sites/isoorg/files/news/magazine/ISOfocus%20(2013-NOW)/sp/isofocus_120.pdf. Accessed 26 Oct 2021
18. ISO 13485:2016: Medical devices—Quality management systems—Requirements for regulatory purposes. International Organization for Standardization (2016). https://www.iso.org/standard/59752.html. Accessed 27 Oct 2021
19. ISO 14001:2015: Environmental management systems–requirements with guidance for use. International Organization for Standardization (2015). https://www.iso.org/standard/60857.html. Accessed 21 Oct 2021
20. ISO 45001:2018: Occupational health and safety management systems–requirements with guidance for use. International Organization for Standardization (2018). https://www.iso.org/standard/63787.html. Accessed 21 Oct 2021
21. ISO 50001:2011: Energy management systems–requirements with guidance for use. International Organization for Standardization (2011). https://www.iso.org/standard/51297.html. Accessed 20 Oct 2021
22. ISO 37101:2016: Sustainable development in communities-management system for sustainable development-requirements with guidance for use. International Organization for Standardization (2016). https://www.iso.org/standard/61885.html. Accessed 23 Oct 2021

Chapter 3
The Fourth Industrial Revolution and the State

Vladimir E. Berezko

Abstract The new technological revolution is of a historic nature in terms of the volume and quality of changes that enter the life of not only the average person but also the society as a whole. Nowadays it is obvious to everyone. According to the researchers, not only labour is changing, but also communication, entertainment and other spheres of life. According to Klaus Martin Schwab, the founder and executive chairman of the World Economic Forum in Davos, revolutionary changes will almost completely transform humanity and change the image of state institutions (Schwab in The Fourth Industrial Revolution, Electronic Book, Moscow, 2016, [19]). The question—to what extent the Fourth Industrial Revolution will affect the fundamentals of state and law, remains open. However, it is quite obvious that the fundamental changes caused by the new technological revolution do not mean the rejection of the substantiated, objective foundations of the state and law, the state understood as the foundation of social peace, stability and prosperity. In this regard, the research is aimed at studying the fundamental aspects of the state, law and public authority under the conditions of the Fourth Industrial Revolution.

3.1 Introduction

The new technological revolution is of a historic nature in terms of the volume and quality of changes that enter the life of not only the average person but also the society as a whole. Nowadays it is obvious to everyone. According to the researchers, not only labour is changing, but also communication, entertainment and other spheres of life. According to Klaus Schwab, the founder and executive chairman of the World Economic Forum, revolutionary changes will almost completely transform humanity and change the image of state institutions [19].

V. E. Berezko (✉)
Moscow State Institute of International Relations MGIMO, Moscow, Russia

A. O. Inshakova and E. I. Inshakova (eds.), *New Technology for Inclusive and Sustainable Growth*, Smart Innovation, Systems and Technologies 288,
https://doi.org/10.1007/978-981-16-9808-8_3

3.2 Methodology

In the course of the study, the system method, the method of comparative law and logical analysis were used. In addition, methods of exegetics and hermeneutics were applied.

3.3 Results

The question—to what extent the Fourth Industrial Revolution will affect the fundamentals of state and law, remains open. However, it is quite obvious that the fundamental changes caused by the new technological revolution do not mean the rejection of the substantiated, objective foundations of the state and law, I would like to emphasize that the state is understood as the basis of social peace and stability. This article is intended to ascertain the impact of the Fourth Industrial Revolution on the state and law.

At the core of the Fourth Industrial Revolution is the digital revolution, which is based on the availability and mobility of the Internet, "tiny" devices for production, artificial intelligence, and machines that are already able to study [19].

And in this regard, the founder of the World Economic Forum worries about the "non-revolutionary" or "linear" thinking of individual leaders. But such "concern" does not seem to be well-founded, since it is "non-revolutionary", but rather evolutionary thinking, which is the most essential characteristic of a good statesman. To destroy everything to the ground is certainly easier, but to build—much more productive, and first of all, for its own people. Indeed, "non-revolutionary" thinking should not be used as a synonym of backwardness, stagnation, etc. In the revolutionary epochs, the romantic belief in a miraculous deliverance from the state and law, in a mystical "liberation" from harassment and oppression, triumphs. And it does not matter that ratio has nothing to do with this naive belief. Understanding comes as always unexpectedly—institutions destroyed in revolutionary storms are reborn—like a Phoenix from the ashes…

The main idea of Klaus Schwab is that new technologies are radically changing the world around us [19].

By the way, it is hard to argue against this fact—just get a new smartphone out of your pocket. But it should be noted that a significant part of the world population does not yet have mobile phones [19].

Indeed, everything will not be the same as it was before. And this is a good thing. But changing the form does not mean that the content will automatically disappear. And what is the essence, the content of the state?

The understanding of the state as certain integrity, an instrument of social harmony, social peace and social prosperity acquires a radically different meaning in the times of the "ubiquitous" Internet. But the classical sense of the state, which entails the management of public, general affairs (res publica), does not disappear at all. The fundamentals of the state remain the same. In this regard, the state should be understood not as a state apparatus, which is typical for the Marxist tradition, but as a "trinity" of territory, population and power [3].

The ancient Greek philosopher Aristotle very figuratively described this legal state—sailors on a ship, seeking to safely complete the voyage. This is their "public affair" (res publica) [2].

And awareness of the fundamental, objective foundations of the state and law will not allow such a ship to crash on the "revolutionary" and "non-linear" reefs. Let us consider, for example, the economic sphere of society. It should be emphasized that the production of goods and services will still be the essence of any economy. The stages of economic activity—production, distribution, exchange and consumption will not disappear. But the problem of distribution of the produced item—goods or services—will not lose its relevance.

And if we imagine that the hard and thankless labour in industry and agriculture will be replaced by the work of robots (which, by the way, can work in the field around the clock, without rest), then, what does the future hold for people?

The answer to this question, in my opinion, depends solely on the way of social organization and the attitude to the accumulation of "human capital"—the greatest value in the modern world. If, for example, "pure" capitalism, based on the postulates of "pure" or "old" liberalism, prevails in such a society, then it will not be difficult to predict the future. Automation and robotization of production will not only fail to solve existing social problems but will exacerbate them to the limit. In particular, the problem of unemployment and social stratification will become acute. The reason for this state of affairs lies in the private way of appropriating the manufactured product and the fulfilment of exclusively private interest by the entrepreneur. It is quite obvious that an entrepreneur who thinks exclusively about daily profits, without thinking about the future, will follow the path of reducing the labour force, minimizing the corresponding payments—wages and various social benefits. But this cannot be considered as a benefit, in terms of the public interest and interests of society as a whole. There are many ways to solve the problem, starting with the establishment of an unconditional basic income and ending with the implementation in practice of an equal classless society. But on the way toward such a society, the same public-legal regulation of economic relations is necessary. Such regulation can be fully and professionally implemented only by the public authorities, the state authorities. Other public relations will undoubtedly require legal regulation. It should be noted that this problem will be highly relevant not only for national states but also for the interplanetary federation [13].

In this regard, it is necessary to resolve the main contradiction of capitalism, which was covered in "Capital" by Marx—between the private character of appropriation and the social nature of labour [15]. In addition, technological progress will exacerbate the contradiction between the private method of appropriation and the declining effective demand. As a result, this can lead to the "death of capitalism. Robots will deliver the final, fatal blow to this economic order" [14].

Essentially, the Fourth Industrial Revolution should lead to a "Golden Age" that has never been seen before in the history of mankind. Perhaps, it will even lead to the "golden millennium". Ideally, the life of humanity should turn into a fairy tale—science, art, endless self-improvement, abandoning the race for survival, because labour productivity due to automation and robotics will lead to the abundance of produced goods and services. But, all this depends on a just social order, which cannot be objectified without a properly organized public authority. In the meantime, Klaus Schwab notes that more than one billion people (more precisely—about 1.3 billion) in the world have no access to … electricity! And these people are utterly divorced from the digital revolution and civilization as a whole! In particular, they are divorced from modern sanitary standards. And if a person does not have modern plumbing, then no digital platform can brighten up his dull life…

The founder of the World Economic Forum is well aware that with new technologies, new problems will arise.

He sees the roots of the problem in the fact that the well-being of those who do not own capital but live at the expense of their own labour, at least, does not increase. And this leads to the "disappointment" of workers who are also worried about the future of their children [19].

The realization of the public interest, which is the prosperity of the whole society, and not just the individual "lucky ones", is the key to solving the arising social problems in the era of the Fourth Industrial Revolution [1].

In the context of the Fourth Industrial Revolution, life expectancy could substantially increase. Klaus Schwab predicts its increase to one hundred years. He also notes the need to revise the retirement age [19].

But in parallel with life expectancy, labour productivity should also grow rapidly. The "Golden Age" of humanity is only possible if people work for pleasure, and not depressingly watch as the life extension coincides with the increase in the retirement age …

Klaus Schwab is fully aware that if the lives of people do not improve, this could lead to "social unrest" and even "armed extremism" [19].

The world of the future is of great concern to the founder of the World Economic Forum. He also wondered: "If the revolution will result in—a world of the precariat, a social class of workers who move from task to task to make ends meet while suffering a loss of labour rights, bargaining rights and job security—would this create a potent source of social unrest and political instability?"[19].

That is a rhetorical question. And the answer to it is known—of course, it will! Especially if the middle class is shrinking, even in developed countries. And even education is now priced as a "luxury" because of the exorbitant cost [19]. Other researchers have also written about the "inequality" and "disadvantaged position" of the middle class, in particular, the futurist Martin Ford [7]. According to the researchers, the middle class will suffer, primarily, due to robotization [14].

Martin Ford, by the way, makes a very accurate remark about the fact that "if the employee is replaced by a machine, the machine will not go shopping and will not consume anything" [7].

Speaking about the possible "inefficiency" of governments in the era of new technologies, Klaus Schwab considers the state not as a "trinity", but identifies it with the concept of "state apparatus". According to his opinion, business and society are practically outside the state. The incorrectness of this point of view was very accurately noted by Professor Atamanchuk [4].

Another key point that Klaus Schwab touches on in his book is that new technologies can disrupt the existing "social contract". The statement is quite controversial since the need for a social contract will not disappear, it can only transform. In any case, there will be a need for "stable legislative frameworks" as well as a legal "ecosystem" [19].

And if there are legislative frameworks, then the legal field in which they will operate is also necessary. And the legal field is unthinkable without a public authority that establishes generally binding legal norms. And the public authorities will be faced with the tasks of "compliance with the laws", "security", etc. It should be noted that various aspects of the legal regulation of digital assets are of serious interest to researchers [21].

New technologies do not eliminate the old ones, for example, the eternal Russian problems—worn-out water pipes. Under these conditions, reprogrammable objects, "WaterNet" and other innovations will not be able to fully prove themselves. Unfortunately, "WaterNet" will not be able to demonstrate its effectiveness in the old, worn-out Moscow five-storey blocks of flats.

Let us consider the next moment. Klaus Schwab writes about how the traditional family can change. He believes that it will turn into a kind of "family network".

But the "family network" does not exclude the traditional family, as it is understood in legal and moral terms!

If members of a large family live in different countries and can communicate indefinitely and without problems through an inexpensive connection, this does not make such a family any less traditional. In this case, the family remains a "unit", and digital technologies only strengthen intra-family ties! For example, it becomes much easier for parents to transfer money to their children studying abroad. And they can easily check their homework!

And the obvious question to ask is—whether it is necessary to rush into the agreement that the family should turn into a "family network"? Although this way of thinking might seem new, it is not really new. It is rather old, like the whole world around us. Even the great Russian philosopher Ivan Aleksandrovich Ilyin warned against the nihilistic attitude and thoughtless denial of the family, which took place in our country after the 1917 Revolution. Ilyin quite rightly pointed out that the "disintegration of the family" is the root cause of all great national catastrophes [11].

And the great humanistic mission of the traditional family, which does not dissolve on the Internet, but becomes even more relevant in the digital society, is, according to Ilyin, "… to pass on from generation to generation … *the spiritual and religious, national and domestic tradition*" [11].

By the way, the need for human, social connections in the digital age does not disappear but becomes increasingly necessary. The reason is that we need to "pause, think, and have a heartfelt conversation". If this does not happen, then people are embarked on frustration and even reading a small article can turn into a "luxury" [19].

The Fourth Industrial Revolution—in addition to making our lives easier—may lead to the Apocalypse. Military robots, new achievements in the development of biological and biochemical weapons—is a matter of great concern. And the role of state, public authority, and law, especially international law, significantly increases under these conditions, because it is the law that embodies the ideals of goodness and justice [8].

Klaus Schwab points out a very important issue—the public authorities should respond to emerging challenges in a timely manner (this is what Boris Nikolayevich Chicherin wrote about). Otherwise, "non-state actors" will handle it [19].

The key point is that the technologies of the Fourth Industrial Revolution should lead to the human freedom expansion, and not to its restriction, for example, as a result of unemployment and poverty. But this requires the creation of adequate social and political systems [18]. And these systems should, among other things, ensure the fair distribution of benefits and the realization of the public interest.

So, what kind of institutions will guarantee the common good? One of the answers seems obvious—the state, but the state which is understood as the "trinity" of territory, population and power. The second answer is a private company that will take over the functions of the state. But in this case, the private nature of the company will immediately change, since, in order to implement public and state functions, it will have to turn into a public authority that can solve res publica. And, in order to solve "public affairs" properly, it will be necessary to establish generally binding legal norms in a certain territory. Thus, a private company will cease to be private and will turn into a public, state power.

In this context, the social contract is subject to revision. And the key point of the social contract revision is the introduction of an unconditional basic income [18]. Moreover, digital platforms contribute to the capital concentration "among an increasingly small number of people" [18]. Martin Ford also speaks about the need to introduce an unconditional basic income [7]. It is noteworthy that he refers to the interesting work of the famous economist Friedrich von Hayek [9].

But it should be emphasized that Hayek, speaking of "uniform minimum income" [9], does not connect it with equitable income distribution. He considered this as a "misfortune". Hayek gives an interesting example that such a system already existed in the form of assistance to the poor. Highlighted throughout the arguments of a well-known economist the idea that the expansion of the social security system and the social state will lead to enormous costs and, as a result—inefficiency. But his reasoning is little different from the maxims of Russian liberals of the nineteenth century, for example, Boris Nikolayevich Chicherin, who also vigorously defended the point of view of the "old", "pure" liberalism, in the style of the classical Russian master, believing that in helping the poor, such a principle as "Love" should prevail!

Meanwhile, there is a paradox in the theoretical reasoning of Hayek, which the supporters of "old" liberalism stubbornly refuse to admit. Matinee performances for the children from orphanages could be held, which in itself is a very right and noble thing—and the principle of "Love" will be fully manifested here. And it is possible to create a social system without poor and needy people. And the role of public authorities in achieving this, not an elusive goal, could not be overemphasized.

The state understood as the "trinity" of territory, population and power in this sense could play a key role. Corporate social responsibility may become the basis of such a system.

The crossroads at which we find ourselves today is a matter of choosing the right road. And what happens if the world chooses the wrong road? The answer is simple: "The experience of political revolutions of the past teaches that inequality consequences do not remain without consequences" [18].

By the way, the New Social Agreement on Values, which was presented by the World Economic Forum's Global Agenda Council on values (2012–2014), contains a key point that has been known since the times of Aristotle, Machiavelli, Hegel, Boris Nikolayevich Chicherin and other philosophers: "The importance of the commonwealth predominates over individual interests." But here it is extremely important to understand another point—public interest implementation, commonwealth achievement is impossible without private interest implementation.

To cope with the possible problems caused by new technologies, to ensure "the interests of consumers and society as a whole" in the new conditions, as well as "the development of innovations", is possible only with the help of flexible management methods–Agile [19].

But here it is important to note the following—Agile is not a panacea and a universal solution to all the problems of modern society. This statement is shared by his unequivocal supporters.

By the way, the ideologists of the Agile approach highlight several points when "Agile should not be used". Among them, in particular, are the purchases for the wedding celebration (unless we are talking about an alcoholic-free wedding) and "designing a space shuttle" [5].

Special attention should be given to the last point. Space projects, as a rule, are carried out by the state, since they require the concentration of a great number of various resources—labour, financial, information, scientific, etc.

Here are two more examples that the ideologists of the Agile approach call "classic" in the sense that Agile will not work there: "unsuitable organization: do not expect the Navy to switch to Agile. Wrong project: the construction of an aircraft carrier is not exactly suitable" [5].

That is why the ideologists of the Agile approach write wonderful words: "Positive reviews about Agile and, in particular, about Scrum—a double-edged sword. The positive thing here is that it is easy to convince people to try Agile. The negative side is that expectations are often overstated. Managers really like these postulates—faster, cheaper, better—and they often forget that there is no such thing as a free lunch" [5].

Let us focus on the next aspect of flexible management. The fundamental principle of the Agile Manifesto is stated as follows: "The highest priority … is to meet the needs of the customer". Can public administration apply this method? It certainly can be applied, since the "needs of the customer" are almost exhaustively formulated in our Constitution. Article 18 could be taken as an example.

The ideologists of the Agile approach emphasize that "Agile starts by defining the necessary minimum and works with it" [5].

The idea of "*minimum viable product* (MVP)" is quite acceptable. And in the case of replacing three suits with one suit and one shirt—for each day, it could be applied. It should be admitted that the world special army forces units also use an Agile approach—at the combat missions, they are all content with the minimum necessary food and clothing.

I deliberately chose a military example as it is a good illustration of the Agile approach. But in the context of the state, such a philosophy may not always be viable. First of all, since the product for the customer—the rights and freedoms of a person and a citizen—it is very difficult to issue it in a "minimum set". It should be provided as a complete package. And it does not matter whether a citizen enjoys the entire set of rights and freedoms, or only the "minimum".

The same applies to the "releases" of the project. Experts in the field of Agile note that "the first release of a project created with Agile is the backbone of the main idea, only the basic and necessary functions" [5].

This idea works perfectly for software. But imagine that an aviation corporation has released a fighter jet, which is gradually updated. And if the "updating" turns out to be technically impossible?

Of course, it is necessary to improve the state, which is understood as a "trinity" and the institution of public service and that is beyond dispute. But they should initially perform efficiently already in the "first release". History shows, that the "second release" may not be forthcoming, being shot down "without idiotic red tape" (*the words of V. I. Lenin*).

Sure, there are projects, including in the field of public administration, where flexible management technologies perfectly demonstrate themselves. But the choice of flexible management application should be made very carefully, so that for the sake of a "flexible worldview" not to part with fundamental values that do not disappear but acquire a new meaning and increased relevance in the twenty first century.

By the way, one of the postulates of the Agile approach is "… cheaper". The author does not call for mindless inflating the budget level and the staff of state bodies. But the profitability of civil servants, persons holding public positions, and of the entire public administration system in a broad sense, should not be measured in mercenary terms, remaining within the worldview of a small shopkeeper-owner—how much have I earned today? The efficiency of civil servants should be measured in a strategic context—the armed forces provide peaceful labour, the internal affairs agencies—public security, law and order. Considering only expenditure budget—yes, the Ministry of Defence and the Ministry of Internal Affairs are very expensive. There is no income.

But let us consider the situation from the public interest perspective. Is it possible to measure peace, law and order in monetary terms? It is entirely possible. A businessman who works in a comfortable environment and feels protected from criminal encroachments pays hefty taxes to the budget. And if we calculate the amount of these financial revenues, they may even significantly exceed the costs for state and public authorities.

And in this sense, the words of Vladimir Vladimirovich Putin, said in one of his Address, are more relevant than ever: "The goal and mission of state service is to serve the people, and those who enter this path must know that by doing this they inseparably connect their lives with Russia and the Russian people without any assumptions and allowances" [20].

The main task of the Agile approach is to ensure the proper result. But in public administration, procedures are no less important—to ensure the rights and freedoms of a person and a citizen. And a civil servant who understands his mission as "service" is, without exaggeration, the key element. This is also pointed out by Agile specialists [5]. The founder of New Public Management, Christopher Hood, quite rightly noted that this concept of public administration "assumes a culture of public service honesty as given" [10]. But it is also clear that following the "service" principle should not remain a pious hope, but should be based on certain guarantees, the ones that New Public Management eliminates.

Of course, the practical application of Kanban boards and brief meetings in public administration is quite possible and can lead to outstanding results. The main thing is that this does not turn into an end in itself: "... the developer is dissatisfied with the fact that he is interrupted several times a day, forced to come to meetings, which makes it difficult for him to complete the work on time" [17]. Reading Stellman makes you remember "Re Conferences" poem by Mayakovsky...

The creation of an effective system of public management in the twenty first century is the subject of numerous scientific studies [12].

One of the concepts is New Public Management, which was mainly developed the British professor Christopher Hood.

The New Public Management concept focuses on results rather than procedures. Hood also noted the need for clear goal-setting and considered "competition to be the key to reducing costs and improving standards" [10]. To do this, according to the authors of the concept, it is necessary to use "proven" management tools in the private sector for public management [10].

Hood noted that the concept of New Public Management includes several provisions from the doctrine of the new institutional economics, as well as the successful application of the "managerialism" experience in the public sector. Another British scholar, Stephen Osborne, noted that "NPM is originated from neoclassical economics and, in particular, rational/public choice theory".

However, Hood also noted the disadvantages of the new approach. He cited the opinion of the critics that New Public Management is like "The Emperor's New Clothes"—a literary folk-tale written by Danish author Hans Christian Andersen—just empty rhetoric and no substance" [10].

The market approach to public management is not always justified, since in this area the public interest is realized, and not the private one, which is manifested in the desire for profit. In no small measure, "NPM, despite its stated claim to promote the "common good" ... is a means of realizing narrow interests" [10].

This, by the way, is the fundamental contradiction between the understanding of the state as a "trinity" of territory, population and power and "managerialism" as an ideology based on private but not the public interest. By the way, speaking about the introduction of market mechanisms in the public management system, it is worth recalling, for example, the experience of the Great Patriotic War—already in August 1941, the orders of the People's Commissariat of Defence of the Soviet Union appeared on monetary and government awards to pilots who distinguished themselves in combat missions.[1]

Understanding the contradictions and shortcomings of New Public Management has led to the need for the development of the new concepts of public administration. In particular, it leads to the development of New Public Governance.

Professor Helen Dickinson writes that if the value foundation of NPM is competition, competition in a market context, then in New Public Governance it is a neo-corporative approach. And the theoretical differences run along the following line—if NPM is economics and management research, then NPG is "organizational sociology and network theory" [6].

Great work edited by Professor Stephen Osborne is dedicated to New Public Governance [16].

New Public Governance, according to Stephen Osborne, involves "… both a *plural state*, where multiple interdependent actors contribute to the delivery of public services and a *pluralist state*, where multiple processes inform the policy-making system" [16]. Consequently, "the NPG is thus both a product of and a response to the increasingly complex, plural and fragmented nature of public policy implementation and service delivery in the twenty-first century" [16].

3.4 Conclusion

The following conclusions are drawn. The evolution of public administration concepts from Public Administration to New Public Management, and then to New Public Governance, is actually dictated by the fundamental features of the state, as a "trinity" of territory, population and power, as well as the law established by the public authority.

The res publica solution will still be necessary in the twenty-first century, the century of the Fourth Industrial Revolution. "Public affair" in any case will have to be solved by the public authorities. It is possible that with the development of corporate social responsibility, some of the functions of state power will be transferred to private companies, especially in the social sphere. But then the nature of corporate activity may change—from private to public.

[1] See, for example, The Order of the National Commissar for the Defence of the Soviet Union "On the procedure for awarding the flight personnel of the Air Force of the Red Army for good combat work and measures to combat hidden desertion among individual pilots" No. 0299 of August 19, 1941.

References

1. Anisimov, A.P., Matytsin, D.E.: The concept and types of environmental entrepreneurship. In: Inshakova, A.O., Frolova, E.E. (eds.) Smart Technologies for the Digitisation of Industry: Entrepreneurial Environment. Smart Innovation, Systems and Technologies, vol. 254. Springer, Singapore (2022).https://doi.org/10.1007/978-981-16-4621-8_25
2. Aristotle: Politics. Moscow (2010).
3. Atamanchuk, G.V.: The Essence of Public Service: history, theory, law, practice. Moscow (2002)
4. Atamanchuk, G.V.: The Theory of Public Administration Moscow (2010)
5. Cole, R., Scotcher, E.: Brilliant Agile. Flexible project management with Agile, Scrum, and Kanban. Electronic book. St. Petersburg (2019)
6. Dickinson, H.: From New Public Management to New Public Governance: The implications for a "new public service". In: Butcher, J., Gilchrist, D. (eds.) The Three Sector Solution: Delivering public policy in collaboration with not-for-profits and business, pp. 41–60. Canberra (2016)
7. Ford, M.: Rise of the Robots. Technology and the Threat of a Jobless Future. Electronic book (2015)
8. Goncharov, A.I., Inshakova, A.O.: Supranational legal mechanism for the use of new-industrial digital technologies 4.0 in foreign trade by Economic Entities of the EAEU and the BRICS member states. In: Inshakova, A.O., Frolova, E.E. (eds.) Smart Technologies for the Digitisation of Industry: Entrepreneurial Environment. Smart Innovation, Systems and Technologies, vol. 254. Springer, Singapore (2022). https://doi.org/10.1007/978-981-16-4621-8_22
9. Hayek, F.A.: Law, Legislation and Freedom. Modern Understanding of the Liberal Principles of Justice and Politics Moscow (2006)
10. Hood, C.: A public management for all seasons? Public Administration, vol. 69, pp. 3–19. Spring (1991)
11. Ilyin, I.A.: The Path of Spiritual Renewal. Why we believe in Russia. Moscow (2006)
12. Inshakova, A.O., Anisimov, A.P.: Legal regulation of the development of renewable energy sources in Russia, the BRICS, and EAEU countries. In: Inshakova, A.O., Frolova, E.E. (eds.) Smart Technologies for the Digitisation of Industry: Entrepreneurial Environment. Smart Innovation, Systems and Technologies, vol. 254. Springer, Singapore (2022). https://doi.org/10.1007/978-981-16-4621-8_30
13. Inshakova, A.O.: "Renaissance" of the category "smart regulation" in domestic law-making theory and practice: preconditions and factors of influence. Legal Concept=Legal Paradigm **20**(3), 6–11 (2021). https://doi.org/10.15688/lc.jvolsu.2021.3.1
14. Katasonov, V.Yu.: Digital Finance. Cryptocurrencies and the electronic economy. Freedom or concentration camp? Moscow (2017)
15. Marx, K.: Capital, vol. 1. Electronic Book (1867)
16. Osborne, S.P. (ed).: The New Public Governance? Emerging perspectives on the theory and practice of public governance. London and New York (2010)
17. Stellman, E., Greene, J.: Learning Agile. Values, principles, and methodologies. Electronic Book. Moscow (2017)
18. Schwab, K., Davis, N.: Technologies of the Fourth Industrial Revolution Electronic Book. Moscow (2018)
19. Schwab, K.: The Fourth Industrial Revolution. Electronic Book, Moscow (2016)
20. The message of the President of the Russian Federation to the Federal Assembly of 15.01.2020 "Message of the President of the Russian Federation to the Federal Assembly". Legal System ConsultantPlus
21. Zavyalova, E.B., Shumskaia, E.I., Kuzmin, M.D.: Smart contracts in the Russian transaction regulation. Lecture Notes in Networks and Systems. 129 LNNS, pp. 205–212 (2020)

Chapter 4
Transformation of Law and Legislation in the Context of Digitalization: Intellectual Property Protection

Mikhail N. Kuznetsov

Abstract One of the primary problems faced by lawyers in Russia and abroad is the lack of a uniform conceptual framework related to the so-called digitalization of law. Russian bylaws as a basis for a future offensive against legal regulation in the civil law environment. Directions of digitalization intrusion into the civil law space: scope of civil law; principles of civil law; breaking ideas about the subject of law; breaking ideas about responsibility in law. Digitalization will not change civil law dramatically until it changes the very person who stands at the center of the civil law environment. Many provisions defining the conceptual framework of this process are reflected in economic terms, which guarantees their stability, since freedom of choice of behavior is assumed, which cannot be said about the legal field.

4.1 Introduction

Without any exaggeration, the twenty-first century can be called the century of the rapidly developing process of digitalization of the social life of the modern family of peoples, in which Russia occupies its rightful place.

Many scientific articles and even several monographs have already been written about the benefits of digitalization, the types and methods of its implementation in specific sectors of the social sphere, both in our country [1] and abroad [7]. Among them, we can note the works within the stated title of the article here and further in the text.

At the same time, at scientific forums, in the speeches of government officials and representatives of the business community, it is noted with concern that the solution of an increasing number of practical issues of digitalization is faced with the lack of its proper legal support both at the internal and domestic, and at the international levels.

M. N. Kuznetsov (✉)
Peoples' Friendship, University of Russia (RUDN University), Moscow, Russia
e-mail: kuznetsov-mn@rudn.ru

A. O. Inshakova and E. I. Inshakova (eds.), *New Technology for Inclusive and Sustainable Growth*, Smart Innovation, Systems and Technologies 288,
https://doi.org/10.1007/978-981-16-9808-8_4

4.2 Methodology

The work used the method of comparative legal analysis, which led to a legal understanding, which began in connection with the widespread introduction of the latest digital information technologies, which is called the fourth industrial revolution, thus becoming one of the main tasks of today's agenda, since the law, by virtue of its very purpose, to regulate social relations that arise in real life, is simply obliged not only "to keep up with changes in these relations, but also to respond to emerging new relations and changing traditional ones."

4.3 Results

Logically, another question arises and is very alarming: will "reacting" lead to hacking the very structure of law and changing the achievements and principles it has developed over the millennia of its history? After all, law, especially civil law, was originally built and perceived in society as the art of good and justice [ius est boni at aequi].

Isn't it dangerous to pass this sacred idea from a human being even to a very advanced artificial intelligence?

The threat of changing and even breaking the basic principles and established civil institutions is felt by every lawyer dealing with the problem of digitalization in law.

A generalized legal analysis of this threat to civil law, and law in general, has not yet been conducted either in our country or abroad. At least, we have not found any relevant publications on this subject, although there are quite a lot of critical and even arguable statements against digitalization of a political and socio-moral nature both in our country and abroad.

However, even more, we can say that there are already too many enthusiastic assessments of digitalization in the banking sector, in transport, in medicine, in the tax service, in street security, and in other areas of modern society.

Meanwhile, taking into account the latter, to crack the law, in our opinion, digitalization still will not succeed. At least as long as it is embedded in the daily life of people, it does not change the person from the inside.

We can name several reasons that objectively prevent digitalization at the present time to radically change modern civil law. Here are some of them, so far, of a general theoretical and methodological nature rather than of an institutional and normative nature.

1. To radically change the law, it is necessary to create new legal categories. Meanwhile, in legal science, there is still no official definition of the concept of digitalization itself, which entails the use of this term everywhere and without proper understanding.

Some lawyers generally consider digitalization to be a technology, shifting this concept toward economic rather than legal categories. Digitalization of various spheres of activity—Prof. A. A. Kartskhiya concludes on the basis of his analysis—implies not so much modern forms and methods of collecting, storing, processing, and transmitting any information in digital format, as methods, techniques, and forms, in other words, technologies of digital regulation of life cycle processes in the economy, politics, law, and in commercial and entrepreneurial relations.

2. It is noteworthy that in the hands of digitalization researchers, not to mention legislators and government managers, there is no such important tool of scientific analysis as "number" and the concept of "numerization" derived from it. Their place, both here and in the West, was taken by "digitalization".

Perhaps this is due to the less extensive statement of individual English words and expressions compared to the richness of meanings of words and expressions in other language cultures. In fact, such variants of "numerical" economics as "numbers economics" would sound clumsy in English. It is more euphonious to put a dead number as a basis and spin the whole civilized world around it, rushing to an "electronic paradise".

Only France initially called the economy of the future "numerical", but then, under the influence of the European Union and pressure from global IT corporations, abandoned this name, allowing the use of the Americanized term "digital" in its regulatory documents.

But the figure cannot embrace and is not intended to describe and express objective reality. "Number" can and is intended for this purpose. It can describe a complex object. A "digit" is an elementary particle. A number is a complex unit filled with energy, an object that can color, reflect, and even transform reality.

"Everything is a number. The number is the measure of everything"—said the great philosopher and outstanding ancient arithmetic Pythagoras (570–490 BC).

The school he created in its own way convincingly proved that the primary in the world is the nature of numbers, which contains and permeates absolutely everything: people, the earth, and the cosmos.

Pythagoras claimed that with the help of a number, you can logically express all the diversity of what surrounds us. Well versed in number and numbers, he considered a number, not a number, to be the origin of the world.

Pythagoras proposed the idea of doubling numbers: even–odd, yin–yang, good–evil, man–woman, etc., rather than discrete manipulations with zero and one.

The main number in the teaching of Pythagoras is one. He did not recognize the number 0 as a reference point, considering it empty, which our world and cosmos are not. The unit corresponds to a point in space through which countless lines can be drawn. Euclid called it the first of the axioms.

The number 2 corresponds to two points through which only one straight line can be invented. Three points, if they are connected, correspond to a plane. Four points—a sphere in which there are four elements: water, air, earth, fire—the interaction of which gives rise to an infinite variety of objects, images, and things.

In our opinion, the term "digit" is legitimate to use only to "digitize" the objects of the external world around us, but not to the inner world of a person, and even more so not to identify and transform the social relations regulated by law.

"Digitization" has no legal content because it means the process of transferring information from physical media to electronic, that is, digital.

This process does not change the quality characteristics and content of the information itself, but makes it more accessible and convenient for further use.

"Digitalization" (more correctly, "numerization") means the creation of a new product in electronic form, which can have new functions. Its goal is to digitize all material and information resources, including the person himself, and to form network interaction platforms based on the obtained data, which represent new business models [2].

3. In addition to the above, it should be pointed out that there is no clear, time-tested, other conceptual and legal apparatus, with the help of which it is only possible to reveal the essence of the process being studied in law (in this case, digitalization).

At the same time, it should be taken into account that "in accordance with the requirements of the theory of law, it is necessary that as the volume side of the concept, that is, the set of objects conceivable in the concept, be adequately represented."

The term "transformation", widely used by both foreign and Russian scientists, does not meet these criteria in the study of digitalization in law and economics. An example of this use is the following expressions that are most commonly found in scientific reports, research, and regulatory documents, such as.

- "the introduction of digital technologies leads to the transformation of all spheres of society, including the economy and social relations."
- "the concept of … digital civil turnover in the context of the transformation of law under the influence of digital technologies" is formulated.
- Transformation of Legal Reality in the Digital age—the title of the collection of scientific works of the Institute of Legislation and Comparative Law under the Government of the Russian Federation.
- "Law in the conditions of digital transformation": materials of the Republican scientific and practical conference.
- Modern technologies challenges and transformation of legal regulation.
- Digital transformation of law [6].

What is the fallacy of using this term in relation to the digitalization process? Transformation is the transition of a substance, energy, or structure to another state, in which the former state ceases to exist.

For example, alternating electric current is transformed into direct current, the heat of the engine is transformed into the movement of the car, feudal relations are transformed into capitalist ones, the energy of wind or water supply is transformed into electrical energy, the industry is transformed into a digital (correctly speaking, numerical) platform by means of so-called disruptive technologies, capturing and crushing the former market, etc.

In the case of digitalization, this does not happen. The norms of civil law remain the same. By giving digitalization the quality of a transformer, we are leading the science of civil law to a dead end. But it will also be able to destroy its integral civilistic structure, which has been honed for thousands of years, with the right methodological approach to the analysis of these social phenomena. Not yet, anyway.

However, a certain destructive impact of digitalization on the law is still possible, and the task of lawyers of various profiles today is to find a balance between such effects and preserving the best that is in the law [3].

4. Earlier it was noted that a number of scientists, including lawyers, consider digitalization to be a technology of digital regulation. However, most legal scholars tend to believe that digitalization is a legal category and that is why it transforms all law, including civil law. But is this really the case? From the point of view of the theory of law, a category can be defined as a well-established form of expression of legal thought that has reached understanding within the framework of the structure of the relevant branch of law, reproducing in a generalized form the real legal reality and/or the deep processes that accompany it.

Digitalization does not fit this definition. This is rather an information and organizational process, which tends not to the law, but to the economy of the post-industrial period. That is why it is called the fourth Industrial Revolution [4].

5. Special attention should be paid to the analysis of the directions of digitalization intrusion into the civil legal space. In our previous publications within the framework of the RFBR project, we drew attention to a number of them. Here and further on, we will focus on other areas of such an invasion. This is primarily the very scope of civil law, that is, the living environment in which social relations arise, in which civil law is called upon to regulate through its norms. Scientists call this invasion "digital formalization of law", believing that in this case digitalization acquires the qualities of a double of law and the quality of a source of law. In our opinion, this point of view is very doubtful. The source of law is a form of objectification of either an already existing or assumed form of human behavior (a rule of law) in connection with this and that. Digitalization does not fit into this design. Acquiring the quality of a double of law, it does not become an independent source of law, orienting people in their social environment.

6. In the countries of the European Union and in the West as a whole, the problem outlined in the title of this article has been discussed for several years. Attention is focused on the fact that the development of the Internet and other digital (or rather numerical) technologies has an impact, and not always positively, on the evolution of some state and public institutions and through their public administration and law in general, including the classical theory of law itself. And this applies not only to the law of the European Union but also to other legal systems.

The military terms "strike on the headquarters" can describe the efforts of those proponents and developers of the invasion of digitalization in law, who propose to introduce such new subjects of law as "robots" or "electronic persons", which in reality should replace a person, endowing virtual personalities with the right of ownership, endowing personal data with the quality of an object of civil law, which, like all objects of classical civil law, can be alienated, introduced into civil circulation, etc.

This idea is already partially included in the program "digital economy of the Russian Federation".

Finally, another direction that strikes at the classical concept of civil law is the transformation (in this case, the use of this term is quite logical) of the very concept of civil liability in law [5].

Who should be held responsible? To solve this problem, the European Parliament offers electronic persons (robots, in particular) an individual number, recorded in a special register accessible to users.

In cases where the robot makes a decision independently and at the same time the counterparty (the party to the transaction or even a third party) is harmed, the owner, the program developer, the insurance company, or even the user of the robot's services may be responsible for the robot's actions. Then where is the classical understanding of guilt or innocent responsibility? And according to some European scientists, private stories, and the very direction of such short stories can lead to the destruction of classical civil law, how wide is the line between mechanical construction and non-human, which is godlike and contradicts the very nature of man as a living being.

4.4 Conclusion

Summing up the above, it is permissible to draw the following conclusions:

1. Objectively taking place in all spheres of the social life of modern society, the process of digitalization (correct—numerization) of law, its features civil law which remains the citadel that keeps the world from lawlessness. This is the sacred power of the system of law as such. The kingdom of lawlessness is already in operation, "the Apostle Paul warned almost 2000 years ago," but it does not require that until the environment is so restraining, the law is able to restrain the objectively existing and necessary process of digitalization for the development of society from sliding into the abyss.
2. Modern civil law, its category, principles, and institutions are under enormous pressure under the onslaught of justified and often unfounded demands for their transformation, that is, essentially drastic changes.
3. A generalized analytical analysis of this threat to the law has not yet been conducted either in Russia or in the West, although many lawyers who study the process of "digitalization" feel alarmed about this.
4. For a fundamental change in the law, it is necessary to create a new categorical conceptual apparatus and, above all, the term digitalization itself, which is used everywhere and everywhere without proper understanding.

   Digitalization is not a legal category, since it does not meet the requirements of this concept. The very term "digitalization" is not a good one; it is more correct to call this "numerization" because not a digit, but a number is a measure of everything.

5. In relation to the process of digitalization in documents, scientific research, and even in civil law, the use of the term "transformation" is unjustifiably widespread; it is a mistake because we lead the science of civil law to a dead end.

   "Transformation" means the transition of a substance, energy, or structure to another state in which the former existence ceases to exist. "Digitalization" cannot and should not destroy rights by replacing them, i.e. transforming them into something else.
6. Scientists should pay special attention to the analysis of the directions of the digital transformation and the civil–legal space:

   - scope of law, the appearance of its digital counterparts that cannot be identified as new sources of law;
   - distortions of the concept of "subject of law", giving robots and electronic persons the qualities inherent in classical subjects of civil law;
   - giving virtual personalities ownership rights;
   - endowing personal data with the quality of civil law objects;
   - transformation of the very concept of civil liability into law.

7. All the "pro and contra" set out above convince the author that you will still not be able to crack civil law, as long as it is introduced into the daily life of people, and does not change the person himself.

Acknowledgements The research was carried out with the financial support of the Russian Foundation for Basic Research in the framework of scientific project No. 20-011-00450.

References

1. Chetvernikova, O.N.: Digital totalitarianism. How it is done in Russia, pp. 320. Book World, Moscow (2019)
2. Dudin, M.N., Frolova, E.E., Protopopova, O.V., Mamedov, A.A., Odintsov, S.V.: Study of innovative technologies in the energy industry: nontraditional and renewable energy sources. Entrep. Sustain. Issues **6**(4), 1704–1713 (2019)
3. Kuznetsov, M.N., Ulvi, O.: Information Communication Technologies and Legal Education Correlation in Russian Federation and European Union. In: Pixel International Conferences the Future of Education, pp. 333–336 (2020). https://conference.pixel-online.net/FOE/acceptedabstracts_scheda.php?id_abs=4757&id_edition=34&mat=ACA&wpage=ped
4. Matytsin, D.E.: Features of the legal infrastructure of the turnover of investment objects in the Russian Federation. In: Inshakova, A.O., Frolova, E.E. (eds.) Smart Technologies for the Digitisation of Industry: Entrepreneurial Environment. Springer Nature Singapore Pte Ltd. 2022 (2022). ISBN 978–981–16–4620–1. https://doi.org/10.1007/978-981-16-4621-8
5. Matytsin, D.E.: Internet investing as a remote algorithm of the retail investment financing. In: Popkova, E.G., Sergi, B.S. (eds.) Modern Global Economic System: Evolutional Development vs. Revolutionary Leap. ISC 2019. Lecture Notes in Networks and Systems, vol 198. Springer, Cham (2021). https://doi.org/10.1007/978-3-030-69415-9_204

6. Rusakova, E.P., Inshakova, A.O.: Industrial and manufacturing engineering in digital legal proceedings in the Asia-Pacific region: a new level of quality based on data, blockchain and ai. Int. J. Qual. Res. **15**(1), 273–289 (2021)
7. Rusakova, E.P., Inshakova, A.O., Frolova, E.E.: Legal regulation of internet courts in China. In: Popkova, E.G., Sergi, B.S. (eds.) Modern Global Economic System: Evolutional Development vs. Revolutionary Leap. ISC 2019. Lecture Notes in Networks and Systems, vol. 198, pp. 1515–1521. Springer, Cham (2021). https://doi.org/10.1007/978-3-030-69415-9_16

Chapter 5
On the Problem of Legal Regulation of Technologies with Elements of Artificial Intelligence

Aleksey V. Barkov, Yana S. Grishina, Aleksandr S. Kiselev, and Aleksey P. Sokolov

Abstract Every year, more and more elaborations capable of self-development appear. Machine learning uses not just a pre-filled algorithm, but also self-tuning of the electronic brain. In simple terms, the computer is training itself without human input. We have become so dependent on new tools that subconsciously we can no longer imagine our lives without the alliance with artificial intelligence. At the same time, the tools we have become accustomed to can cause irreparable harm. This is why it is important to define the status of artificial intelligence from the ligal point of view, as well as define the range of persons who can and should be held liable as a result of the incorrect and illegal use of AI.

5.1 Introduction

From an ideological point of view, the question of fixing the legal status of artificial intelligence did not arise by chance. As far back as in the twentieth century, science-fiction writer Isaac Asimov formulated "The Three Laws of Robotics" in his story "Roundabout". A few would have thought at the time that half a century later, scientific and technological advances would allow the application of these laws to become a reality. Asimov's laws are the principles by which any artificially created object, endowed with self-awareness, must act. Mankind has already got used to the fact that phenomena from science fiction (wireless communication, flights into space, submarines and many other things) are becoming an integral part of our life

A. S. Kiselev (✉)
Moscow State Regional University, Moscow, Russia

A. V. Barkov · A. S. Kiselev
Financial University Under the Government of the Russian Federation, Moscow, Russia

A. P. Sokolov
Military University of the Ministry of Defense of the Russian Federation, Moscow, Russia

Y. S. Grishina
Russian State University of Justice Moscow, Moscow, Russia

A. O. Inshakova and E. I. Inshakova (eds.), *New Technology for Inclusive and Sustainable Growth*, Smart Innovation, Systems and Technologies 288,
https://doi.org/10.1007/978-981-16-9808-8_5

and the rules for their treatment are subject to legal regulation with time. Now it is the time to define what Artificial Intelligence is from the ligal point of view.

5.2 Methodology

According to the purpose of this study, the most significant methods are general scientific methods (analysis and synthesis, dialectics and metaphysics, induction and deduction, description and modelling) as well as special scientific methods (comparative-legal, formal-logical) of research.

5.3 Results

5.3.1 On the Influence of Artificial Intelligence on the Spheres of Human Life

Legal regulation is necessary when objects of the world around us are inseparable from daily human activities and professional life [1]. To begin with pointing out in which spheres AI is applied today.

Artificial intelligence is already widely used in healthcare both in Russia and abroad. Google (DeepMind Health) and IBM (Watson Health) offer "smart" developments for patient assessment and preliminary diagnostics. Google's solution is already working in several hospitals, including the British Moorfields Eye Hospital, but Artificial Intelligence serves only as a supporting tool [2]. Google DeepMind Health analyses the available information about the patient's symptoms and provides a list of recommendations [3].

IBM Watson Health also helps diagnose: it recognises venous thrombosis, cardiomyopathy and heart attacks. The company's main mission is to support digital transformation in the medical field [4]. In a famous case, IBM Watson identified a rare form of leukaemia in a 60-year-old patient who was initially misdiagnosed. To do this, the system "studied" 20 million scientific articles on cancer in 10 min [2]. In addition, artificial intelligence is helping to improve the lives of people with incurable illnesses. For example, a wheelchair has been developed for those whose bodies are completely non-functional, which is controlled by the power of thought and gaze. The know-how was presented by Russian scientists from the Kurchatov Institute Research Center [5].

And the first Israeli company has developed a device called MyEye that allows visually impaired people to learn to recognise their surroundings. The device identifies other people and objects and transmits this information to the person using discrete audio. Such technologies have been used as an experiment by domestic researchers at Saratov State University to help students in the educational process [6].

Meanwhile, scientists are working on the safety of our planet's inhabitants and introducing artificial intelligence into the road environment. For example, the company Pony.ai has created the PonyAlpha self-driving car system; the system is a network of sensors that continuously reads information to assess objects in the environment and their distance, providing key information necessary for safe driving [7]. In addition, artificial intelligence is being introduced into human office life [8]. Today, such old-timers as Automation Anywhere and UiPath are developing technologies in this direction.

5.3.2 On the Application of Artificial Intelligence in the Humanities Professions

On February, 1st 2019, the Australian version of The Guardian published its first article written by a robot on its website. E. N. Seitzhelilova points out "About a billion news stories were generated by artificial intelligence in the media. Many foreign media outlets have robot journalists working alongside their regular employees to relieve people of their routine and allow them to do more skilled work without being distracted by monotonous tasks". The motto is "To help, not to leave unemployed!" [9].

The robot writer has also been actively worked on by the non-profit research organisation OpenAI and has created an artificial intelligence, GPT2, capable of writing fake news almost perfectly. The AI is based on the use of pre-trained deep neural language models which have improved the artificial intelligence's overall understanding of human speech construction [10]. For this purpose, a neural network was loaded with about 100 million English articles and news with a volume of data of 40 Gb. The system processes the information and writes articles based on it. In doing so, it is able to recognise stylistics and avoid the traditional mistakes of generators—loss of focus and the essence of the material.

5.3.3 Experience of the Application and Legal Regulation of Artificial Intelligence

In China, a Zero Trust pilot project was able to expose more than 9,000 officials for corruption. Artificial intelligence detected irregularities such as illegal transfer of ownership, manipulation of assets and inadequate infrastructure. However, the intelligence was only able to provide the final results, and it did not show the whole chain of data, i.e. it did not have an evidence base in fact [5].

The Russian President's Decree No. 490 of October, 10th 2019 "On development of artificial intelligence in the Russian Federation" contains the following definition:

"Artificial intelligence is a complex of technological solutions that allows imitation of human cognitive functions including self-learning decision-making without a predetermined algorithm and obtaining results at least comparable with the results of human intellectual activity when performing specific tasks" [11]. A few months earlier, on February, 11th 2019, US President D. Trump signed an executive order called the American AI Initiative, which aims to set US AI priorities for the future. Prior to this initiative, the US lacked a comprehensive national plan for the development of artificial intelligence. The order follows a number of government initiatives undertaken by various countries over the past few years, including Singapore, Canada, the European Union and China. The initiative listed the main areas of action as (1) investing in AI research; releasing AI resources; setting standards for AI management; creating an AI workforce; and international cooperation in AI [12]. The decree can broadly be divided into two wide groups. The first group includes specific planning and funding for AI development programmers, while the second group consists of guidance documents that outline the general direction in which AI should be developed.

The US Decree does not contain a definition of AI and this is an important distinguishing feature compared to the domestic regulation. To date, both Russian and foreign doctrine understand the term "Artificial Intelligence" ambiguously and there is no consensus among experts on which technologies should be considered as AI and which should not. The Russian Presidential Decree sets the ideology for the development of Artificial Intelligence in the country. The American President's Decree only defines the direction of its development.

5.3.4 Discussion About the Status of Artificial Intelligence in Civil Relations: Subject or Object of Relations?

Many people today are concerned with the question: can artificial intelligence integrate into society in such a way that it can perform legally significant actions generating certain legal consequences? In fact, this would represent an unprecedented evolution from a machine that appeared as an object of social relations to an autonomous object with legal capacity and tortuousness.

In practice, machine learning has the following difficulties: in order for a computing system to find a specific solution, it is necessary for a human to specify a certain set of data; in fact, AI does not think for a human, but helps to perform auxiliary tasks within clearly defined boundaries [13]. In this connection, we will partly agree with the position of V. P. Kamyshansky and A. V. Koretsky that "legal legitimization of an AI subject as a subject of law may take years and even decades as any integration of new subjects into society will require reconstruction of consciousness and formation of new communications in society during several generations" [14]. On the other hand, humans are already enabling AI to make decisions on which people's lives depend; supercomputers capable of surpassing human thinking are being developed [8].

Accordingly, today it is necessary to determine whether AI can be equal or even superior to humans in terms of thinking fictions, have rights and bear responsibilities. An AI cannot now be recognised as a natural person because, firstly, it is a product of human intellectual activity and, secondly, it is not endowed with morality, which fundamentally affects its understanding and direction of its actions. Accordingly, the subjective side of legal relations involving AI is devoid of a primary basis. However, this did not stop Hasselt (Belgium) from officially including the robot Fran Pepper in the population register in 2017. The robot Fran could go to school and then continue his studies at a local university [15]. The relevant document was signed by the mayor of the city. It is noted that the robot is able to communicate with people, understand the emotions of the interlocutor and has a unique self-learning system. It is the first recorded case in the world when an AI is recognised as an independent entity of social relations. The inclusion of a robot as a citizen implies constitutional rights and obligations. In our opinion, it is obvious that such measures were taken only for the purpose of a successful PR-campaign to improve the city's attractiveness, and there was no intention to give the robot a real constitutional legal status.

M. N. Petrenko classifies AI according to "what the simulation of cognitive processes is based on:

- based on rigid algorithms in AI, which cannot be changed and independently improved by the AI itself, but only used for processing various information, including that unknown at the time of AI creation, about the surrounding reality;
- on human-assembled algorithms which can be changed and independently improved by AI itself" [16].

Obviously, in the first case, all the responsibility for the unlawful acts committed by means of AI must be borne by the user and/or right holder, as algorithms cannot "think" without external influence. In the second case, both the definition of a liability and the question of the legal capacity of AI are debatable. If in the near future an AI will be created whose capacity limits cannot be established, it will not be quite correct to attribute all responsibility for wrongful actions of the AI to its creator.

In the same case, self-regulation cannot be resorted to because it is not known whether AI will obey laws and morality. For example, as in the scenario of the 2004 film I, Robot, the program may deem human actions counterproductive and turn all Internet and technological resources against humanity. Accordingly, responsibility for such actions should lie with those individuals who knew or should have known of the hypothetical possibilities of AI getting out of control and allowing AI to be used in practice without imposing restrictions on it. Researcher O. V. Makarov believes that "AI cannot be a subject of civil law relations" [17].

F. A. Novikov is quite categorical in his position: "A program using artificial intelligence uses some algorithms peculiar to human thinking, but in real life a person will solve this problem differently, engaging including mental structures, which have not yet been studied enough and therefore cannot be translated into mathematical formulas and program code" [18]. It is difficult to argue with Y. S. Kharitonova's position that "Artificial intelligence systems, most of which apply statistical-based learning methods to find patterns in large data sets and make predictions based on these patterns, are used in a variety of applications and software products" [19]. The author defines AI as a tool, a set of technical tools that help people in various spheres of life. Other scientists—Gurko [20]—admit the possibility of giving AI a special competence.

Researchers V. P. Kamyshinsky, A. V. Koretsky propose "three types of legal capacity:

- special capacity—having a significant neural network, ability to create and create principally new objects of law;
- limited capacity—capable of autonomously performing clearly defined main task within a single competence (android-assistant, android-nurse, android-surgeon, android-operator);
- lack of competence (incompetence)—AI carriers that have no human-like embodiment (graphic editors, aggregators and other programs)" [14].

This gradation, in our opinion, is quite fair, taking into account the functional features of the entire array of existing technologies, which can be understood as artificial intelligence.

5.4 Conclusion

It is possible that the emergence of an AI that surpasses human consciousness and is equal to it will require the legal status of an electronic entity. Although such an AI will de facto be considered a man-made thing, i.e. an object, it will be indistinguishable from humans by its cognitive functions and worldview. This issue should be studied from the point of view of morality and ethics; public opinion should be sought, and only then should direct legal regulation be applied. However, attempts to confer a special status on AI at this point in time do not seem advisable.

At the same time, civil liability for the damage caused by technologies with elements of artificial intelligence should already be introduced today by analogy with the liability for the damage caused by a source of increased danger. The most effective and optimal mechanism for holding the creators of technologies liable by means of compensation for harm from their stabilisation funds, which could operate by analogy with the mechanism of property liability of self-regulatory organisations of which the developers of artificial intelligence could be members. In turn, owners would be required to insure liability for harm caused by artificial intelligence technologies. At the same time, the producers and users of AI systems will be held administratively, criminally and financially liable, and the risks of using neural networks in their activities will be specified in user agreements.

Acknowledgements This research was supported by the Russian Scientific Foundation within the framework of the grant "Mission-oriented corporate governance as a factor in ensuring high standards of social and environmental well-being".

References

1. Inshakova, A.O.: Information technologies for the formation of the European space of freedom, security and justice. In: Inshakova, A.O., Inshakova, E.I., Deryugina, T.V. (eds.) Advances in Intelligent Systems and Computing, vol. 1100. AISC, pp. 444–453 (2020). https://doi.org/10.1007/978-3-030-39319-9_51
2. Kahn, S.V., Titova, J.I., Sergazykyzy, A.: Artificial intelligence in medicine. Actual Issues Sci. **78**, 34–38 (2021)
3. DeepMind's health team joins Google Health. https://deepmind.com/blog/announcements/deepmind-health-joins-google-health (2019). Accessed 17 Oct. 2021
4. IBM Watson Health. https://www.ibm.com/uk-en/watson-health. Accessed 17 Oct. 2021
5. Hubr site: China will abandon Zero Trust anti-corruption AI system because of its high efficiency. https://habr.com/ru/post/439070/ (2019). Accessed 18 Oct. 2021
6. Venig, S.B., Solovyova, V.A.: Eye-tracking: regularities of educational information searching. Int. Annu. Ed. Appl. Psychol. Theory Res. Pract., 3 (2016)
7. Pony.ai. https://www.pony.ai/tech?lang=en. Accessed 17 Oct. 2021
8. Matytsin, D.E.: Neoindustrial tools for the circulation of uncertified securities: digital technologies for the implementation and protection of the rights of investors and issuers. Legal Concept = Legal paradigm, vol. 19, 73–83 (2020)
9. Seytzhelilova, E.N.: Artificial intelligence and the future of journalism: forecasts of media experts. In: Intercultural Communications. Collection of Scientific Works of the International Scientific Conference, pp. 67–72 (2019)
10. Das, A., Verma, R.M.: Can machines tell stories? A comparative study of deep neural language models and metrics. IEEE Access **8**, 181258–181292 (2020)
11. Presidential Decree 490: On Development of Artificial Intelligence in the Russian Federation. Collected Legislation of the Russian Federation (2019)
12. The American AI Initiative: What Could Implementation Look Like? https://www.project-disco.org/innovation/030619-the-american-ai-initiative-what-could-implementation-look-like/. Accessed 18 Oct. 2021
13. Grishina, A.V.: Application of artificial intelligence in economics and finance and development prospects. Econ. Manag. Innov. Technol. **12**(99), 18 (2019)
14. Kamyshansky, V.P., Koretsky, A.V.: Concept and legal status of the carrier of artificial intelligence. Law Power **1**(37), 42–50 (2019)
15. Robot obtained citizenship. https://cumgeek.com/articles/v-belgii-robot-poluchil-grazhdanstvo-i-ofitsialnyj-status-usynovlyonnogo/ (2017). Accessed 18 Oct. 2021
16. Petrenko, M.N.: Artificial intelligence: about the problems of legal status. Alley Sci. **1.1**(17), 491–494 (2018)
17. Makarov, O.B.: Subjects of civil law: the present and the future. Arbitr. Civ. Process **3**, 39–43 (2017)
18. Novikov, F.A. (ed.): Symbolic Artificial Intelligence: Mathematical Foundations of Knowledge Representation. Yurite, Moscow (2018)
19. Kharitonova, Y.S.: Legal regulation of artificial intelligence technology application in military affairs: experience of Russia and China. J. Appl. Res., 1 (2021)
20. Gurko, A.: Artificial Intelligence and Copyright: A Look into the Future. Intellectual Property. Copyright and Allied Rights, vol. 12, pp. 7–18 (2017)

Chapter 6
Criminal Procedure Aspects of Fight Against Economic Extremism Among Young People

Valentina A. Lazareva, Pavel P. Fantrov, and Vladimir M. Shinkaruk

Abstract The purpose of this study is to analyze the possibilities of criminal procedure activities as a factor of influence on the stability of development and security of society and the economy through the identification, suppression, and investigation of economic crimes, including those with an extremist orientation. In this regard, the question of the correlation of the goals and objectives of criminal procedure activity, as provided for in Article 2 of the Criminal Procedure Code of the Russian Soviet Federative Socialist Republic (CPC of the RSFSR) of 1960, with the purpose of criminal proceedings in the interpretation of the Code of Criminal Procedure of the Russian Federation (Article 6), which has been discussed for a long time, is very controversial and ambiguously solved in the theory of criminal procedure. To achieve this goal, the authors used the method of systematic research of social, economic, and legal phenomena, which allowed us to characterize the criminal procedure as a developing phenomenon and one of the factors of effective influence on the control and prevention of economic extremism among young people. The article shows that criminal procedure activity has an impact on all the processes taking place in society, and therefore its regulatory and legal organization and the implementation quality can and should be considered as factors of the effectiveness of the fight against economic extremism among young people. The study sets the directions, ways, and options for the development of the theory, and practice of criminal proceedings formulates practice-oriented recommendations. Originality/value of research is determined both by the subject of the study, the formulated and justified conclusions, and the author's

V. A. Lazareva
Department of Criminal Procedure and Criminalistics, Samara National Research University Named After Academician S. P. Korolev, Akademika Pavlova Str., 1, 443011 Samara, Russian Federation

P. P. Fantrov (✉) · V. M. Shinkaruk
Department of Criminal Procedure and Criminalistics, Volgograd State University, Universitetsky Prospect, 100, 400062 Volgograd, Russian Federation
e-mail: pavelfantrov@volsu.ru

V. M. Shinkaruk
e-mail: shinkaruk@volsu.ru

A. O. Inshakova and E. I. Inshakova (eds.), *New Technology for Inclusive and Sustainable Growth*, Smart Innovation, Systems and Technologies 288,
https://doi.org/10.1007/978-981-16-9808-8_6

interpretation of the results of previous research in the field of criminal procedure, criminalistics, the theory of criminal intelligence operations, and forensic inquiry.

6.1 Introduction

Extremism as a social phenomenon among young people, which destabilizes the social, political, and economic situation in the country, poses a serious threat to its unity and territorial integrity, and therefore to the interests of society, the rights, and freedoms of citizens. Extremist manifestations among young people in the context of the development of the modern Russian economy cause great economic damage to the Russian state, destabilizing the community and hurting inclusive growth [8]. The growing trends toward the growth of extremist sentiments, their penetration into the youth environment, the active use by members of extremist organizations of new information technologies, including social networks [5], for the dissemination of extremist materials, attracting new members to their ranks, and coordinating illegal activities, noted in the Concept of Public Security in the Russian Federation, approved by the President of the Russian Federation on November 20, 2013, require the most serious attention and involvement of various methods of identifying, suppressing, and preventing negative manifestations. The legal mechanism formation for prevention of extremist crimes, including economic crimes, is becoming an important and relevant issue of the state's criminal policy, as evidenced by the intensification of scientific, including dissertation research in this area, devoted mainly to the criminal-legal and criminological characteristics of extremist crimes [9, 17] or the identification of traces of extremist activity, private methods of investigating extremist crimes, problems of interaction between law enforcement agencies [23]. The impact issues of criminal procedure activities on the suppression and prevention of economic extremism and terrorism have not yet been sufficiently developed, although some problems have been covered in dissertations [18, 19]. Meanwhile, research in this direction is very relevant, since penal prohibition, as is known, is implemented through criminal procedural activities, and forensic recommendations for the identification and investigation of economic and extremist crimes must comply with the requirements of the criminal procedure law.

Based on this, the authors set out to show the impact of criminal procedure activities on the stability of development and security of society and the economy by identifying and suppressing extremist crimes among young people, to consider, in this regard, the debatable issue of the goals and objectives of criminal procedure activities and the purpose of criminal proceedings, to identify and analyze, if available, the procedural features of the proceedings in this category of criminal cases.

6.2 Methodology

To achieve this goal, the authors used general scientific dialectical approaches of cognition of social and legal phenomena and processes, current and previously existing criminal procedure legislation, and law enforcement practice. The applied methodology of systematic research of social and legal phenomena allowed us to characterize the criminal procedure as a developing system and a social institution that is interconnected with other social institutions, and, consequently, as a factor of effective influence on the control and prevention of economic extremism among young people. The use of formal-logical and comparative-legal private-scientific methods of cognition made it possible to comprehensively study the research subject, formulate proposals for countering economic extremism among young people by investigating and considering cases of crimes of this orientation. The research is based on the modern provisions of the theory of criminal law, criminal procedure, criminalistics, criminal intelligence, and expert activity.

6.3 Results

The increase in the number of crimes committed on the grounds of destabilizing the social and economic foundations of society makes it necessary to use effective means not only to prevent extremist sentiments among young people but also to actively counteract them by suppressing criminal actions and bringing those responsible for their commitment to criminal liability. Great importance is attached to the legal qualification of actions for the commission of which the state establishes such responsibility and work organization of the criminal justice system, without which cannot be realized criminal liability.

Crimes of extremist orientation among young people, as is known, are heterogeneous. Among them, there are justifiably distinguished, first, such special elements of crimes as public calls for the implementation of extremist activities; for the destruction of the territorial integrity of the state, the organization of an extremist community, the organization of the activities of an extremist organization, the financing of extremist activities (art. 280, 280.1, 282, 282.1, 282.2, 282.3 Criminal Code of the Russian Federation); secondly, other crimes provided for by the Criminal Code of the Russian Federation committed for reasons inherent in extremist activities; thirdly, terrorist activities that represent an extreme form of extremism [20]. The criminalistic characteristics of each group of crimes, including economic and extremist ones, undoubtedly affect the forms and methods of their detection and disclosure [21], tactics of some investigative measures. However, the initiation of a criminal case, investigation, and court proceedings of economic crimes, having an extremist connotation, is carried out according to the general rules established by the Criminal Procedure Code of the Russian Federation, obeys its principles, is provided with appropriate guarantees, and is accompanied by the application of various procedural measures, that is, state coercion [15]. The functioning of the criminal justice system is designed to ensure the achievement of an outcome that meets the needs of society [12].

This allows us to consider criminal procedure activities as a factor of influence on the fight against extremist manifestations among young people and the prevention of economic crimes of extremist orientation [22]. At the same time, the effectiveness of the impact of criminal procedure activities on the level of extremism in society is determined not only by the effectiveness of the activity itself to initiate and investigate extremist crimes in the economic sphere but also by the effectiveness of criminal proceedings in general [11]. Therefore, any problems of criminal proceedings caused by imperfection of legal regulation or shortcomings of law enforcement practice affect the efficiency of proceedings in cases of crimes, including extremist ones, which have their own specific problems.

It may seem that the formulated thesis about criminal procedure activity as a factor in the fight against extremist and economic crime is in a certain contradiction with the widely spread in modern times concept of criminal procedure (criminal proceedings) as a way to protect the rights and freedoms of the individual, which was also justified by one of the authors [13]. Among the issues that are important for determining the essence of the criminal procedure, we can mention the protracted, and according to some estimates, fruitless, discussion about the relationship between the concepts of purpose, purpose, and objectives of criminal proceedings [2]. The theoretical nature of the problems chosen for the study does not detract from their importance for solving practical problems, including legal regulation of criminal procedure activity.

According to I. B. Mikhaylovskaya, fixing the defense of the rights and legitimate interests of an individual as the purpose of criminal proceedings, including against unjustified charges, convictions, restrictions on their rights and freedoms, and changing the system of principles of criminal proceedings, "deprives the normative basis of the thesis about the fight against crime as the goal of procedural regulation of the functioning of criminal justice" [14].

The view on the purpose of the criminal procedure, which was presented at the turn of the centuries and millennia and was widely spread later, allowed us to assert that a new, protective type of process was being formed in Russia [7].

It should also be noted that the difference in theoretical approaches to each of these concepts-task, purpose, including in application to criminal proceedings [4, 6], is largely predetermined, both by the lack of a clear distinction in theory between goals and objectives, and by the ambiguity of the concept of the criminal procedure itself, which denotes both criminal procedure law, and the proceedings regulated by it in a specific criminal case of extremist orientation, and the functioning of the entire criminal justice system. In this regard, bringing the perpetrators to criminal responsibility, assigning them a fair punishment, and compensating for damages are called by some authors the goals of the criminal procedure [16], while others are called the tasks [1, 3, 24]. It is easy to find a justification for each point of view, if desired, which is not necessary for the terms of our research. Therefore, as a compromise, we can assume the possibility of considering the tasks solved by the participants in the criminal procedure as their immediate goals. Being in close interaction, the goals and objectives of the criminal procedure do not completely coincide when we talk about the proceedings on a specific case and the criminal procedure as a general concept.

In particular, Article 2 of the Code of Criminal Procedure of the RSFSR quite precisely defined the tasks of the proceedings in a particular criminal case—to solve the crime, establish the guilt of the offenders; this task should be solved today by the bodies carrying out criminal prosecution, the purpose of which is paragraph 55 of Article 5 of the Code of Criminal Procedure of the Russian Federation calls the establish the guilt of the offender, and therefore the disclosure of a crime, including extremist and economic orientation. The solution of these tasks ensures the achievement of the goal outlined in Part 2 of Article 6 of the Criminal Procedure Code of the Russian Federation—the conviction and punishment of the guilty and the prevention of the conviction of the innocent. Dualism in determining the purpose of the criminal procedure already indicates the limited means and methods of solving the problems of solving crimes of extremist and economic orientation and establishing the guilt of the offender, as evidenced by the fact of legal regulation of criminal proceedings, although in Part 1 of Article 6 of the Criminal Procedure Code of the Russian Federation, the nature of this restriction is more clearly expressed. In other words, the legal regulation of criminal proceedings on crimes of economic and extremist orientation is aimed not only at solving the crime and thereby protecting the rights and legitimate interests of persons and organizations who have suffered from the crime, but also at preventing the restriction of the rights and interests of the person accused of the crime, his unjustified conviction. Only if these requirements are met, criminal proceedings can contribute to strengthening the rule of law, preventing and eliminating crimes of economic and extremist orientation, educating citizens in the spirit of unswerving enforcement of laws and respect for the rules of human society, that is, to meet the public demand or, to put it another way, to meet the social purpose of criminal proceedings as the most important type of state activity.

6.4 Conclusion

Answering the question from the above positions, whether it is possible to consider the criminal procedure, the purpose of which is to protect the rights and freedoms of legal entities and individuals, as a factor of influence on the crime of extremist and economic orientation, we state:

As social processes, crime and criminal procedure activities are, of course, interrelated and mutually dependent. The criminal procedure activity ensures the realization by the state of the right to punish the person who committed the crime, contributes to the solution of the tasks of general and special prevention, which, although it does not give grounds to consider the criminal procedure as a means of combating extremist and economic crime, cannot be ignored as a factor of influence on it. Hence, the conclusion is that effective activities for the solution of crimes of economic and extremist orientation, exposing the perpetrators, ensuring just punishment of the guilty, and protecting the interests of the victims, have a positive effect on the state and dynamics of crimes, including extremist orientation, and harmonizes public relations. Accordingly, inefficient activities that do not ensure the implementation of the purpose of criminal proceedings hurt public processes and contribute to the growth of anti-social, illegal, including extremist manifestations. The search for legal means and ways to improve the effectiveness of criminal procedure activity remains an important area of scientific research.

Acknowledgements The reported study was funded by RFBR, project number 20-011-00688 (A) "Model for the prevention of extremism among the youth: integration of the activities of civic associations and law enforcement agencies in the context of legal socialization".

References

1. Alekseev, A.I., Ovchinsky, V.S., Pobegailo, E.F.: Russian Criminal Policy: Overcoming the Crisis. Norma, Moscow (2006)
2. Alexandrov, A.S., Alexandrova, I.A., Kruglov, I.V.: Appointment of Criminal Proceedings and Punishment. Nizhny Novgorod Law Academy, N. Novgorod (2006)
3. Azarov, V.A., Revenko, N.I., Kuzembaeva, M.M. (2006). Functions of Preliminary Investigation in the History, Theory and Practice of the Criminal Procedure in Russia. Publishing House of OmSU, Omsk
4. Barabash, A.S.: Public Initiation of Russian Criminal Proceedings. Legal Center-Press, Saint Petersburg (2009)
5. Bokov, Y.A.: Smart technologies in electoral fundraising. In: Book Series: "Lecture Notes in Networks and Systems", vol. 155, pp. 1167–1175. Springer (2021). https://doi.org/10.1007/978-3-030-59126-7_127
6. Davletov, A.A., Azarenok, N.V.: Criminal Justice Program. M.: Jurlitinform (2009)
7. Fantrov, P.P., Shinkaruk, V.M.: Smart technologies and prevention of extremism among young people. In: Book Series: "Lecture Notes in Networks and Systems", vol. 155, pp. 665–673. Springer (2021). https://doi.org/10.1007/978-3-030-59126-7_74
8. Fantrov, P.P., Bobovkin, M.V., Medentsov, S.A.: Improving the activities of the prosecutor's office for preventing crimes of extremism among minors. J. Legal Concept **20**(2), 74–81 (2021). https://doi.org/10.15688/lc.jvolsu.2021.2.10
9. Fridinsky, S.N.: Criminal law and criminological characteristics of extremist crimes. Ph.D. in Law thesis. Rostov-on-Don (2003)
10. Inshakova, A.O.: Extremism as a threat to national security: yesterday, today, and tomorrow. J. Legal Concept **19**(2), 6–12 (2021). https://doi.org/10.15688/lc.jvolsu.2020.2.1
11. Inshakova, A.O.: Coordination of the activities of law enforcement agencies in the fight against crime on guard of the national security of the Russian Federation. Legal Concept **20**(2), 6–11 (2021). https://doi.org/10.15688/lc.jvolsu.2021.2.1
12. Kozyavin, A.A.: Social Purpose and Functions of Criminal Proceedings. Yurlitinform, Moscow (2010)
13. Lazareva, V.A.: Theory and Practice of Judicial Protection in Criminal Proceedings. Samara University, Samara (2000)
14. Mikhailovskaya, I.B.: Purposes, Functions and Principles of Criminal Proceedings (Criminal Procedure Form). Prospect, Moscow (2003)
15. Mikhailovskaya, I.B.: The social purpose of criminal justice and the purpose of the criminal procedure. J. State Law **5**, 111–118 (2005)
16. Orlov, Yu.K.: Problems of the Theory of Evidence in Criminal Proceedings. Yurist, Moscow (2009)

17. Pavlinov, V.I.: Criminalistic anti-state extremism: criminal law and criminological aspects. Doctor of Law thesis Moscow (2008)
18. Pogodin, I.V.: Extremist Crimes: Methods of Evidence. Yurlitinform, Moscow (2012)
19. Sergeeva, A.V.: Features of pre-trial proceedings in criminal cases of extremist crimes. Ph.D. in Law thesis. Moscow (2012)
20. Shinkaruk, V.M., Biryukov, S.Y., Rezvan, A.P.: The interaction of the investigator with the body of inquiry during the investigation of crimes in criminal cases of extremism: the problems and ways of their solving. J. Legal Concept **20**(2), 65–73 (2021). https://doi.org/10.15688/lc.jvolsu.2021.2.9
21. Shinkaruk, V.M., Fantrov, P.P., Solovyova, N.A.: The role of network technologies in preventing youth extremism in the context of integration of government and civil society's activities. Materials of Conference ISC 2020 "Smart Technologies" for Society, State and Economy". In Book Series: "Lecture Notes in Networks and Systems", vol. 155, pp. 507–513. Springer (2014). https://doi.org/10.1007/978-3-030-59126-7_57
22. Shinkaruk, V.M., Frantsiforov, Yu.V.: Investigation of extremist crimes. J. Legal Paradigm **2**, 57–63 (2020)
23. Skorikov, D.G.: Investigation of Extremist Crimes. Diss ... Candidate of Legal Sciences. Volgograd (2014)
24. Zinatullin, T.Z.: Hierarchy of functions of the Russian criminal process. J. Rus. Judge **4**, 46–47 (2001)

Chapter 7
Consumers' Right to Information in the Remote Sale of Goods as a Factor of Inclusive Growth

Andrey N. Sadkov and Nikolay V. Kotelnikov

Abstract The authors claim that remote sale reduces the costs of consumers for the purchase of socially significant goods and increases the availability of such goods for all groups of the population, regardless of the location of the consumer. This is interpreted by the authors as one of the factors of inclusive growth. It is proposed to consider the right of consumers to receive reliable and complete information about the product as a subjective right that provides everyone with equal opportunities to meet household needs and improve well-being. The key role of the consumers' right to information in the system of measures is emphasized that ensure satisfaction of their legitimate interests in the remote sale of goods. General proposals are formulated to improve the Russian legislation regulating relations in the field of consumer protection.

7.1 Introduction

The need to ensure the effective comprehensive long-term development of society leads to the search for new criteria that could become guidelines that allow choosing the right direction for optimal achievement of the intended goals. Expressing concern about the greatly increasing stratification of society, economists and politicians of developed countries, based on the results of social research, concluded that such an indicator of the development of the national economy as the share of gross domestic product per capita does not reflect the real state of the level of development of society. This is largely due to the fact that when calculating gross domestic product, both indicators of private production and the public sector of the economy are taken into account. However, high profitability in the public sector does not always indicate

A. N. Sadkov (✉)
Volgograd State University, Volgograd, Russia
e-mail: sadkov@volsu.ru

N. V. Kotelnikov
Volgograd Academy of the Ministry of Internal Affairs of Russia, Volgograd, Russia
e-mail: nivako26@yandex.ru

A. O. Inshakova and E. I. Inshakova (eds.), *New Technology for Inclusive and Sustainable Growth*, Smart Innovation, Systems and Technologies 288,
https://doi.org/10.1007/978-981-16-9808-8_7

a high median income of the population. It often happens that the volume of the national gross domestic product is growing, and the population really lives worse. In addition, there are enough examples in the modern world of how profits (especially from extractive industries) go abroad, and with all the wealth of the country's natural resources, people live very poorly. The lack of objective information reflecting the level of complex development of society cannot ensure the adoption of effective decisions that contribute to the sustainable development of mankind in the long term. Back in 2009, the commission, headed by the Nobel Laureate in Economics, the authoritative American economist Joseph Stiglitz, responsibly stated that it was time to shift the emphasis to measuring the level of public welfare from measurements of economic production (primarily by estimating gross domestic product (GDP)) to measure the well-being of people [11, P. 55].

During the World Economic Forum in Davos in 2017, a proposal was made on the need to use such an indicator as the Inclusive Development Index (IDI) as an alternative indicator of gross domestic product [1]. According to modern researchers, the Inclusive Development Index should be considered as a modern analogue of GDP, devoid of many disadvantages of the latter [9, P. 79].

The inclusive development index is directly related to the category of "inclusive growth", which, according to R. Hasmath, is characterized by a focus on improving the well-being of the population, reducing inequality due to the uniform distribution of wealth and benefits among the population, characteristic of the social economy [2].

Undoubtedly, this approach deserves the closest attention. It seems that the slogan "the economy is for a person, not a person for the economy" [3] is more relevant than ever before at the present stage of society's development.

Undoubtedly, the degree of satisfaction of the median consumer interests should be recognized as one of the indicators of sustainable inclusive development. This is primarily determined by the availability of socially significant goods for him. The remote sale of such goods reduces the costs for both producers of goods and consumers of these goods. The popularity of remote purchase and sale of goods is evidenced by the steady growth in the volume of online commerce [4]. In addition, distance trading makes it possible to ensure the availability of a particular product for all groups of the population, regardless of the location of the consumer, which undoubtedly should be considered as one of the factors of inclusive development and the Internet of Things [5].

The problem of proper and reliable informing the consumer about the quality and properties of the goods during its remote purchase and sale is more acute than ever before. It seems that each of us has come across either personally, or such cases have taken place in his immediate environment, when the goods purchased in the order of online trade did not meet the buyer's idea of its quality and properties. Often, It does not often only generate a feeling of deep dissatisfaction, but is also associated with additional property and temporary losses for the consumer.

7.2 Methodology

The research work is based on the general scientific method of materialistic dialectics. In addition, such methods of scientific cognition as analysis, synthesis, deduction, induction, method of system research, method of structuring material, formal legal method, and comparative legal method were used.

7.3 Results

In the modern legal environment, effective protection of consumer rights is considered as one of the most important factors in creating a competitive, transparent, and fair market environment. This fully applies to the problems of distance trading, which uses modern digital technologies that ensure the interaction of the seller and the buyer. That is why, as one of the fundamental goals of the implementation of the Strategy of the State Policy of the Russian Federation in the field of consumer protection for the period up to 2030 (approved by the decree of the Government of the Russian Federation dated August 28, 2017, No. 1837-r [6]) is to improve the level and quality of life of the population of the Russian Federation. As can be seen, the Government of the Russian Federation sees a direct link between a high degree of consumer protection and an increase in the standard of living of the population.

Forming a list of basic consumer rights, the Russian legislator emphasizes the consumer's right to receive objective information concerning the purchased goods, as well as about those persons who created such goods, about the seller, and about the performer of the consumed works and services. The norms of paragraph 1 of Article 8 of the Law of the Russian Federation dated July 2, 1992, No. 2300–1 "On Consumer Rights' Protection" [7] (hereinafter—the Law on Consumer Rights Protection) determine that the consumer has the right to demand the provision of necessary and reliable information about the manufacturer and seller of the goods, the person performing the work and the person performing the services, the working hours of these persons, and the qualities of the goods (works, services) sold by these persons.

The importance of the right to information in the legal status of the consumer is due to a number of reasons.

Firstly, this right is one of the conditions for consumer protection, since it allows the consumer to orient himself about whom to make claims against and whom to demand compensation from for damage caused as a result of the sale (delivery) of substandard and unsafe goods (work, services).

Secondly, guided by reliable and complete information about the properties of the product, the consumer can thoughtfully and consciously choose a product (work, service) that has the qualities he needs. The proper fulfillment of the obligation to provide information by the seller and the manufacturer of the goods to the consumer is designed to ensure that the consumer can competently choose the goods, works, or services he needs. In the legal literature, the increased attention paid to the consumer's right to information about goods (works, services) is conditioned by the presumption that the consumer does not have special knowledge about the properties and characteristics of goods (works, services), which is highlighted in the norms of paragraph 4 of Article 12 of the Law on Consumer Protection [8].

Making the right choice and safe use of many modern goods requires the consumer to have the appropriate knowledge, which an ordinary citizen does not possess as a rule. As an example of such goods, technically complex goods, computer equipment and software, household chemicals, and other similar goods can be cited. A similar statement is true for some works and services, for example, banking and insurance services, and paid medical services. Citizens who do not have the necessary and reliable information about the goods, works, or services consumed often attract the attention of fraudsters selling expensive goods that either do not meet the stated characteristics, or such goods are not really needed by the consumer (first of all, this applies to medical products, biologically active food additives, hygiene products, and cosmetics).

Of particular importance for the proper choice by the consumer of those goods that he really needs, the right to information has a remote interaction between the seller and the buyer. The norms of Article 26.1 of the Law on Consumer Protection recognize the possibility of concluding a retail sale agreement in the absence of direct familiarization of the consumer with the product (or a sample of such a product) at the conclusion of the contract. Interacting with the seller in the framework of distance trading, the consumer chooses the product based on familiarization with the description of the product offered by the seller. The description of the product, its image, and photos are most often placed in catalogs, brochures, booklets, on the websites of sellers or manufacturers. It is obvious that the consumer in this situation largely depends on the reliability and completeness of the information provided by the seller.

Thirdly, the consumer's right to receive reliable information about the product, its manufacturer, and seller is also important because it ensures the proper exercise of another consumer right—the right to the safety of goods, works, and services. It seems that this is why the provisions formulated in paragraph 3 of Article 12 of the Law on Consumer Rights Protection grant the consumer the right to demand full compensation for the damage caused to his life, health, and property, if the infliction of such damage is causally related to the failure of the seller and the manufacturer of the goods to fulfill the obligation to provide the consumer with complete and reliable information about the goods being sold or work or service.

The construction of "informational" legal relations arising between entrepreneurs and consumers in remote trading assumes that the consumer's rights to receive complete and reliable information about the product correspond to the obligation of the seller of the goods (manufacturer, contractor) to provide this information. In this regard, the Law on Consumer Protection establishes a number of requirements that determine the specifics of the performance of such a duty.

As already noted, in accordance with the rules of Article 8 of the Law on Consumer Protection, the consumer has the right to demand the provision of necessary and reliable information about the manufacturer of the goods and the seller of such goods, about the mode of its operation and about the goods being sold (works, services). The Law on Consumer Protection establishes a number of requirements for the information provided to the consumer. Such information must be reliable, necessary, timely, and accessible.

Information that contains data that exactly corresponds to the real properties and qualities of the product (work, service) is recognized as reliable. Thus, reliable information should not contain errors, distortions, and deliberately false information. According to the antimonopoly authorities, the unreliability of the translation of information about the product (work, service) into Russian should be regarded as the provision of inappropriate information [9].

The amount of the information about the product, which the legislator calls "necessary information", forms a set of information sufficient to form a clear idea of the consumer properties and qualities of the goods to an ordinary citizen. A certain amount of information about the product, characterized by the term "necessary information", helps the consumer, who does not have special knowledge about the product, to make an accurate choice of the product he needs in accordance with his preferences.

The list of the necessary minimum information about the manufacturer of the goods (the executor of the goods, the seller of the goods), which the consumer has the right to rely on, is clearly reflected in the Law on Consumer Protection. According to the instructions of the legislator, information about the seller (performer, manufacturer) should reflect:

- the name of the organization or the name of an individual entrepreneur (in accordance with the norms of Article 54 of the Civil Code of the Russian Federation [10], it is established that the name of a legal entity must necessarily include an indication of its organizational and legal form. The names of non-profit organizations, and in cases stipulated by law, the names of commercial organizations, must contain an indication of the nature of the legal entity's activities. An individual entrepreneur who is a seller of goods (a performer of works and services, a manufacturer of goods) is obliged to provide the consumer with information about the passage of the state registration procedure as an individual entrepreneur and information about which body carried out such registration);

  - the location of the seller of the goods (the manufacturer of the goods, the contractor of services and works);
  - the seller's working hours;
  - information about the license and (or) accreditation, if the entrepreneur's activity is subject to licensing and (or) state accreditation (in this case, the consumer has the right to request information about the type of activity permitted to a potential counterparty, about the license number and (or) the number of the certificate of state accreditation, about the validity period of the license and (or) the certificate of state accreditation, as well as about the body that issued the license and (or) the certificate of accreditation).

The establishment of a list of mandatory information about the product to be specified during its sale protects the consumer from an unscrupulous seller who, by hiding a part of the information about the product from the consumer, could mislead the consumer about the consumer properties of the product and its quality. The minimum amount of information about a product (work, service) is fixed in the norms of paragraph 2 of Article 10 of the Law on Consumer Protection and includes an extensive list of information components. For certain types of goods, such a list (as a rule, a wider one) is established by special legal acts.

However, it cannot be assumed that the provision of the minimum mandatory amount of information about the product, established by the legislator, will be recognized as the proper fulfillment by the seller of his duties to inform the buyer of the goods. This interpretation of the norms of Article 10 of the Law on Consumer Protection, in our opinion, unreasonably reduces the content of the information obligation of the entrepreneur to his clients. Based on a systematic interpretation of the norms of Article 8 and Article 10 of the Law on Consumer Protection, we believe that, depending on the type and nature of the goods (services, work), the entrepreneur is obliged to provide additional information necessary to ensure the correct choice of the consumer. It is necessary to agree with the statement that if the consumer needs other or additional information that is not prescribed by a regulatory legal act, then there is no reason to believe that the contractor has the right to refuse to provide such information to the consumer [11].

The requirements concerning the availability of information about a product (service, work) to the consumer follow from the provisions of paragraph 2 of Article 8 and paragraph 1 of Article 10 of the Law on Consumer Protection. In these provisions, the legislator prescribes the manufacturer of the goods and its seller to bring the appropriate information to the consumer's attention in a clear and accessible form when concluding a consumer contract by the methods adopted in the relevant areas of trade and service. As a general rule, all the information that is required must be communicated to the Russian consumer in Russian. At the discretion of the manufacturer of the goods (contractor of works and services, seller of goods), the information prescribed by law about the goods (work, service) may be brought to the consumer in the state languages of the subjects of the Russian Federation and the native languages of the peoples of the Russian Federation. The manufacturer and the seller can place the information addressed to the consumer on the packaging or label of the goods, state it in the technical documentation attached to the goods, leaflet inserts for each unit of goods, or in any other way accepted for certain types of goods.

The peculiarity of the exercise of the consumer's right to information during the remote sale of goods is that the law provides for step-by-step informing of the consumer. Prior to the conclusion of the contract, the consumer should be provided with the opportunity to get acquainted with the main consumer properties of the goods (services, works), with information about the place of manufacture of the goods, the address (location) of the seller of the goods (manufacturer of the goods, contractor of works and services) and its full brand name (name), the price and conditions of purchase of the goods, its delivery, service life, expiration date and warranty period, the procedure for payment for the goods, as well as the period during which the offer to conclude the contract is valid. This is quite clearly stated in the norms of paragraph 2 of Article 26.1 of the Law on Consumer Protection. Subsequently (that is, at the time of delivery of the goods), the consumer must be provided in writing with information about the goods provided for in Article 10 of the Law on Consumer Protection, and information about the procedure and timing of the return of the goods provided for in Article 26.1 of the Law on Consumer Protection.

7.4 Conclusion

Distance trading allows you to reduce the seller's costs for organizing the sale of goods (there is no need to maintain a large number of warehouses, hire workers to work in trade pavilions, etc.). This extends the discretion of the seller to set the price of the goods. Reducing the price of the product ensures its availability to the public.

The use of effective technologies of remote interaction between counterparties in the purchase and sale of goods through online commerce contributes to the expansion of the buyer's capabilities both to choose the product itself and to choose the seller. This has a positive impact on the level of competition in a particular commodity market, stimulating manufacturers and sellers of goods to search for the most optimal models of interaction with buyers. We believe that achieving an optimal level of interaction between sellers (producers) of goods and consumers of such goods is one of the factors that have a positive impact on the pace of inclusive growth.

It should be remembered that the effectiveness of distance trading largely depends on the level of balance of the guarantee system and ways to protect consumer rights. Determining the content of the consumer's right to information during the remote sale of goods, the legislator must take into account the specifics of the procedure for concluding and executing a retail sale agreement, the specifics of using technologies for remote interaction of counterparties. We believe that it is necessary to establish a simplified procedure for the return of goods and compensation for losses caused to the consumer if the properties of the purchased goods do not fully correspond to the information provided about them. The need for such additional measures is due to the special significance for the consumer of information about the product during remote interaction between the seller and the buyer, since the information provided by the seller about the product is the main (and often the only) guideline for choosing the goods needed by the consumer.

References

1. Davos came up with an alternative to the GDP indicator. And who is the world leader now? http://news.mail.ru/card/117/ (2021)
2. Hasmath, R.: Inclusive Growth, Development and Welfare Policy: A Critical Assessment, New York (2015)

3. Dugin, A.: The economy is for a person, not a person for the economy. https://www.geopolitica.ru/news/aleksandr-dugin (2021)
4. Mammadli, ZA.: З. A. The features of consumer rights violations in online trading. Legal Concept. **2**(20), 149–154 (2021)
5. Inshakova, A.O.: Law as the basis of infrastructure support of the digital economy and technology of the internet of things. Legal Concept. **3**(18), 6–11 (2019)
6. "On the Strategy of the state policy of the Russian Federation in the field of consumer protection for the period up to 2030": Order of the Government of the Russian Federation No. 1837-r of 28.08.2017 (2021). http://www.consultant.ru/document/cons_doc_LAW_256217/
7. "On consumer protection": The Law of the Russian Federation No. 2300-1 of 07.02.1992. RossiyskayaGazeta. 07.04.1992. No. 80
8. Kirillovyh, A.A.: Information obligations of an entrepreneur: some problems of realization of consumers rights to information. https://www.jusinf.ru/internet_mag/article3925/ (2021)
9. About some issues related to the application of the Law of the Russian Federation. "On Consumer Protection": Order of the Ministry of Antimonopoly Policy of the Russian Federation No. 160 of 20.05.1998 (2021). http://www.consultant.ru/document/cons_doc_LAW_21455/
10. "Part One of the Civil Code of the Russian Federation": Federal Law No. 51-FZ of 30.11.1994. RossiyskayaGazeta. 08.12.1994. No.238-239
11. Stiglitz, D., Sen, A., Fitussi, J.-P.: Misjudging our life: why does GDP make no sense? Report of the Commission on Measuring the Effectiveness of the Economy and Social Progress, Moscow (2016)

Chapter 8
Protecting the Rights of Indigenous Peoples as an Essential Element of Inclusive Economic Growth

Fridon R. Ananidze and Naser A. R. Al Ali

Abstract The notion that the fruits of economic growth should be enjoyed by wide sections of society has not always received universal support. Traditional approaches to economics have long focused on the role of market forces as an engine of economic growth rather than on ensuring equitable development outcomes driven by market forces. In practice, however, as a result of a policy of total non-government intervention in the economy, there is growing evidence of unwanted outcomes, such as increased income inequality and persistent poverty levels, which weaken social cohesion. Over time, the number of people who understand that economic development and growth must bring prosperity and well-being for all increases, including among scientists and policy-makers at the national and international levels. This increase in public awareness has led to increased recognition and understanding that growth must be inclusive, including the growth of prosperity and the protection of the rights of all vulnerable groups. Within the framework of this article, special attention is focused on protecting the rights of such a vulnerable category of the population as indigenous peoples.

8.1 Introduction

Indigenous peoples live compactly on all continents inhabited by humans. In North America, these are Indian tribes (Iowa, Apache, and Roquez, Eskimos, Aleuts, and others); in Latin Amerika—Aimara, Aruaki, Aztecs, Maya, Inca, Yanomami, and others; in Australia—the Australian Bushmen; in Africa—Maasai, Enderois, San (bushmen), Mbati (pygmies), and others; in Europe—the Sami (Lapps); in Asia—the Ainu and Ryukyu peoples (Japan), the Nanai and some other peoples of China; in Oceania—Chamorro (Guam), Hawaiians

F. R. Ananidze (✉)
Peoples' Friendship, University of Russia, Moscow, Russia
e-mail: ananidze_fr@rudn.ru

N. A. R. Al Ali
Law Institute of the Russian University of Transport, Moscow, Russia

A. O. Inshakova and E. I. Inshakova (eds.), *New Technology for Inclusive and Sustainable Growth*, Smart Innovation, Systems and Technologies 288,
https://doi.org/10.1007/978-981-16-9808-8_8

(Hawaii), Maori (New Zealand), Papuas (Papua New Guinea and some regions of Indonesia), Naura (Nauru), and others [16]. A large group of indigenous peoples is made up of the small indigenous peoples of the Russian Federation living in a vast area from Karelia to Kamchatka and Sakhalin—Vepsians, Sami, Koryaks, Khanty, Mansi, Nanais, Yukagirs, Nivkhs, Nenets, Chukchi, Shors, and many others [16]. It is approximately from 200 to 360 million people who share common needs and problems [8].

Indigenous peoples as peoples with all their own characteristics (cultures, customs, traditions, languages, ancestral lands, traditional way of life, etc.) have been existing for centuries. Nevertheless, it was only in the past century when the world community started to recognize the vulnerability of millions of representatives of aboriginal communities and realize special politics to increase the level of social and legal aspects of their life.

However, neither the international law science nor international legal documents still are able to define what members of the Earth population should be included in the concept of "indigenous people" and what criteria the concept of "indigenous people" do consist of [1]. Nowadays, in international law practice, it is possible to find how the concept functions in various kinds of its definitions.

The issue is further complicated by the fact that a whole host of distinctive features characterizes indigenous peoples: first, they represent different ethnic groups, ethnicity, and races; secondly, they profess different religions; thirdly, they speak different languages; fourthly, they are representatives of different cultures, traditions and customs, and alike.

The only thing that is a standard feature for them is that almost all of them live on their ancestral (traditional) lands and lead a traditional way of life [16].

Thus, based on such a variety of races, languages, cultures, religions, and the mere fact that indigenous peoples are at different stages of their socio-economic development, and they have different interests and aspirations, it is not very easy to develop internationally one adequate definition which would be a common name for all aborigines.

8.2 Methodology

The methodological basis of the study involves a combination of general scientific (dialectical, historical, inductive, deductive, analytical, synthetic) and private scientific methods (formal-legal, comparative-legal, interpretative, statistical, procedural, and dynamic).

8.3 Results

As Russian researcher, R. Garipov states, "the development of the concept of "indigenous people" should be based on the approach of comprehensive coverage of indigenous peoples, which would include all their (indigenous peoples) diversity and, at the same time, be based on some general criteria" [12]. We agree with this statement and believe that starting point here is a criterion of determining indigenous peoples as "descendants of the population, that lived in this certain area before the arrival of people of other racial and ethnic origin, with another religion, culture, language, customs" [13].

Another acceptable, in our opinion, criterion for defining the term "indigenous people" is their non-dominant position in the nationwide collective. According to professors I. P. Blishchenko and A. Kh. Abashidze, "at the national level indigenous peoples usually do not occupy a dominant and sometimes re in a discriminatory provision" [11]. Here we completely agree with them, and cannot agree with the opinion of R. Sh. Garipov, who claims that "this is not a sign of the indigenous people, but a consequence of the unfair state policy towards them. In addition, in a number of states, the indigenous population constitutes the majority in relation to non-indigenous nations (countries of South America) and in some of them state policy is changing for the better before our eyes (for example, Bolivia, where a new constitution was adopted)" [12].

We proceed from the fact that "the result of the unfair state policy toward indigenous peoples", and such was the colonization of the ancestral lands of the aborigines, the above first criterion can be considered the same success. At the same time, it should be specified that indigenous peoples around the world constitute a minority of the population of states with the exception of three Latin American states—Bolivia (49%), Ecuador (40%), and Guatemala (40%), and note that only in Bolivia, the Quechua and Aymara have status from the state language along with Spanish [5].

The essence of the next criterion for defining the term is the complexity diversity of languages, cultures, traditions, customs, and beliefs of these peoples.

When defining the concept of "indigenous people", it should also be noted that the overwhelming majority of indigenous people avoid the creation of overly centralized political institutions and organize their life mainly at the community level. Decisions are made after a common opinion has been reached within the community, which, in terms of organization and management of society, is very different from most modern states [4].

One more circumstance should be specially emphasized: the consciousness of belonging to an indigenous people (self-identification) is an integral part of their culture, beliefs, and existence as a separate ethnic group. In general, it is expressed in spiritual closeness to their origins, roots, identity, and unites almost the entire world's indigenous peoples [10].

This circumstance was manifested in the "Lovelace Case" [24], considered by the Human Rights Committee. The crux of the matter was this: A Canadian Indian woman named Sandra Lovelace, registered as a Malecite Indian, married a non-Indian. If an

Indian woman did not marry an Indian, then, according to the Canadian Indian Act of 1970, she would lose her Indian status. This meant that such a woman could not enjoy all the benefits and preferences enjoyed by members of the Indian community (the right to inherit family property, the right to be buried on a reservation, take part in ceremonies and rituals, and others). Simultaneously, when the male Indians married non-Indian women, they did not lose this status.

Thus, Sandra Lovelace after marriage lost the status of an Indian. When she separated from her husband and decided to return to the reservation, attempts were made by the Canadian authorities to prevent her from living with her fellow tribesmen on her former reservation. In the end, having exhausted all domestic remedies of their violated rights Sandra Lovelace's individual petition addressed to the Committee of the United Nations for Human Rights, whose jurisdiction is recognized by the state. In her appliance, she claimed that the Canadian Indian Act of 1970 disrupted her rights contained in art. 2 and 3 of the International Covenant on Civil and Political Rights of 1966 (equal right of men and women), as well as refusing to her «the right, in community with other members of their group, to enjoy their own culture» on the basis of art. 27 of the Covenant. The Committee decided that Sandra Lovelace IME la eligible to return to the reservation and to live together with other members of her tribe.

Later, in 1985, the Canadian Indian Act was amended to remove provisions that discriminated against women. It is noteworthy that, based on the amended law, all women who suffered like Sandra Lovelace were restored to their rights.

Thus, the criterion of self-identification is an important component of the concept of indigenous people, especially since it found its consolidation in paragraph 2 of art. 1 of the ILO Convention No. 169, which says that «the indication of the peoples themselves as belonging to the indigenous or tribal way of life is considered as a fundamental criterion for determining the groups to which the provisions of this Convention apply» [18].

The actual content of another criterion for defining the concept of «indigenous people» is their attachment to the original (traditional, paternal) lands. The overwhelming majority of indigenous peoples consider the land to be alive and holy, and any economic activity (mining, and others) is considered a great sin. For example, the indigenous people of Haruku, living in eastern Indonesia on the island of Haruku, part of the Moluccas, believe that theiunr lands' commercial use has devastating consequences for their sacred sites [5]. Also, Australia's aborigines believe that their ancestors' spirits inhabit the land around them, so any activity related to the extraction of minerals can hurt them. In a word, the world around and the land for indigenous peoples is a sacred, living creature, and "owning or treating the land as a commodity to be used and then abandoned is considered by them to be the worst crime deserving of damnation" [7].

Traditional lands for indigenous peoples are the foundation of their existence and development. These lands give them food (hunting, fishing, and gathering), on them they lead their traditional way of life, here they perform their cult rites, and here are their holy places. Over the centuries, the indigenous peoples have developed "a certain set of rules on the ownership and preservation of these lands as the main factor of their existence" [19]. On the din from the representative of the Indigenous Council on this occasion, he said: "execution of indigenous people is the surest way to separate them on the part of the earth" [18]. It should be emphasized that this «territorial» feature is the main distinguishing feature of the indigenous peoples from national minorities.

If you trace the history of the emergence and evolutionary development of the concept of "indigenous people" in the international legal literature, you can find that the term "indigenous people" originally appeared. This was in 1921 in connection with a study by the ILO on the problems of indigenous workers [12]. Then it was included in the text of international legal acts developed and adopted within the framework of the ILO, and to one extent or another, regulating certain rights of the indigenous population. On the basis of analysis and the relevant provisions of these acts, it can be concluded that "they mostly regulate labor relations between indigenous to the village and representatives of metropolises" [9].

ILO Convention 1957 No. 107 "On the protection and integration of indigenous and other communities, Tribal and Semi-Tribal Peoples in Independent States" (hereinafter—the Convention No.107) summarized existing in international law, the first period and the fragmented components of the concept of "indigenous" (the term "indigenous people" appeared in international legal documents later) and in Art. 1 held that it applies:

1. (a) to persons who are part of the population, leading a tribal or semi-tribal lifestyle in independent countries and who are at a lower socioeconomic stage of development than the rest of the population of the state, and whose legal status is regulated partially or completely by their own customs, traditions or special legislation;
   (b) to persons who are part of the population leading a tribal or semi-tribal lifestyle in independent countries, and are considered as an indigenous population due to the fact that they are the descendants of the inhabitants who inhabited the country or geographical area of which this country is a part, during its time conquest or colonization, regardless of their legal status, leading a lifestyle that is more consistent with the socio-economic and cultural system of those times than the system of the country they are part of.
2. For the purposes of this Convention, the expression "semi-tribal" includes groups or individuals who, although they are close to losing their tribal characteristics, are not yet integrated into the national collective" [2].

A few decades later, Convention No.107 has been revised, and on its basis in 1989, the ILO adopted liters and Convention No.169 "About the Indigenous and Tribal Peoples in Independent Countries" (hereinafter—the Convention No.169). There were certain objective reasons for this, the essence of which is briefly as follows:

In a first, the Convention No.107 actually been focused on "integration" of the indigenous population lives in the dominant society was preceded by the United

Nations in the form of the Indian question, adopted by the UN system organizations more in the year 1953. Also, in 1953, the ILO published its first comprehensive study on the working and living conditions of indigenous peoples [7].

In a second, in this period the provisions of the universal legal instruments such as the International Convention on Human Rights of 1966, the Optional Protocol to the International Convention on Civil and Political Rights of 1966, and others are reflected in the national legislation in a number of states and by the ethnic and cultural features of indigenous peoples and their protection, and that are regulated by the national legislation of a raw of countries.

In a third, in a given period the increase of voluntary integration started. This means that indigenous peoples could be integrated into the national community only voluntarily, i.e., only if there is a clearly defined will of themselves. According to par. 1 of art. 5 of the UNESCO Declaration on Race and Racial Prejudice of 1978, "each people is free to decide for itself the issue of preserving and, if necessary, adapting or enriching the values that it considers fundamental to its identity" [15].

Fourth, indigenous peoples have a deliberative vote in certain international intergovernmental organizations in solving their problems at the highest international level. For example, in 1951, the Scandinavian states established the inter-parliamentary international organization called the Scandinavian Council, which included members of 5 Scandinavian countries (Norway, Sweden, Finland, Denmark, and Iceland) and consisted of 20 representatives from each parliament. Among other deputies in the Nordic Council, indigenous people were presented by representatives of the Greenland Eskimos (Denmark) and the Sami people's (Norway, Sweden, Finland)—the indigenous peoples of these member states of the Organization. The main thing here is that they (representatives of indigenous peoples) are fully-fledged MPs of this significant inter-parliamentary forum of the region's states [19].

Fifth, in 1982, ECOSOC established the Working Group on Indigenous Issues to address the promotion and protection of human rights and fundamental freedoms of indigenous populations through the development of and the rules in this regard.

All of these new aspects are reflected in the provisions Co. Mr. Wentz ILO number 169, according to art. 1 which it applies:

a. "to peoples leading a tribal way of life in independent countries, the social, economic and cultural conditions of which distinguish them from the rest of the population of the country and whose legal status is regulated partially or completely by their own customs, traditions or special legislation;
b. to peoples in independent countries, which are considered as indigenous peoples due to the fact that they are the descendants of the inhabitants who inhabited the country or geographical area of which this country is a part, at the time of its conquest or colonization, or the establishment of the current borders of states, and who regardless of their legal status, they preserve partially or completely their own social, economic, cultural and political institutions" [14].

Thus, based on the analysis of the above convention definitions, the following main criteria for defining the concept of "indigenous people" can be identified:

Indigenous peoples are the descendants of the population that lived in this particular territory before the arrival of people of a different racial and ethnic origin, with a different religion, culture, language, customs; non-dominant position of indigenous peoples in the national communities of states; linguistic, cultural, ethnic, racial and other differences of indigenous peoples from the rest of the population; self-identification is the voluntary awareness of belonging to an indigenous community.

On September 13, 2007, the UNGA adopted the Declaration on the Rights of Indigenous Peoples [3] (Resolution 61/295) [19]; however, this document only regulates the legal status of indigenous peoples (rights, freedoms, guarantees, obligations) and does not say anything about the definition of the term "indigenous people".

In such a situation, a starting point for the definition of indigenous people can become a working definition of the United Nations, formulated by the Special Rapporteur on the issue of discrimination against indigenous populations for the UN Sub-Commission on Prevention of Discrimination and Protection of Minorities Jose Martinez Cobo, according to whom "indigenous people is from the current descendants of peoples who lived in the present territory of all or part of a country at the time when persons of another culture and ethnic origin came to it from other parts of the world, who conquered them and put them in a dependent and colonial position through conquest, colonization, and other means; at present, these peoples live more in accordance with their special customs and social, economic and cultural traditions than with the institutions of the country of which they are a part, with a state structure that is based mainly on the national, social and cultural characteristics of other dominant segments of the population" [20].

José Martinez Cobo also did not ignore the problematic issue of the marginalization of certain indigenous peoples. He states that "Although the isolated marginal groups that exist in the country have not been subdued or colonized, they should also be extended to include the concept of 'indigenous people" and names the following factors as evidence for his claim:

- "they are descendants of groups that were on the territory of the country at the moment when groups of another culture or ethnic origin arrived there;
- precisely because of their isolation from other parts of the country's population, they managed to keep the customs and traditions of their ancestors practically intact, which are similar to the customs and traditions characteristic of the indigenous population;
- they are the odds subordinated to the state structure, which is based on alien they national, social and cultural characteristics" [21].

In his final report of 1982, J. M. Cobo proposed the formulation of the investigated term: "indigenous people are indigenous communities, peoples and nations that maintain a historical continuity with societies that existed before the invasion of the conquerors and the introduction of the colonial system and developed in their own territories who consider themselves to be different from other strata of society currently prevailing in these territories or in part of these territories. They constitute non-dominant strata of society and want to preserve, develop and pass on to future generations the territory of their ancestors and their ethnic identity as the basis for the continuation of their existence as a people in accordance with their own cultural characteristics, social institutions and legal systems" [22, 23].

Working Group on the OH on Indigenous Populations repeatedly discussed the issue of defining the concept of indigenous people, and set aside time to consider this issue and decided to use the definitions and criteria set forth g wasps hearths J.M. Cobo. As we have noted above, adopted in the subsequent Declaration of Indigenous Peoples' Rights does not contain a definition of indigenous people.

8.4 Conclusion

On the basis of the analysis carried out, the following general criteria and components of the concept of the term «indigenous people» can be distinguished:

Firstly, indigenous peoples are the descendants of the population that lived in this particular territory before the arrival of people of a different racial and ethnic origin, with a different religion, culture, language, customs;

Secondly, at the national level, indigenous peoples occupy a non-dominant position.

Thirdly, the presence of a variety of languages, cultures, traditions, customs, and beliefs among the indigenous peoples of the spruce complex.

Fourth, self-identification is a voluntary awareness of belonging to an indigenous community.

Fifth, the main feature of indigenous people from other ways of life is a wide adherence to traditional ("father's") lands, nature, and the environment in general, resulting in great love and respect for her, in the interconnectedness and interdependence of these two concepts (the root people and land), since all of this is an integral part of the culture, religion, life, and the mainstay of survival, existence, and development of indigenous peoples [6].

References

1. Abashidze, A.K., Inshakova, A.O., Dementev, A.A.: New challenges of international law in the digital age. In: Popkova, E.G., Sergi, B.S. (eds.) Modern Global Economic System: Evolutional Development vs. Revolutionary Leap, pp. 1125–1132. Springer Nature, Cham (2021)
2. Abashidze, A.K., Ananidze, F.R., Gugunsky, D.A., Kruglov, D.A., Solntsev, A.M.: ILO Convention No. 107 "On the Protection and Integration of Indigenous and Other Populations Leading Tribal and Semipribal Life in Independent States" 1957. Ensuring the rights of indigenous peoples in accordance with international and national law. Collection of acts. RUDN, Moscow 445 p. (2017)
3. Abashidze, A.K., Ananidze, F.R., Gugunsky, D.A., Kruglov, D.A., Solntsev, A.M.: United Nations Declaration on the Rights of Indigenous Peoples of 13 September 2007. Ensuring the

rights of indigenous peoples in accordance with international and national law. Collection of acts. RUDN, pp. 203–204 (2017)

4. Abashidze, A.K., Ananidze, F.R., Solntsev, A.M.: Mezhdunarodno pravoviye osnovy zashity menshinstv y korennyh narodov (International legal framework for the protection of minorities and indigenous peoples. Textbook. RUDN), 572 p., RUDN, Moscow (2015)
5. Abashidze, A.K., Ananidze, F.R., Solntsev, A.M.: The rights of indigenous peoples in a globalizing world (ensuring in accordance with international and national law. A course of lectures delivered within the framework of the program "The UN system and mechanisms for protecting human rights for representatives of indigenous peoples". RUDN, 330 p. (2017)
6. Aga Khan, S., Ben Talal, H.: Dying Forests: Human Impacts of Deforestation. Report of the Independent Commission on International Humanitarian Affairs. 110 p. (1987)
7. Aga Khan, S., Ben Talal, H., Blishchenko, I.P.: Indigenous Peoples: A Global Pursuit of Justice. Report to the Independent Commission on International Humanitarian Affairs, 226 p. (1988)
8. Aga Khan, S., Ben Talal, H., Kudryavtsev, H.M.: Will Humanity Preserve Humanity? Report of the Independent Commission on International Humanitarian Affairs (1988)
9. Ananidze, F.R.: On the question of the definition of "indigenous people". Vestnik RUDN. Series: Legal Sciences, vol. 1, p. 110, Moscow (1997)
10. Anisimov, A.P., Matytsin, D.E.: The concept and types of environmental entrepreneurship. In: Inshakova, A.O., Frolova, E.E. (eds.) Smart Technologies for the Digitisation of Industry: Entrepreneurial Environment. Smart Innovation, Systems and Technologies, vol. 254. Springer, Singapore (2022). https://doi.org/10.1007/978-981-16-4621-8_25
11. Blishchenko, I.P., Abashidze, A.K.: For Indigenous peoples and the international law. Russian justice, 3 p., Moscow (1994)
12. Garipov R.S.: The concept of "indigenous people" and their status in international and domestic law. International law and international organizations, 3 p. (2013)
13. James, S.A.: Indigenous Peoples in International Law, 2nd edn. Oxford University Press, Oxford (2004)
14. Melkov, G.M.: ILO Convention No. 169 "On Indigenous and Tribal Peoples in Independent Countries" 1989. Human rights. Collection of international documents (1998)
15. Nikulin, I.D.: UNESCO Declaration on Race and Racial Prejudice 1978. International normative acts of UNESCO (Conventions, agreements, protocols, recommendations, declarations) (1993)
16. Prokhorov, A.M.: Peoples and religions of the world. Encyclopedia. Institute of Ethnology and Anthropology of Miklukho-Maklay, 928 p. (1998)
17. Prokhorov, A.M.: The peoples of Russia. Encyclopedia. Institute of Ethnology and Anthropology of Miklukho-Maklay, Moscow, 479 p. (1994)
18. Rights of the Indigenous Peoples to the Earth-Paper submitted to the UN Working Group of Indigenous Populations–UN Doc. E/CN4/Sub.2/AC4/1985/WP4 (1985). Accessed 28 January 2021
19. Sandvik, G.: Legal status of the Sami. State and law, 9 p. (1992)
20. UN documents. E/CN.4/Sub.2/1972/566/L. Accessed 28 January 2021
21. UN documents. E/CN.4/Sub.2/1986/7/Add.4, paragraph 379. Accessed 28 January 2021
22. UN documents. E/CN/4/Sub.2/1983/21/Add. 8, paragraph 379. Accessed 28 January 2021
23. UN Documents E/CN4/Sub.2/1983/Add4. Accessed 28 January 2021
24. Views of the Human Rights Committee under Art. 5 (4) of the Optional Protocol concerning Communication. NR 6/24. 30 July 1981. Report of the Human Rights Committee, GAOR 36th Session. Supp. 40 (A/36/40). Ann. XVIII, 166 p.

Chapter 9
Creative Industries in the Context of the Global Digitalization of Copyright

Rimma S. Rahmatulina and Galina N. Chernichkina

Abstract The transition to an intensive path of development of States in the modern world is not possible without investment in human beings. The absence of such investments can lead to a humanitarian crisis to preserve a single cultural space. All this dictates the need to develop traditional culture, intangible cultural heritage, historical identity, connections in the transfer of cultural, ethnic traditions and knowledge. At the same time, the result of this task in the context of widespread digitalization is the solution to such issues as the formation of a single register of cultural heritage objects. The registry should contain account information about the object in an online form. The creation of an interactive form of various cultural heritage objects is associated with the process of forming creative industries based on them. This article is devoted to solving the problem related to the protection of intellectual property, which also accompanies the development of digital creative industries while preserving historical heritage.

9.1 Introduction

Currently, in many countries of the world, a "creative economy" has begun to develop, which is increasingly influenced by digital technologies. This term was first introduced by John Hawkins. According to John Hawkins, "creative economics deals with ideas and money. This is the first kind of economy where imagination and ingenuity decide what people want to do and do. And what they want to buy" [1]. Russian creative economy is "an economy that is associated with intellectual property in various fields of activity" (The order of the Government of the Russian Federation of 20.09.2021 No. 2613-r about the development of the concept of the creative

R. S. Rahmatulina (✉)
Financial University Under the Government of the Russian Federation, Moscow, Russia
e-mail: RSHRahmatulina@fa.ru

G. N. Chernichkina
Russian State University of Justice, Moscow, Russia
e-mail: Chernichkinagn@rsuj.ru

A. O. Inshakova and E. I. Inshakova (eds.), *New Technology for Inclusive and Sustainable Growth*, Smart Innovation, Systems and Technologies 288,
https://doi.org/10.1007/978-981-16-9808-8_9

industries) [2]. At the moment, the creative industry in the Russian Federation does not have sufficient legal regulation, and in this regard, it needs to be developed and improved.

In the United States, creative industries focus on a sector more focused on works of art, while excluding the field of computer programming and scientific research not related to art. In the UK, the first place in the creative industry is put by digital technologies and use the concept of "creative and digital industry", which allows the wide use of various copyright objects in digital form, online services. In Singapore, as a result of the digital revolution, digital marketing, the design and media arts industry are successfully developing.

The concept of "creative industries" is associated with tourism, with such areas of economics and culture as folk art crafts, design, theater, cinema, architecture, the fashion industry, television, their use and commercialization.

9.2 Methodology

The potential of creative industries is high, as noted in many Russian and foreign studies. Every year, the sphere of creative industries includes both new types of activities and various objects of intellectual property are used. As the main sector of the economy in many countries of the world, creative industries are associated with the formation of new creative ideas, their implementation and commercialization of results. Creative industries originated in the early nineties of the last century and in many countries are already a priority area of public policy. In countries such as Indonesia and the United Arab Emirates, the creative economy has shown rapid growth in the years 2017–2018. However, only 44 settlements on the territory of the Russian Federation have the status of a historical settlement of federal significance. Creative economy and creative industries are new terms that have existed for about twenty years around the world, but are widely used, established, and constantly developing.

9.3 Results

9.3.1 Creative Industries as a Legal Phenomenon

The creative industry is an activity in which subjects realize various economic and creative tasks. The term creative industries is multifunctional and interdisciplinary in nature. In article 3 of the Fundamentals of the legislation of the Russian Federation on culture, cultural activities are "activities for the preservation, creation, dissemination and development of cultural property" [3].

It should be noted that "cultural industries" and "creative industries" are replacing each other's concepts, although the concept of "cultural industries" is more correlated with historical heritage, folk traditions, folklore, and the work of different nationalities. Creative industries are creative technologies, digital art, design projects, the result of which is profit and job creation through the distribution of copyright objects. Copyright objects are the basis of the creative economy, its main component, since creativity is increasingly becoming the engine of interesting ideas and innovations, the cultural development of society.

"The development of digital technologies will lead to a reduction in jobs, and thanks to creative industries, new ones will be created that are inextricably connected with creativity and cannot be automated" [3, 4]—as indicated in the concept of the development of creative industries in Russia. In this regard, a large role in the field of tourism, show entertainment, festivals, concerts is assigned to copyright objects, which are transformed into a massive market product of creative industries.

The concept of the development of creative industries in Russia to achieve national goals contributes not only to "opportunities for the self-realization and development of talents; decent, efficient work and successful entrepreneurship, but also digital transformation" [3, 3].

The global integration of digital technologies [5, 239] has especially influenced copyright objects, their creation and wide use, thereby developing different types of individual and folk art. New products, goods, services in digital form are emerging. A new phenomenon in cultural life was digital services and platforms, social networks, on which objects of museum funds, theater projects, entertainment television programs. Spectacular online projects and events of participants of many countries of the world increase every year and issues and problems of protection of copyright objects arise. The digital revolution has not just changed the creative potential and objects created in digital form, it has created opportunities for business development in this area. In the Resolution adopted by the UN General Assembly on December 19, 2019, 2021 is the International Year of the Creative Economy and for the Development of the Creative Economy, noted the need for "the development of digital technologies, an innovative and digital economy and electronic commerce, Establishing appropriate digital infrastructure and communication networks to promote sustainable development, Increasing public and private investment in creative industries and establishing appropriate normative base".[1]

In the recommended manual of the World Intellectual Property Organization (WIPO), the creative industries "include a diverse group of activities, all of which rely to a greater or lesser degree on the contribution of original work and its protection through various IP rights" [6].

In the Russian concept of the development of creative industries, creative industries are associated with companies whose sphere of activity is intellectual property, manufactured goods and services with economic value, providing economic growth, cultural development, and art.

[1] https://undocs.org/pdf?symbol=ru/A/RES/74/198.

9.3.2 On the Identification of Objects of Creative Industries and the Register

Currently, a significant part of the small towns of the Russian Federation is the center of unique monuments of cultural and natural heritage, which can be centers of cultural and cognitive tourism. In many countries of the world, as in Russia, a state register of cultural heritage objects (historical and cultural monuments) is maintained. Registers of objects [7] provide interaction of information systems. Therefore, the entry of the object into the register can be accompanied by an interactive link to its media characteristics in order to familiarize itself with it when taking an online tour, participating in certain projects. The development of the status of a historical settlement inevitably entails the creation on its basis of a creative economy based on the capitalization of intellectual property obtained during the design of an object of historical and cultural heritage for their involvement in scientific, tourist, and educational content. The uniform brand of the region which basis is formed by the trademark, the geographical indication or the name of the place goods origin has to unite development of creative clusters of historical objects (crafts, the gastronomic industry and the like). In addition, the brand can be based on such types of intellectual property that complement the brand with a unique combination of colors, original graphics of objects, a set of phrases and sounds. The above may be an integral part of the digital content of a historical object, and may also be protected as an industrial design or trademark. The name of the place of origin of the product and geographical indication are fixed assets indicating the unique properties of the product, determined by the natural or human factors of the geographical object. The trademark must also have a unifying meaning and be associated with the historical region and object.

The state is a key investor in the development of the creative industry of historical territories and regions. Extrabudgetary funds are the source of funding, and the mechanism for its creation is the trust capital fund (endowment funds). Endowment—this property, formed from donations, constitutes the target capital, which remains inviolable ("eternal capital"), and only income from the target capital is spent. This tool was most widely used in the countries of the Anglo-Saxon system: in the UK and the USA. It combines foreign models of endowment fund institutions with characteristic tax benefits that create attractiveness for investors and philanthropists. In the UK, the minimum period for which the fund is formed is not determined, as well as the minimum amount of fixed assets of the fund [8]. In the USA, since 2006, regulation has been carried out by the states on the basis of the Uniform Prudent Management of Institutional Funds Act (UPMIFA) [9]. Success of funding endowments is confirmed by examples of existing funds: Bolshoi Theater Fund (founded since 2002)[2]; A.S. Pushkin State Museum of Fine Arts (founded since 2015)[3]; the first in the regions of

[2] https://bolshoi.ru/partners/fund (Accessed: 21.10.2021).

[3] https://www.pushkinmuseum.art/museum/support_us/development_fund/index.php (Accessed: 21.10.2021).

Russia Museum Development Fund named after M.A. Vrubel, established by Omsk entrepreneurs and public figures (founded since2016).[4]

The basis for creating targeted capital for the development of historical objects and the creative industry can not only be charity, but also co-investment in the framework of public–private partnership. At the same time, digital technologies allow not only to raise funds online, as noted in the scientific work on the financial algorithm of Internet investment [10, 1850], but also allow the investor to control the targeted use of investments. In the context of digital technologies, the blockchain of systems (distributed databases) can be a good basis for building a co-financing system, as well as accounting for created intellectual property objects as part of the development of the creative industry of a historical object, as this is shown by the example of the use of accounting for complex goods [11, 328].

Digital platforms In creative industries, digital platforms are not only ways to promote the results of creative activity, but also their monetization. Using digital platforms helps capture new, only emerging intellectual property objects. Digital content helps not only promote the creative industry, but also contributes to the growth of e-commerce. So, for example, in the UAE, various creative platforms are developing very quickly using historical heritage and modern cultural objects. The DED Trader license in Dubai allows you to start a business on the Internet that is open to entrepreneurs of different nationalities. The world leader of creative industries—Great Britain is a country with a developed system of licenses and advanced creative technologies. The basis of such contractual models for attracting online funds in the conditions of digitalization, as noted in the literature, can be a smart-contract [12, 459]. Global digitalization of processes can serve as the development of creative entrepreneurship platforms, both in Russia and in other countries. Noting the development of creative digital platforms, attention should be paid to the achievement of national IT companies [13, 273] in building an intelligent judicial system. Such advanced technologies will help to quickly respond to copyright violations in the field of digital creative industries.

9.4 Conclusion

The legal regulation of new institutions related to creative industries will solve many complex issues. For example, access to international digital creative platforms will expand international communications, help to conclude licenses faster, while reducing costs and expenses. We believe that global digitalization will be the connecting basis for building creative industries to restore historical heritage and create new creative results. In addition, this will not only strengthen the economy and provide work for the population, but will also attract international investors to small regions.

[4] https://fond.vrubel.ru/ (Accessed: 21.10.2021).

References

1. Hawkins, J.: Creative Economy. Publishing House “Classics-XXI”, Moscow. http://www.johnhowkins.com/wordpress (2011)
2. Order of the Government of the Russian Federation No. 2613-r dated 09/20/2021. http://publication.pravo.gov.ru/Document/View/0001202109270012
3. Fundamentals of the legislation of the Russian Federation on culture of 09.10.1992 N 3612–1. http://pravo.gov.ru/
4. The concept of the development of creative industries and mechanisms for the implementation of their state support in large and largest urban agglomerations until 2030, 2. http://publication.pravo.gov.ru/File/GetImage?documentId=86c2f205-5e40-4c47-ba73-a1d913d52a52&pngIndex=2
5. Inshakova, E.I., Ryzhenkov, A.Y., Inshakova, A.O.: Neo-industrialization of the Russian economy: technological and digital development. In: Popkova, E.G. (ed.) Ubiquitous Computing and the Internet of Things: Prerequisites for the Development of ICT. Studies in Computational Intelligence, vol. 826, pp. 239–250. Springer Science + Business Media, Cham (2019)
6. How to Make a Living in the Creative Industries. WIPO. https://www.wipo.int/edocs/pubdocs/en/wipo_pub_cr_2017_1.pdf (2017)
7. Regulations on the Unified State Register of Cultural Heritage Objects (Historical and Cultural Monuments) of the Peoples of the Russian Federation. Approved by Order of the Ministry of Culture of the Russian Federation No. 954 dated October 03, 2011. Bulletin of normative acts of Federal executive authorities. No. 4 (2012)
8. Endowments in the arts. Arts council England. https://www.artscouncil.org.uk/publication/endowments-arts (2010). Accessed 21 Oct 2021
9. The Uniform Prudent Management of Institutional Funds Act 2006. (UPMIFA). https://www.uniformlaws.org/HigherLogic/System/DownloadDocumentFile.ashx?DocumentFileKey=d7b95667-ae72-0a3f-c293-cd8621ad1e44&forceDialog=0. Accessed 21 Oct 2021
10. Matytsin, D.E.: Internet-investing as a remote algorithm of the retail investment financing. In: Modern Global Economic System: Evolutional Development vs. Revolutionary Leap. Springer book series Lecture Notes in Networks and Systems, pp. 1850–1857. Springer Nature, Cham (2021)
11. Inshakova, A.O., Goncharov, A.I., and Ershova, I.V.: Digital blockchain registration of legally significant stagesof complex good’s export-import supplies by business entities of the EAEU and BRICS jurisdictions. In: Inshakova, A.O., Inshakova, E.I. (eds.) Competitive Russia: Foresight Model of Economic and Legal Development in the Digital Age: Proceedings of the International Scientific Conference in Memory of Oleg Inshakov (1952–2018), pp. 328–336. Springer Nature, Cham (2020)
12. Inshakova, A.O., Deryugina, T.V., Malikov, E.Y., Ponka, V.F.: Smart contracts as a legal instrument of developingthe export potential of intellectual property rights in international economic activity. In: Inshakova, A.O., Inshakova, E.I. (eds). Competitive Russia: Foresight Model of Economic and Legal Development in the Digital Age: Proceedings of the International Scientific Conference in Memory of Oleg Inshakov (1952–2018), pp. 459–465. Springer Nature, Cham (2020)
13. Rusakova, E.P., Inshakova, A.O.: Industrial and manufacturing engineering in digital legal proceedings in the asia-pacific region: a new level of quality based on Data, Blockchain and AI. Int. J. Qual. Res. **15**(1), 273–290 (2021). https://doi.org/10.24874/IJQR15.01-16 (Q 2)

Chapter 10
Some Issues of Proof in Insurance Disputes in the Conditions of Digital Transformation of Law in Russia

Anna L. Polina-Stashevskaya

Abstract The objective of the research in the article below is to identify the features of methods of proof in insurance disputes in the context of digital transformation of law. The active introduction of online technologies, due to the development of the economy, digitalization, and the pandemic, affected the insurance industry and initiated the implementation of a new model for concluding insurance contracts and exchanging documents in the process of their execution. The adaptation period for creating optimal legal structures of electronic insurance contracts and the sequence of actions has formulated a pool of issues requiring legal analysis at the doctrinal and law enforcement levels. The analysis of judicial practice in 2020–2021 shows that the subject of proof in insurance disputes has expanded and included the conditions for the conclusion of contracts. Also, the number of disputes on the invalidity of insurance contracts as a way to protect against forms of abuse of rights by insured people has increased greatly. The development of a standard of proof for this category contributes to the unification of judicial practice and improves the current legal regulation mechanism.

10.1 Introduction

The procedural legislation of the Russian Federation does not contain standards of proof, in contrast to English and American law, which provide for the criterion according to which the judge makes a decision based on the balance of simple probability. The absence of such regulation created some legal uncertainty and raised the subjective assessment over the search for objective truth. Insurance disputes arising from aleatory obligation always touch upon issues of honesty and abuse of law, predetermining the uniqueness of the methods of proof due to the fact that these institutions have an evaluative legal nature. The introduction of online technologies,

A. L. Polina-Stashevskaya (✉)
Department of Civil Law and Procedure and International Private Law, RUDN University, 6 Miklukho-Maklaya str., 117198 Moscow, Russia
e-mail: zaytseva.al@mail.ru

A. O. Inshakova and E. I. Inshakova (eds.), *New Technology for Inclusive and Sustainable Growth*, Smart Innovation, Systems and Technologies 288,
https://doi.org/10.1007/978-981-16-9808-8_10

which have changed the process of concluding a contract and identifying customers, on the one hand, increased the availability of insurance, and on the other hand, created the preconditions for the use of digital innovations by dishonest insures [1].

The framework legal regulation of the information obligations of the insuree, enshrined in Article 944 of the Civil Code of the Russian Federation, does not contain a classified system of consequences. The Russian legislator considered it optimal when establishing the fact that the insurer was provided with deliberately false information about significant circumstances, to recognize the insurance contract as invalid and apply the consequences provided for in paragraph 2 of Article 179 of the Civil Code of the Russian Federation. At the same time, the method of proving the fact of presenting deliberately false information is not fixed in any regulatory legal act but can be formulated on the basis of judicial acts. It can be presented in the form of a mathematical formula in which the fact will be the sum of the insuree's unfair actions aimed at misleading the insurer with the obligatory establishment of advance knowledge.

10.2 Methodology

Based on the method of comparative analysis, it was possible to conduct a comparative analysis of the best models of proof, to identify common features of such definitions in insurance disputes as deliberately false information, abuse of law. As a result, the research allows to formulate an assessment of the current model of proof for insurance disputes on the conclusion of an insurance contract and invalidity, the impact of digitalization on the process of proof.

10.3 Results

10.3.1 The Procedure for Providing Insurance in Electronic Form: Issues of Proving the Conclusion of the Contract

In 2019, amendments were made to Article 940 of the Civil Code of the Russian Federation, according to which the legislator allowed the conclusion of an "electronic insurance contract". The legalization of electronic document exchange required the expansion of boxed insurance products, modeled on the presumption of the insuree's honesty, since the risk assessment at the pre-contractual stage is minimally realizable.

Of fundamental importance for the conclusion of a property insurance contract is the fact that the parties to the contract agree on its essential terms, which is provided for in Articles 432, 942 of the Civil Code of the Russian Federation. The requirement to comply with the written form of the transaction also remains a prerequisite for the emergence of insurance legal relations. In the era of digital transformation of law, the proof of the conclusion of an insurance contract by the courts is considered when presenting evidence that allows the content of the transaction to be reproduced unchanged on a tangible medium with the signature of the contract by parties and the possibility of reliable identification of the person who expressed the will [2].

The change in the form of contract conclusion has expanded the range of legally significant circumstances to be proved in insurance disputes about the conclusion of contracts. Pre-digital realities required the submission of the original insurance contract and confirmation of the fact of payment of the insurance premium, the introduction of the electronic form of contracts determined the relevance of means of proof of the conclusion of contracts in the form of correspondence between the parties. The courts have formed as one of the requirements the certification of such correspondence of the parties by a notary. However, such an approach to the relevance of this evidence is possible if the other party disputes the evidence in form or content, claims its forgery (Article 186 of the Civil Procedure Code of the Russian Federation) or falsification (Article 161 of the Arbitration Procedure Code of the Russian Federation). Otherwise, the party representing the correspondence has the right to substantiate the position with reference to Article 102 of the Fundamentals of the Legislation of the Russian Federation on Notaries, designed to provide evidence and containing a condition on the possibility of certification, if there are grounds to believe that the presentation of evidence will subsequently become impossible or difficult.

In the judicial practice of the Supreme Court of the Russian Federation, a standard of proof is formed in cases of the conclusion of insurance contracts, which includes checking the consistency of the essential terms of the insurance contract. Thus, one of the interesting precedents in this category is the civil case of LLC «IC Soglasie» vs Ogannesyan M. on the recognition of the insurance contract as not concluded, invalid. The judicial instances (first and appeal) disagreed on the legal qualification of facts: non-conclusion or invalidity. The plot of the case is based on the conclusion of an electronic insurance contract with online filling in of a questionnaire and the choice of an insurance product, in which, it is impossible to insure an object under construction. When filling in the questionnaire, Ogannesyan M. indicated that he intended to insure the object of completed construction, specifying its characteristics. The court of the First instance concluded that the parties did not agree on the essential terms of the contract, the list of which included the object of insurance (Article 942 of the Civil Code of the Russian Federation), and the court of Appeal pointed to the invalidity of the contract since the potential insuree had deliberately provided the insurer with deliberately false information on the essential conditions. Such a difference in the positions of the law enforcement officer is due to the specifics of the case and the prioritization of established facts.

It is also worth highlighting the approach of the law enforcement officer, which contains a literal interpretation of the provisions of the law on signing a contract concluded using online technologies. In one of the cases, the court refused to recognize the insurance contract as concluded, since the parties exchanged electronic images of documents that did not have originals, and in violation of the norms of the current legislation, were not signed either by a simple or enhanced electronic signature of the insurer, indicating during the assessment of the evidence that silence, according to Art. 438 of the Civil Code of the Russian Federation, is not an acceptance.

One of the features of proof in cases of this category is the imposition of the burden of proof on the plaintiff, who claims the conclusion or the non-conclusion of the insurance contract. Such a procedural ideology allows one to follow the principles of legality, competitiveness, and equality of parties in insurance disputes, since one of the parties is such a scientific category as a "weak side" or a consumer. However, in cases of contract conclusion, a professional dishonest insuree often hides under the guise of a weak side [3].

Each component of the mechanism of proof in insurance disputes constitutes a social and legal phenomenon and is placed in the framework of a procedural form.

10.3.2 Features of the Legal Regulation of the Procedure for Proving the Invalidity of Insurance Contracts

The adoption of the Concept for the development of the provisions of Part 2 of the Civil Code of the Russian Federation on the insurance contract dated September 25, 2020, led to a new stage in understanding the need to finalize the mechanism of legal regulation of the institution of invalidity of the insurance contract. The analysis of foreign experience in legal regulation has raised the question of the expediency of using the experience of diversified consequences of violation of informational pre-contractual obligations for the Russian legislator. The existing definition of the invalidity of insurance contracts in Part 3 of Article 944 of the Civil Code of the Russian Federation in conjunction with the provisions of Article 179 of the Civil Code of the Russian Federation has become in demand in recent years as a tool for countering insurance fraud. For a long time, the courts have been forming a uniform approach to assessing the evidence presented by insurers to mislead them at the stage of pre-contractual legal relations, indicating that the insurer is a professional market participant and is obliged to assess the risk at the stage of concluding a contract. This concept has influenced the expansion of forms of abuse of rights by dishonest insurers.

However, in 2019–2021, insurers changed the procedure of proof and filled the gaps in legal regulation in this area. Thus, in judicial practice, precedents have appeared on the invalidity of insurance contracts, the analysis of which allows us to conclude about using the presentation of evidence based on the chronology of the emergence and development of legal relations in the method of proving, while insurers were able to refract the approach of law enforcement officers to the interpretation of "deliberately false information" as a criminal legal definition with the requirement to provide relevant evidence—a court verdict that has entered legal force, and deliberately delimited and filled the gap in the legal regulation of this category.

The most interesting is the approach of the courts on recognizing insurance contracts as invalid using the institution of abuse of rights on the part of the insured. The legal symbiosis of Articles 10, 179, and 944 of the Civil Code of the Russian Federation has created a stable form of counteraction against dishonesty on the part of insurers and exposed pseudo-interest in insurance [4].

One of the interesting court cases is the civil case on the claim of Ustyuzhanin A.Yu. to LLC "IC Soglasie" on the recovery of insurance compensation and on the counterclaim of LLC "IC Soglasie" to Ustyuzhanin A.Yu. on the recognition of the insurance contract as invalid. Only in the second round of consideration of the case, the courts recognized the insurance contract as invalid with the application of the provisions of Article 10 of the Civil Code of the Russian Federation, establishing the entire chain of actions of the parties to the legal relationship. The fact was taken into account that the history of insurance acquires a fantastic aspect associated with deliberately misleading the insurer by submitting a contract for the construction of a house worth 36 million rubles, transferred in cash by the insuree to unknown builders. At the same time, the millionaire-insurer has numerous enforcement proceedings for debts for utilities, artificial jurisdiction, etc.

The largest number of court disputes on the invalidity of insurance contracts in 2019–2021 is connected to the insurance of residential buildings and personal insurance. The regulation of issues of the beginning of the limitation period when claiming the invalidity of insurance contracts still remains an acute problem.

The specific nature of disputes on the invalidity of contracts is based on the unique evaluative nature of the legal definitions themselves, which act as the basis for regulating the invalidity of an insurance contract. The experience of legal regulation in England is interesting, where the consequences depend on the intention to violate (deliberate or reckless) the information obligation by the insured. It is the time of digital decisions that contribute to making decisions on the classification of requirements for the information obligations of the insured, the subject composition, and the implementation of positive foreign experience.

10.3.3 Impact of Digitalization on the Proof Process

The new realities of the digital economy, Big Data, blockchain have created the prerequisites for changing the structure of the evidence, the emergence of a new legal category "electronic proof". In some disputes, the list of evidence of data enrichment at the stage of execution of legal relations between the parties was reduced. For example, the use of biometric data in legal relations between a client and a bank directly affects the list of evidence, optimizing the process of proof [5].

Since 2020, the possibility of obtaining financial services, including insurance, is realized in insurance in Russia using a financial platform (the Marketplace system), which will provide round-the-clock access to financial services regardless of your location (p. 1 of art. 1, par. 7 p. 1 of art. 2 of the Law of 20.07.2020 N 211-Federal Law). The creation of Internet platforms for the emergence of a legal relationship changes the mechanics of the process, making it more secure, provided that competent legal structures are used, embedded in the software with a prerequisite for the correct sequence of actions [6].

A similar procedural platform is the module “Electronic justice” (formerly State Automated System “Justice”), which allows using a browser to implement electronic exchange of documents within the framework of legal proceedings.

The change in the proof process is also associated with the improvement of the procedural codes of the Russian Federation in terms of the introduction of video conferencing and online meetings. One of the new problems of proof is the verification of written evidence during an online meeting; a legislative solution has not yet been found [7].

In the structure of evidence of the digital age, the correspondence of the parties in messengers and e-mail takes an important place. If earlier the courts refused to accept correspondence as evidence, then in recent years they have formed a uniform approach to the inclusion of such evidence in compliance with the conditions of assurance.

10.4 Conclusion

The improvement of proof models in insurance disputes and the creation of standards for cases of non-conclusion and invalidity of insurance contracts was a natural consequence of the introduction of digitalization into the life of society and the court. Legal regulation gaps are filled by law enforcement officers in specific cases, and individual precedents also have the effect of filling the legal gap.

The specific legal nature of insurance legal relations requires the process of proving to have the same uniqueness in the “chronological evidence base”.

The change in the format of the emergence of legal relations and the transition to the digital plane makes the need for the development of new digital standards of proof and their legalization in procedural legislation inevitable. Discussions about the development of artificial intelligence and its applicability in the process of proving in Russian courts appear to be an indicator of the time of changes in legal theories of procedural legislation and the transformation of legal norms.

The information society and big data environment, increasing the availability of information, form the complexity of the architecture of the evidence base, changing the process of collecting, presenting, and examining evidence.

References

1. Ermakova, E.P., Rusakova, E.P., Sitkareva, E.V., Frolova, E.E.: Main components of protecting consumers of financial products in Asian-oceanic (APAC) counties. Int. J. Eng. Technol. (UAE) **7**(4), 157–162 (2018). https://doi.org/10.14419/ijet.v7i4.38.24343
2. Matytsin, D.E.: Digital technologies of the bank of russia for regulating investment relations. In: Inshakova A.O., Frolova E.E. (eds.) Smart Technologies for the Digitisation of Industry: Entrepreneurial Environment. Smart Innovation, Systems and Technologies, vol. 254. Springer, Singapore (2022). https://doi.org/10.1007/978-981-16-4621-8_19
3. Inshakova, A.O., Inshakova, E.I., Deryugina, T.V.: Information technologies for the formation of the European space of freedom, security and justice. In: Artificial Intelligence: Anthropogenic Nature vs. Social Origin. 1100 AISC, 444–453 (2020). https://doi.org/10.1007/978-3-030-39319-9_51
4. Frolova, E.E., Inshakova, A.O., Dolinskaya, V.V.: The concept and legal framework for judging corporate conflicts on the financial market. In: Conflict-Free Socio-Economic Systems: Perspectives and Contradictions, pp. 101–108. Bingley, West Yorkshire (2019). https://doi.org/10.1108/978-1-78769-993-920191012
5. Inshakova, A.O., Frolova, E.E.: Preface. In: Inshakova A.O., Frolova E.E. (eds.) Smart Technologies for the Digitisation of Industry: Entrepreneurial Environment. Smart Innovation, Systems and Technologies, vol. 254. Springer, Singapore (2022)
6. Frolova, E.E., Ermakova, E.P., Protopopova, O.V.: Consumer protecrion of digital financial services in Russia and abroad. In: Popkova, E.G., Sergi, B.S. (eds.) ISC Conference—Volgograd: 13th International Scientific and Practical Conference Artificial Intelligence: Anthropogenic Nature vs. Social Origin, pp. 76–87. 1100 AISC (2020). https://doi.org/10.1007/978-3-030-39319-9_8
7. Rusakova, E.P., Frolova, E.E., Gorbacheva, A.I.: Digital rights as a new object of civil rights issues of substantive and procedural law. In: Advances in Intelligent Systems and Computing. 1100 AISC, pp. 665–673 (2020). https://doi.org/10.1007/978-3-030-39319-9_74

Chapter 11
Legal Personality of Credit Institutions in the Civil Legislation of the Republic of Tajikistan

Alexander A. Pukhart and Shukhratpur Mukhammadyusuf

Abstract The issue of legal personality of credit institutions in civil legal relations is not new, however, there is no single or the most popular and sufficiently justified position corresponding to modern scientific and legal achievements. The article examines the issue of the legal personality of a credit institution as a participant in civil legal relations. Analyzing modern scientific achievements, the unacceptability of using the category of legal capacity to all types of legal entities in general and to credit organizations, in particular, is stated. The existing opinion about the legal personality of credit institutions on the basis of just one legal capacity is refuted, and the widespread view that credit organizations acquire legal personality after obtaining a license, and not after creation, as is the case with other legal entities, for which it is not necessary to obtain a license, is questioned.

11.1 Introduction

Every jurist knows that in order to participate in civil legal relations, a person must possess legal personality, which, in turn, is formed from legal capacity, meaning that a person has civil rights and obligations, and legal capacity, meaning that a person has the ability to independently acquire and exercise civil rights and obligations. We would like to note right away that this traditional, very common, and legally defined view is not always supported by the authors. For example, Polich S. B. says that: "Legal personality as a decisive characteristic is determined precisely by the status, position, condition of a person, and legal capacity and capacity—his legal capabilities and their practical implementation" [16]. In addition to the traditional view on this issue, the authors also emphasize transactionality, delictability, and transdealability as constituent elements of the legal personality of persons [1], however, most

A. A. Pukhart (✉) · S. Mukhammadyusuf
Law Institute, RUDN, Moscow, Russia
e-mail: Pukhart_aa@pfur.ru

S. Mukhammadyusuf
e-mail: mshukhratpur@list.ru

A. O. Inshakova and E. I. Inshakova (eds.), *New Technology for Inclusive and Sustainable Growth*, Smart Innovation, Systems and Technologies 288,
https://doi.org/10.1007/978-981-16-9808-8_11

consider them as an integral part of legal capacity. Mark A. J. correctly notes that legal personality provides individuals with a set of rights and obligations recognized by a certain legal system and, depending on the historical period, its volume increases or decreases [13]. The above two elements together form the legal personality of persons in civil legal relations and endow them with the ability to be the subject of these relations. To participate in legal relations, credit organizations must also possess legal personality, which has a special character.

It should be noted that in the civil legislation of the Republic of Tajikistan (RT) [23], the legal personality of credit institutions is regulated by the rules on the legal personality of legal entities. The real legal reality is confirmed by many authors [9].

Due to the fact that the legal personality of credit institutions follows from the legal personality established by law for legal entities, it is necessary to consider this problem comprehensively.

Emphasizing the participation of credit organizations in civil legal relations, most authors speak only about their legal capacity, without mentioning the legal capacity of credit organizations as the second element of its legal personality [24]. For example, A. V. Mul notes that the legal personality of legal entities has a specific character and the possession of only one legal capacity is sufficient for a legal entity to be able to participate independently in legal relations [14]. You can look at this question from a different angle. If an individual having only legal capacity, for example, a person with mental disabilities recognized by a court as incapacitated, does not have the right to acquire civil rights and duties for himself and fulfill them, i.e. is not considered an independent entity, so why is a legal entity having only legal capacity without having legal capacity, can acquire civil rights and duties for himself and fulfill them by his actions? It turns out that an incapacitated individual does not have the right by his actions to acquire civil rights and obligations for himself and fulfill them, and an incapacitated legal entity, i.e., a legal entity with only legal capacity has the right by his actions to acquire civil rights and obligations for himself and fulfill them. A legal entity with only legal capacity can be considered an independent and full-fledged subject of civil legal relations, but an individual does not, the latter needs legal capacity for this, and the former does not. A natural person who does not have legal capacity is considered incapacitated, and a legal person who does not have legal capacity is not considered incapacitated. An individual without legal capacity is not considered a full-fledged and independent subject of civil legal relations, and a legal entity does not need the legal capacity to confirm its status as an independent and full-fledged subject of civil legal relations. Undoubtedly, such reasoning, at least, has a debatable character.

At the moment, the authors consider various types of legal capacity of legal entities: general legal, sectoral, and special [21], general (universal) or limited (special) [12], in terms of credit organizations directly, some authors tend to the fact that credit organizations have special legal capacity [15], others insist on general legal capacity [19], others support exclusive legal capacity [20], they also talk about partially limited legal capacity [22].

A legal entity with the purpose of carrying out its daily activities should have the same rights and obligations as an individual. However, the first concept is much more meaningful and includes collective interest, concerted actions, etc. [1] In the civil legislation of the Republic of Tajikistan, the issue of legal capacity and legal capacity of individuals has found its detailed consolidation, and as for a legal entity, Article 49 of the Civil Code of the Republic of Tajikistan (Civil Code of the Republic of Tajikistan) establishes only its legal capacity, and nothing is provided about its legal capacity either in the said act or in others. A possible reason for such a provision in the legislation is that a legal entity carries out its activities through its bodies, which ultimately consist of individuals making decisions. This argument emphasizes that a legal entity is a continuation of the will of an individual, which is why the legal capacity and legal capacity of a legal entity has not found detailed consolidation in legal norms, as in the cases of an individual, the author emphasizes. De facto, legal entities as an "artificial' entity personifies a union of people. This view of the issue also comes from the theory of fiction. The question of whether legal entities have legal capacity is complicated [10], but there are countries (Switzerland, Ukraine, Moldova) where the concept of legal capacity of legal entities is enshrined in law [17].

Thus, the question of the legal personality of credit institutions and the related category of legal capacity and the ratio of legal entities and credit institutions in this matter requires a comprehensive analysis.

As you know, by virtue of civil legislation, an individual acquires legal capacity from the moment of birth, and his full legal capacity arises gradually from age, taking into account his mental state, and from the moment of acquisition of the latter, he becomes a legal entity. Again, I would like to refer to a different view of S. B. Polich: "We believe that it is legal capacity that arises after legal personality, but not vice versa! [16]. Since the emergence of civil legal capacity, its "happy owner" is a legal entity because the corresponding right and corresponding duty are already recognized for him. Legal personality can arise even when legal capacity has not yet arisen, and take place in the absence of it." The moment of acquisition of legal capacity and legal capacity is established in the Civil Code of the Republic of Tajikistan, and the question of the moment of acquisition of legal personality is not established either in the above or in other legal acts. However, the authors tend to believe that the categories of legal capacity and legal personality are categories of general and private, where the former is considered private, and the latter is general. Below are additional arguments in favor of this position.

In the case of legal entities, as well as with credit organizations, the onset of legal personality has its own specifics, and concepts such as birth, death, mental state, age, guardianship and guardianship, partial, incomplete, and limited legal capacity, and its deprivation do not apply to them. Although we should not forget about the restriction of the rights of legal entities, as well as credit institutions, within the framework of the norms provided for by law. In accordance with art. 49 of the Civil Code of the Republic of Tajikistan, a credit organization as a legal entity has civil rights corresponding to the types of activities on the basis of the license obtained and bears obligations related to this activity, i.e., credit organizations have civil rights and obligations and this means that they have legal capacity, and this, as we know, is half of what is necessary for credit organizations to participate in civil legal relations as an independent entity.

In order to participate directly in civil legal relations, a credit institution, in addition to having civil rights and obligations, must also have the ability to acquire, create, exercise, and perform civil rights and obligations by its actions. For an individual, such an ability is provided by his legal capacity, which a legal entity does not have, at least the law does not establish his legal capacity for the latter. The Civil Code of the Republic of Tajikistan does not establish a norm on the legal capacity of legal entities that would be applicable to credit institutions. This issue is considered only in the doctrine. The question of whether the concept of legal capacity is applicable to legal entities in general, and to credit organizations in particular, on the one hand, is indisputable, and on the other relatively doubtful. The reason is that the legal capacity applied to an individual is related to his age and mental state, and these indicators are completely not applicable to legal entities, as well as to credit organizations. Maybe the legal capacity of legal entities is a special kind of legal capacity? In the current position of the law and doctrine, there is no sufficient and reliable argument in favor of this hypothesis, and there are enough facts to refute it.

With one legal capacity established by law, it is impossible for these entities to cover, evaluate and ensure their legal status as subjects of civil legal relations. Civil legal capacity grants them civil rights and obligations, but not the possibility of independent exercise of these rights and obligations [7]. For individuals, such an opportunity was provided by their legal capacity, and for legal entities in general, and credit institutions in particular, the law does not provide for a category of legal capacity and, due to the above argument (age and mental state), it is not considered permissible to legislate such a category for these entities.

In this regard, the question arises, how can a legal entity and a credit institution, having no legal capacity, i.e., the ability to create civil rights and obligations for themselves and fulfill them, by their actions, take part in a civil legal relationship as an independent entity?

It is stated that legal entities acquire legal capacity and legal capacity simultaneously from the moment of state registration [17]. Due to the fact that the legal capacity and legal capacity of legal entities are constantly and inextricably linked throughout their existence, there is a widespread opinion that there is no need to separate these elements [12]. The authors emphasize one exception to this theory. The fact is that some authors, referring to the content of the law, note that legal entities whose activities are subject to licensing acquire legal capacity after obtaining this license and, accordingly, the legal capacity of a credit institution as a legal entity whose activities are licensed should occur after obtaining a license, and not after state registration, as is the case with legal entities engaged in activities that do not require a license. For example, A. V. Mul cites the opinion of many authors that the moment of acquisition of legal capacity and legal capacity of legal entities may not coincide [14]. This may be the case when a license is required to carry out certain types of activities. From this theory, it turns out that the legal personality of legal entities operating under a license is formed in two stages: first, from the moment of state registration, they acquire legal capacity, and after obtaining a license, they acquire legal capacity and eventually become legal entities.

Credit organizations receive the right to carry out banking activities from the moment of obtaining a license. Credit organizations receive a license after they are created and pass state registration. The above opinion suggests that after the creation of a credit institution, it acquires civil rights and bears obligations, i.e., for the time being, it acquires only legal capacity, but if a credit institution has not yet received a license, it does not have the right to carry out banking activities, as a result of which it cannot acquire civil rights for itself and exercise them, create civil duties for itself and fulfill them, so it is still incapacitated and therefore non-legal [4].

This view of the issue on the part of the authors suggests that the doctrine discusses two types of legal personality of legal entities.

Legal personality of legal entities whose activities are not subject to licensing. Legal personality of legal entities whose activities are subject to licensing.

1. Legal entities acquire legal capacity after creation, i.e., after state registration and entry of relevant information into the register. Legal entities acquire legal capacity not after creation, as is the case in the first case, but after they receive a license.
2. Legal personality comes immediately. Legal personality is acquired in two stages.
3. There is no need to separate the categories of legal capacity and legal capacity. There is a need to separate the categories of legal capacity and legal capacity.

Given their arguments, the authors tend to support the above theory about the two-stage formation of the legal personality of legal entities operating under a license. First of all, it should be emphasized that the most important indicator of legal capacity, as an integral element of the legal personality of each participant in civil legal relations, is the ability to exercise their civil rights and obligations directly by their actions. The composition of the capacity consists of four main elements forming its structure:

- acquire civil rights by their actions;
- to exercise civil rights by their actions;
- create civil duties by their actions;
- to fulfill civil duties by their actions.

From the moment of its creation, even before obtaining a license, a credit institution has the right to acquire civil rights and obligations for itself and fulfill them by its actions. In order to equip workplaces in non-residential premises, a credit institution may conclude civil contracts with other entities, for example, with a furniture company, manufacturers, suppliers, contractors, etc. This fact indicates that the credit institution has not yet received a license, already has legal personality, participates in civil legal relations, acquires civil rights and obligations for itself by its actions, and fulfills them. From the moment of creation, before obtaining a license, a credit institution enters into an NBT with respect to what is characterized as administrative and legal, organizing working personnel enters into labor contracts with employees thereby entering into labor relations, concluding various transactions enters into civil law relations, etc.

After the credit institution has been established and information about it has been entered into the register by virtue of Article 10 of the Law of the Republic of Tajikistan "On Banking Activities", the National Bank of Tajikistan (NBT) and the credit institution participate in the process of obtaining a license, i.e. after the creation of a credit institution before obtaining a license, legal relations already arise between the credit institution and the NBT. The parties to this legal relationship are the credit institution and the NBT [2]. For example, by virtue of Chap. 2 of the said article in order to prepare a conclusion when issuing a license, if necessary, the NBT requests additional information from the credit institution, which is a direct confirmation that the NBT participates in a legal relationship with the credit institution and thereby the legal personality of the credit institution is certified.

Thus, it turns out that there is no two-stage acquisition of legal personality. Those legal entities that carry out their activities on the basis of a license acquire legal personality from the moment of creation, as well as those legal entities whose activities are not subject to licensing. The moment of acquisition of legal personality from all legal entities, regardless of the need to obtain a license, is the same and is associated with the moment of creation, i.e., passing state registration and entering the relevant information into the register [6].

In this context, it is necessary to agree with the fair remark of Frolova E. E. that state registration is sufficient for a credit institution to engage in certain activities within the framework of the law and make transactions not prohibited by law, except for the one for which the presence of a special permit (license) is considered mandatory [3, 5].

As for the category of legal personality of credit organizations, it should be noted that a credit organization, as a legal entity, acquires legal personality from the moment of creation, and even if their activities are subject to licensing, this does not mean that their legal personality comes after the acquisition of a license, not to mention the unacceptability of using the category of legal capacity in relation to any legal entities, including credit organizations [9]. The above argument sufficiently justifies this position. The license only grants the credit institution the right to carry out certain operations or transactions, i.e., to engage in activities for which obtaining a license is mandatory, but not as it does not endow the credit institution with legal capacity. This is like a diploma that entitles an individual to engage in a certain profession or hold a certain position, without affecting his legal capacity. A license does not make a credit institution legally capable or capable, as does a diploma for an individual.

Thus, from the moment of creation of a credit institution, i.e., after passing state registration and entering information about it into the register before obtaining an NBT license, a credit institution acquires civil rights for itself by its actions, exercises acquired civil rights by its actions, creates civil duties for itself by its actions, fulfills these duties by its actions, therefore, from that moment it is considered a legal entity and has the right to participate independently and fully in civil legal relations. All legal entities, regardless of whether their activities are subject to licensing or not, acquire legal personality from the moment of creation, i.e., passing state registration and entering information about it into the register.

The current wording of the civil legislation of the Republic of Tajikistan, where only the term "legal capacity" is established in relation to legal entities, is not quite able to express the true essence of the legal personality of legal entities, as well as credit organizations. Polich S. B. rightly noted that: "Using simple logical rules, admitting and applying a formal analysis of the verbal phrases and expressions used, it is necessary to recall the rule formulated long ago in civil law that subjects of law and legal personality are not equivalent concepts, as well as in no way should legal personality be identified with legal capacity" [16]. Legal capacity implies the ability to have civil rights and duties that everyone owns, and not everyone has the ability to acquire civil rights and duties by their actions and fulfill them. One term "legal capacity" established in the civil legislation of the Republic of Tajikistan for legal entities is impossible to cover the entire scope of its legal status and its capabilities as a subject of civil legal relations. Different categories are different concepts.

The present provision of the legislation, on the issue under consideration, can be characterized as an attempt to cover the entire scope of the concept providing the legal status of the subject by means of one term. The term "legal capacity" does not endow a legal entity and, together with it, a credit institution with the opportunity to acquire civil rights and obligations for themselves and fulfill them by their actions. The concept of legal capacity is not comprehensive and does not cover the content of legal personality. However, the term legal personality may well cover the whole essence of the legal status of the subject.

It is possible to speak about a separate allocation of legal capacity and legal capacity only when it comes to the legal personality of individuals. A legal entity is an inanimate entity, not a person born as a person (an individual), but a fictional one. However, this question is not quite unambiguous as it may seem at first glance. The fact is that there has been a discussion in the scientific community for a long time about the fictitiousness and unreality of the term "natural person". It is noted that this term is artificially created to denote human participation in legal relations [11]. Such a judgment emphasizes that an individual, as well as a legal entity, is a fictitious category. If both terms are fictional and unreal, then why does one (individual) have legal capacity and the other (legal entity) does not? The opinion of Ramalho [8] based on the legislation of Portugal gives the most reasonable answer to this question. The legal personality is divided into a single one for individuals and a collective one for legal entities, and at the same time, the legal personality of the former is considered ultralegal and supra-legal, and the latter legal and lawful. This view of the issue is completely new in our legal doctrine and very undeveloped on the part of the authors, but due to its relevance deserves attention, however, we will not unreasonably go so far.

Continuing our own reasoning, we take the liberty to believe that in the matter of legal entities, including credit institutions, it is necessary to talk about their legal personality, without dividing its elements into legal capacity and legal capacity. The legal personality of an individual consists of two elements (legal capacity and legal capacity), but there is no legal entity, the first becomes legal personality from the acquisition of full legal capacity, and the latter from the moment of formation. Taking into account the specifics of this subject, the separation of the elements of its legal personality is unreasonable. Legal entities, as well as credit organizations, do not possess legal capacity and legal capacity, they possess legal personality as a single and whole category, covering the entire scope of their legal capabilities and legal status as a participant in civil law relations.

Due to the lack of a legal definition, the concept of legal personality, it is possible to give it a certain doctrinal formulation. Legal personality is a broad and comprehensive concept that covers the categories of legal capacity and legal capacity, provides a person with the opportunity to act as an independent subject of civil rights and obligations, have these civil rights and obligations, and exercise them by their actions, bear civil liability.

References

1. Arcelia, E., Adriano, Q.: Natural persons, juridical persons and legal personhood. Mex. Law Rev. **8**(1), 101–118 (2015)
2. Belkova, E.G.: On the categories of legal personality and legal capacity. Bull. Irkutsk. State Econ. Acad. **6**, 55–58 (2006)
3. Ermakova, E.P., Frolova, E.E.: Pravovoe regulirovanie tsifrovogo bankinga v Rosii I zarubezhnykh stranah (Evropeyskiy souz, Ssha, Knr). (Legal regulation of digital banking in Russia and foreign countries (European Union, USA, China). Bull. Perm Univ. Leg. Sci. **46**, 606–625 (2019)
4. Ermakova, E.P., Rusakova, E.P., Sitkareva, E.V., Frolova, E.E.: Main components of protecting consumers of financial products in Asian-Oceanic (APAC) countries. Int. J. Eng. Technol. (UAE) **7**(4.38), 157–162 (2018)
5. Frolova, E.E.: Licensing of banking operations as a form of state regulation of banking activities (theory, practice, problems), p. 189. Ph.D. in Law thesis. Mocscow (2004)
6. Idrisova, E.A.: Features of the legal status of a credit institution. Theory Pract. Mod. Sci. **10**(28), 74–80 (2017)
7. Inshakova, A.O., Goncharov, A.I., Mineev, O.A., Sevostyanov, M.V.: Amendments to the civil code of the Russian Federation: contradictions of theory and practice. Contrib. Econ. 147–153 (2017). https://ideas.repec.org/h/spr/conchp/978-3-319-55257-6_20.html
8. Joaquim, R. (2019). A personalidade jurídica das pessoas coletivas: evolução dogmatic. Vargas revista direito GV. São Paulo, 15(3). Retrieved from https://www.scielo.br/j/rdgv/a/FxBkjb5DN4tvycCbgGzJ9ZR/?lang=pt
9. Kashirin, S.V.: Legal regulation of the activities of credit institutions in the banking services market, p. 216. Ph.D. in Law thesis. Moscow (2011)
10. Kirsanova, A.V.: The concept of personality, capacity, capacity of legal entities. Yuridicheskiy Vestnik SamGU **1**(2), 134–140 (2015)
11. Komzolova, Y.A.: Fiction individuals in the Russian legal science. Monit. Law Enforc. **3**(12), 68–70 (2014)
12. Lomaev, O.Y.: A civil capacity and legal capacity of legal entities. Int. Sci. J. Innov. Sci. **6**, 152–154 (2016)
13. Mark, A.G.: Legal Personality. International Encyclopedia of the Social & Behavioral Sciences, pp. 8646–8650 (2001)
14. Mul, A.V.: The moment of the emergence of legal capacity and legal capacity of legal entities. Issues Mod. Jurisprud. **7**(68), 33–38 (2017)

15. Ovechkin, A.P.: The legal status of a credit institution. Bull. Law Inst. MIIT **2**(10), 104–111 (2015)
16. Polich, S.B.: Nekotorye osobennosti pravosub"ektnosti lits—uchastnikov grazhdanskikh i semeynykh otnosheniy [some peculiar features of legal personality of persons participating in civil and family relations]. bulletin of Perm University. Law–Perm University Herald. Juridical Sci. **4**, 664–684 (2018)
17. Proshin, S.N.: The capacity of a legal entity is one of the components of its legal. Notary. Moscow Yurist **2**, 26–29 (2012)
18. Pyatkov, D.V.: The position of domestic of the jurists on the definition of natural person: the past and the present. News ASU Leg. Sci. **6**(104), 187–191 (2018)
19. Rakhimzoda, M.Z.: Problems of legal regulation of credit institutions in the Republic of Tajikistan. Legislation **1**(25), 25–32 (2017)
20. Rybakova, S.V.: Financial legal personality of credit organizations. J. Russ. Law **9**(249), 47–54 (2017)
21. Solovtsova, E.A.: On certain issues of the legal status of credit institutions in the Russian Federation. In: Actual problems of modern science: V International Scientific and Practical Conference. Stavropol University. Stavropol, Feb. 16–20, pp. 211–214 (2016)
22. Tarasenko, O.A.: Studying the category legal capacity of a credit institution through the prism of law, doctrine and practice. Bull. O.E. Kutafin Univ. (MSAL) **3**(43), 24–32 (2018)
23. The Law of the Republic of Tajikistan On Banking Activity dated May 19, 2009, No. 524; [Electronic resource]. http://mmk.tj. Last accessed 16 Jan 2021
24. Trofimov, K.T.: Credit organizations in the banking system of the Russian Federation (civil law problems), p. 392. Doctor of Law thesis. Moscow (2005)

Part II
Legal and Technological Support for the Sustainable Development of Ecopreneurship and Environmental Protection Measures in Economic Activity

Chapter 12
Modern Problems of Digitalization of Environmental Protection in the Russian Federation

Agnessa O. Inshakova, Denis E. Matytsin, and Svetlana Yu. Kazachenok

Abstract The chapter notes that the digitalization of the economy carried out in Russia has given rise to a new phenomenon—the digitalization of law, including environmental law. Environmental digitalization provides an opportunity to cope with the unprecedented environmental problems that are faced with the environment today, including climate change, increased pollution of air, water bodies, and soil, which poses threats to human life and health. It is digitalization technologies that can help change the situation, providing humanity with a secure and sustainable future. The use of digital environmental technologies for national environmental monitoring, improving the safety of industrial enterprises, in agriculture, and also in order to improve the comfort of life in an urban environment seems to be very promising.

12.1 Introduction

Today, the global community, including the Russian Federation, is in transition to a new technological paradigm, referred to in various policy documents as the "digital economy' and the digital revolution". The main feature of this new way of life is not

A. O. Inshakova (✉)
Head of Department of Civil and International Private Law, Volgograd State University, Volgograd, Russia
e-mail: ainshakova@list.ru; gimchp@volsu.ru

Department of Law, Volga Branch of Volgograd State University, Volzhsky, Russia

D. E. Matytsin
Insitute of Law, Volgograd State University, Volgograd, Russia
e-mail: Dmatytsin@ya.ru; gimchp@volsu.ru

Department of Civil Law and Procedure, Volga Branch of the International Law Institute, Volzhsky, Russia

S. Yu. Kazachenok
Department of Civil and International Private Law, Volgograd State University, Volgograd, Russia

A. O. Inshakova and E. I. Inshakova (eds.), *New Technology for Inclusive and Sustainable Growth*, Smart Innovation, Systems and Technologies 288,
https://doi.org/10.1007/978-981-16-9808-8_12

only the emergence of fundamentally new information technologies but also changes in the state of public institutions, the mechanism of interaction between civil society and the state, and the emergence of new values and ideologies. These processes affect all sectors of social life, including environmental protection. Considering the transboundary nature of environmental problems, one of today's tasks is holding a new UN international conference on "digitalization and sustainable development" in 2022, which could summarize the progress of the world community toward the goals and values of sustainable development, launched at the UN international conference on sustainable development in Rio de Janeiro in 1992, and discuss strategies and plans for the world community in the future.

12.2 Methodology

In the process of research there were used general scientific methods, such as formal-logical, dialectical, system-structural, and critical knowledge. When interpreting the results of the study methods of synthesis, classification and generalization were used. Particular scientific methods were also used in the work: formal-legal, the principle of evaluation of legal processes, the method of comparative analysis, etc.

12.3 Results

In modern science there is no generally recognized definition of digitalization. Some authors believe that digitalization is "not only the use of digital technology and the creation of new opportunities, including the realization of interests in profit, it is a process of transition to digital business, which requires detailed regulation by law" [1]. Other authors note that "digitalization can be considered as a process of application of digital technologies in various spheres, while digital transformation involves the occurrence of specific results from their application" [2]. Thus, in Russia, the main emphasis is placed on the technical and/or economic aspects of digitalization, which can be explained by the special attention of the state to the problems of the digital economy.

Not all areas of digitalization of the economy and new technological solutions are applicable to the goals and objectives of digitalization of environmental protection (for example, electronic identification technologies have no environmental specifics). It seems that the main tools of the digitalization of environmental policy and law can be the use of specialized software at production facilities, contributing to environmental protection and optimization of the use of natural resources; maintaining an open environmental policy of companies, with the ability to track its execution through the timely publication of reports and statistical information (at present, an open environmental policy is conducted by a very low percentage of companies); development of automated some of these tools have already been tested abroad. For example, the German chemical-industrial company "Henkel" has been using technology in its activities for several years, which greatly facilitates the work with large amounts of statistical information [3].

A special place among the areas of digitalization of the economy can be taken by the digitalization of banking, within which a new direction, "green banking," is already being developed. This strategy means that the bank finances environmentally friendly projects related to the development of renewable energy sources, as well as the implementation of environmental technologies. Along with this, many banks are already assessing the environmental risks of projects in lending. Among the equally important areas of digital technology's impact on environmental protection is the Internet of Things, as the 2017 International Declaration on the Internet of Things for Sustainable Development was adopted in Geneva, which addressed the use of Internet of Things technologies to combat climate change, supporting the implementation of the Internet of Things in urban and rural areas to create more "smart" and sustainable cities and communities. Environmental control sensors help measure temperature, humidity and air composition, radiation levels, and can help save energy [4].

Thus, the modern legal regulation of issues of digitalization of the economy affects a very wide range of areas, many of which are not relevant to environmental protection. At the same time, there is no doubt about the possibility of using the achievements of digitalization for environmental purposes, related to the automation of production, the use of sensors, the Internet of things, risk analysis and accounting, etc.

In Russia, the achievements of digitalization can be used in the field of environmental protection in a variety of sectors. For example, there are clear prospects for the use of digitalization in the environmental assessment of projects, the chipping of plants and wild animals (which are directly related to Big Data), the emergence of new algorithms for data collection and storage, and the opportunities for their integration and algorithmic management of the environment. It is possible that new technologies for collecting data on environmental pollution can lead to automatic management of technological processes and improve the economic mechanism of environmental management. To see the full picture of changes occurring in the environment is to seriously improve environmental safety. However, as noted in the scientific literature, the anticipated results of digitalization will inevitably entail essential changes in economic, social, production, cultural, and psychological relations, thereby affecting the place and role of humans as "homo sapiens" in the world. Therefore, one of the main dangers of the ideas and technologies of Industry 4.0 may be too much (and perhaps even one-sided) reference to technology and economics while ignoring the social, value, moral, environmental, and energy aspects of its consequences [5]. There is no place for real ecological development in the paradigm of continuous economic growth. The prescription of the growth of consumption as a civilizational factor inevitably leads and will lead to the degradation of nature. The only question is to what extent and with what consequences. While agreeing with these fears, let us note that to date, the development of digital technologies in the field of environmental protection does not give grounds for such pessimistic predictions. The experience of the use of digital technologies in Russia and many other countries around the world allows us to conclude that it is possible to drastically reduce the negative impact of humans on nature in the following important sectors of public life:

(1) With the help of digital technologies it is possible to increase the rationality of the use of certain types of natural resources. For example, they can be used to ensure more rational use of forest resources, to reduce water loss through the creation of online networks, and the use of sensors that transmit information about leaks and water consumption.
(2) Digitalization of environmental management. Digital technologies open up new opportunities both for environmental monitoring (for example, in terms of protecting forests from fires or protecting agricultural land from overgrazing) and for state environmental oversight (automating the system of control over emissions and discharges of harmful substances into the environment, waste disposal). Further development of digital workflow can be used in the issuance of environmental licenses, calculation of emissions, and environmental payments.
(3) Digitalization and circular economy. In its most general form, the latter is a scientific concept that proposes, through a series of economic measures, to reduce the production of waste, as well as to ensure its reuse. For example, the digitalization of management and education reduces the consumption of paper, which reduces its production and deforestation. This will also be facilitated by the digitalization (in the long term) of court proceedings, in particular the filing of lawsuits (including environmental cases) via the Internet, as well as the development of mediation technologies in environmental dispute resolution. And this is by no means the only positive consequence of the introduction of the ideas of the circular economy. The implementation of the concept of circular economy will allow the reuse of renewable, recyclable, and biodegradable materials in subsequent life cycles to reduce the negative impact of waste on the environment. Waste recovery and recycling will enable the recovery or use of materials, substances, and products whose useful life has expired or whose ability to be used (e.g., due to breakage) has decreased. However, such materials and products contain valuable substances, materials, energy, or other components, allowing them to be used for other purposes or to profit from their recycling.

A separate direction within the circular economy is the creation of an exchange or sharing platform (lease) of various objects of movable and immovable property, which reduces the volume of their production (and consumption of natural resources), as well as the amount of waste produced. If we talk about the practical side of this activity, Brightstar, for example, has resold about 15 million used digital devices. It refurbishes more than 22 metric tons (Mt) of printed circuit boards, 42 Mt of screens, and 35 Mt of batteries annually. Their Buyback and Exchange program takes used devices from developed countries and sells them in developing countries, targeting retailers and wireless carriers [6].

(4) Logistics. Logistics accounts for 13% of global emissions, and its influence continues to grow as more and more goods are exchanged around the world. In 2016, the U.S. Environmental Protection Agency released a report indicating that greenhouse gas emissions from trucking in the U.S. have increased by 76.3% since 1990. And while trucking remains the largest contributor to global warming in the transportation sector, the negative environmental impact of aviation and maritime transport is also growing. Meanwhile, digital technology can reduce emissions from the logistics industry by as much as 3.6 billion metric tons by 2025. That is a 10% reduction over a ten-year period. However, it is worth emphasizing that we have not yet found a way to break the link between economic growth, emissions growth, and resource use. Failure to update current business practices will result in a global gap of 8 billion tons between the supply and demand of natural resources by 2030. Financially, this would mean a loss of $4.5 trillion in economic growth by 2030 [7].

(5) The utility sector. The use of digital technology is necessary to connect residents' homes to electricity. At the same time, thanks to developments in mobile money, households wishing to connect to the generation unit pay with smart meters and mobile payment methods. In addition, we can highlight the "green benefits" of mobile money (it can be used to pay for a variety of services, including the installation of solar power generators).

(6) Education. Effective environmental protection requires qualified personnel, including not only state and municipal environmental management officials but also employees of businesses, environmental public organizations, and all other citizens. In the conditions of COVID-19 pandemic their training can be carried out using remote digital technologies.

12.4 Conclusion

The digitalization of the economy carried out in Russia has generated a new phenomenon—the digitalization of law, including environmental law. It is ecological digitalization that can help cope with the unprecedented challenges facing humanity in the twenty-first century, including climate change and increasing pollution that threaten human life and health. Digitalization technologies can help improve our lives by providing humanity with a secure and sustainable future. Digital environmental technologies are becoming very promising in global and national environmental monitoring, in improving the safety of industrial plants, in agriculture, and in making life more comfortable in urban environments. Programs of digital environmental protection could have the greatest effect in logistics, transportation, natural resource use, production, and consumption waste management (especially with the introduction of circular economy standards), as well as in a number of other sectors. However, in order to increase the efficiency of these activities, separate chapters should be added to the environmental legislation of the Russian Federation and its constituent entities, providing for specific areas in which digital technologies can be used (environmental oversight, monitoring, accounting, interaction between authorities and the public, etc.).

Acknowledgements The study was carried out with the financial support of the Russian Science Foundation (project No. 20-18-00314).

References

1. Cherevko, V.V., Shepelev, D.V.: Digitalization and legal interests of the Russian Federation. Actual Probl. Mod. Sci. Pract. **2**, 132–137 (2019)
2. Zubarev, S.M.: Legal risks of digitalization of public administration. Actual Probl. Russ. Law **6**, 23–32 (2020)
3. Zherebtsova, O.V.: Digitalization of the ecological industry of the region. In: Actual Problems of Science and Practice: Gatchina Readings 2019. Collection of Scientific Papers on the Materials of the VI International Scientific-Practical Conference, pp. 458–460. State Institute of Economics, Finance, Law and Technology, Gatchina (2019)
4. Kozlova, M.A., Fedosov, P.V., Izmailov, I.Y.: The impact of digital technology on the transition to a green economy. Mod. Econ. Success **3**, 164–171 (2020)
5. Inshakova, A., Frolova, E., Rusakova, E., Kovalev, S.: The model of distribution of human and machine labor at intellectual production in industry 4.0. J. Intellect. Cap. **2**(4), 601–622 (2020). https://doi.org/10.1108/JIC-11-2019-0257
6. World Economic Forum (2021). How can digital enable the transition to a more sustainable world? http://reports.weforum.org/digital-transformation/enabling-the-transition-to-a-sustainable-world/. Last accessed 22 Sept 2021
7. Dallemand, J.: Is digital transformation the key to environmental sustainability in logistics? (2021). https://blog.datumize.com/digital-transformation-and-environmental-sustainability-in-logistics. Last accessed 22 Sept 2021

Chapter 13
Role and Importance of Modern Technologies in Combating Global Climate Change: Legal Aspect

Denis E. Matytsin, Anatoly Ya. Ryzhenkov, Olesya P. Kazachenok, and Viktor F. Pon'ka

Abstract This Chapter argues in favor of the conclusion that climate change is of an objective nature and consists in the growth of average annual temperatures registered for several decades in many countries, including Russia. Undoubtedly, the causes of this phenomenon are complex; however, the growth of anthropogenic impact on nature, which manifests itself in the increase in greenhouse gas emissions, makes its contribution to the development of the dynamics of this process. This led to the development of international cooperation in mitigating the impact on climate and the elaboration of measures for the adaptation to climate change that has already occurred. To resolve these issues the 1992 United Nations Framework Convention on Climate Change was adopted as well as two international instruments developing and supplementing its provisions—the Kyoto Protocol and the Paris Climate Agreement. These international instruments stipulate a list of measures that must be taken by the countries joining these treaties, which will reduce anthropogenic impact on nature and slow the growth of average annual temperatures. Each country adopted a

D. E. Matytsin (✉)
Insitute of Law, Volgograd State University, Volgograd, Russia
e-mail: Dmatytsin@ya.ru; gimchp@volsu.ru

Department of Civil Law and Procedure, Volga Branch of the International Law Institute, Volzhsky, Russia

A. Ya. Ryzhenkov
Department of Civil and Private International Law (The Basic Department of the Southern Scientific Center of the Russian Academy of Sciences), Volgograd State University, Prospekt Universitetsky, 100, 400062 Volgograd, Russia
e-mail: gimchp@volsu.ru

O. P. Kazachenok
Department of Civil and Private International Law, Volgograd State University, Prospekt Universitetsky, 100, Volgograd 400062, Russia
e-mail: o.kazachenok@yandex.ru

V. F. Pon'ka
Head of the Department of Legal Support of the Russian Academy of Sciences, Honorary Worker of Higher Professional Education, Moscow, Russia
e-mail: ponka@pran.ru

A. O. Inshakova and E. I. Inshakova (eds.), *New Technology for Inclusive and Sustainable Growth*, Smart Innovation, Systems and Technologies 288,
https://doi.org/10.1007/978-981-16-9808-8_13

number of regulatory technical measures to fulfill the assumed international obligations, and Russia is not an exception in this case. Meanwhile, despite the adoption of a range of bylaws dedicated to the specific questions of fulfilling the assumed obligations, there is no systematic approach to the issue of climate protection at the level of federal laws in Russia, which requires correction (for example, through the supplementation of the Federal Law "On Environmental Protection" of January 7, 2002 with a special chapter). In this Chapter, it is also noted that some universal issues in the area of counteraction to global climate change are still relevant, they are typical for all countries and are to be resolved, in particular, the issue of climate refugees, compensation for harm caused by climate change and some other issues.

13.1 Materials

The regulatory framework of the study consists of instruments of international and national law. Among international treaties covering the main issues of counteraction to global climate change, the provisions of the 1992 UN Framework Convention on Climate Change, the 1997 Kyoto Protocol thereto, the 2015 Paris Climate Agreement as well as numerous Russian bylaws regulating Russia's fulfillment of the assumed international obligations were used in the Chapter. The latter include the following bylaws which are of the most interest: Decree of the Government of the Russian Federation of September 15, 2011 "On Measures to Implement Article 6 of the Kyoto Protocol to the United Nations Framework Convention on Climate Change" (with the Provision on the Implementation of Article 6 of the Kyoto Protocol to the United Nations Framework Convention on Climate Change), the Executive Order of the Government of the Russian Federation of December 25, 2019 "On the Approval of the National Action Plan of the First Stage of Climate Change Adaptation until 2022", the Directive of the President of the Russian Federation of December 17, 2009 "On the Climate Doctrine of the Russian Federation" and a number of other bylaws.

Works by Russian and foreign scholars dedicated to certain aspects of climate protection were used in the course of the study, in particular, the works by Brinchuk [2], Grinin [9], Druzin [6], Zolotova [24], Ivanova [11], Jaffe [12], Knox [13], Matveeva [14], Ovchinsky [16], Parker-Flynn [17], Sokolova [19, 20], and Falileev [8]. The works by the said authors made it possible to make a number of conclusions and suggestions aimed at the development of the legal support of technical measures for further climate protection (in the context of Russia).

13.2 Methods

A number of general scientific methods, in particular, formal logical, dialectical, systemic structural methods, and the method of critical cognition, were used in the course of the study. Synthesis, classification, and generalization techniques were applied in the interpretation of the results of the study. Specific scientific methods were used in this Chapter as well: formal legal method, the principle of legal process assessment, the method of comparative analysis and so forth.

13.3 Introduction

According to instrumental data on global surface temperature, the last three decades have been the warmest since the middle of the nineteenth century. The average global surface temperature of the first decade of the twenty-first century was 0.5 °C higher than in the period from 1961 to 1990 and 0.2 °C higher than in the period from 1990 to 2000. In its turn, the last decade of the twentieth century was warmer than the previous decades. Regional climate anomalies are also associated with the occurring global changes. In the territory of the Russian Federation, the warming is much more intense than for the Earth on the whole. In recent decades, the rate of warming in the territory of Russia on the whole has been more than twice the global rate, and more than four times in some regions such as the Arctic Zone of the Russian Federation. In addition, there is large territorial and seasonal heterogeneity of the temperature changes.

Against the background of climate warming, many regions of Russia experience the growing frequency and intensity of dangerous hydrometeorological phenomena, including floods, forest fires, squalls, whirlwinds, hurricanes, cloudbursts with thunderstorms, hail and squally wind, heat waves, severe droughts, etc. An important peculiarity of climate change and its effects in Russia is that they began significantly affect many economic industries—the power industry, agriculture and forestry, transport, construction, environmental protection, housing, and utilities. Threats to human health increase, especially because of atmospheric pollution. Moreover, the frequency and intensity of these threats in Russia have a very clear tendency to increase [18].

The said tendencies are reflected in a number of program policy documents adopted in the Russian Federation. For example, in the Climate Doctrine of the Russian Federation [5], it is stated that climate change is one of the most important international problems of the twenty-first century, which goes beyond scientific discussions, it is a complex interdisciplinary issue covering environmental, economic, and social aspects of sustainable development of the Russian Federation. The unprecedentedly high rate of global warming observed in recent decades is of particular concern to the Russian authorities. Modern science provides increasingly convincing evidence that human economic activity, associated primarily with greenhouse gas emissions, has a significant impact on the climate as a result of fossil fuel burning. In their turn, the authors of the Environmental Security Strategy of the Russian Federation until 2025 approved by Decree of the President of the Russian Federation No. 176 of April 19, 2017 draw attention to the fact that on average about 950 dangerous hydrometeorological phenomena (floods, drought, strong wind, heavy precipitation, etc.) causing significant damage to economic industries and the life of the population are registered in the territory of the Russian Federation every year. According to expert estimates, the material damage from dangerous hydrometeorological phenomena can reach one percent of the gross domestic product in certain years [21].

Hence it follows that state authorities of the Russian Federation attach great importance to the issue of global climate change, which cannot be solved on the scale and with the resources of a single country. This results in the increasing participation of Russia in the international cooperation in coping with global climate change and sets the objective at the national level to elaborate effective measures mitigating the effects of climate change or ensuring adaptation to them. The effects of climate change become a constant factor affecting the development of modern society; they are a big challenge for it. The development and adoption of the United Nations Framework Convention on Climate Change and two protocols to it (the Kyoto Protocol, which expired, and the 2015 Paris Climate Agreement) were an international response to this challenge [1], [22]. These instruments are aimed at decreasing the growth of global surface temperature by reducing greenhouse gas emissions.

Meanwhile, despite Russia's active participation in the international cooperation and development of a number of bylaws aimed at fulfilling the assumed international obligations, the climate is not an object of protection from the perspective of environmental legislation of the Russian Federation. Measures organized for its protection are not systematic. Undoubtedly, the development of a legal mechanism to combat global climate change (with the introduction of amendments to environmental, financial, administrative, and other legislation) is only an external reflection of solutions to this strategic task requiring new knowledge, scientifically grounded proposals and recommendations of representatives of various natural, technical, and other sciences. The development of an affective mechanism to combat climate change can be implemented only in case of a set of interdisciplinary, basic, exploratory, and applied research in the study of climate change, climate engineering, assessment of socio-economic and environmental losses and damage as well as potential benefits from occurring and expected climate change. In this regard, the purpose of this Chapter is to make a contribution to the settlement of this complex issue from the perspective of legal science.

13.3.1 International Cooperation in Combating Global Climate Change: Legal Aspect

The development of international climate cooperation has passed several stages associated with the gradual awareness of the depth of the problem of climate change by the most part of the international community.

(1) The starting point of international climate cooperation is the adoption of the United Nations Framework Convention on Climate Change of May 9, 1992

[23]. According to Article 2 of this Convention, its ultimate objective is "to achieve, in accordance with the relevant provisions of the Convention, stabilization of greenhouse gas concentrations in the atmosphere at a level that would prevent dangerous anthropogenic interference with the climate system. Such a level should be achieved within a time-frame sufficient to allow ecosystems to adapt naturally to climate change, to ensure that food production is not threatened, and to enable economic development to proceed in a sustainable manner". To achieve the said objectives, the convention stipulates a very wide list of obligations which must be assumed by the states that have signed this document and the protocols to it.

(2) The Kyoto Protocol to the 1992 Framework Convention on Climate Change. The main provisions of the Kyoto Protocol included the following areas of activity: determination of the permissible volume of greenhouse gas emissions from 2008 to 2012 for all industrially developed countries that signed the document; development of the mechanism for the correction of quotas for certain countries (international trade in quotas, implementation of joint projects for the introduction of technologies ensuring emission reduction); development of the mechanisms of control over the levels of emissions (the need to create national anthropogenic emission assessment systems, monitoring of emissions and discharges). The Protocol stipulated a system of quotas for greenhouse gas emissions. The essence of this system consisted in the fact that every country was allowed to emit a particular amount of greenhouse gases. In addition, it was supposed that some countries or companies would exceed the quota of emissions. In such cases these countries or companies could buy the right to additional emissions from the countries or companies the emissions of which were less than the assigned quota [14].

It appears that the significance of the Kyoto Protocol is that, on the one hand, it was intended to create legal conditions to ensure the limitation and reduction of greenhouse gas emissions into the atmosphere, consequently, the main purpose of the adoption of the Kyoto Protocol is to reduce the anthropogenic load on one of the components of the environment—the air. On the other hand, the said document contained legal regulations due to which "the international market mechanism of the settlement of global environmental issues began to form". First of all, legal conditions for the trade in quotas for greenhouse gas emissions into the air were created, which predetermines the considerable economic interest of states in the reduction of greenhouse gas emissions [24].

(3) The Climate Change Conference (Copenhagen, December 7–19, 2009). It was held like "the last stand" to resolve issues related to the formation of the climate regime after 2012. This expected importance was reflected in the fact that the Conference was held at a higher level than usually. The result of the Copenhagen Conference was a political agreement, in particular, the Copenhagen Accord. It was not adopted by all the governments but raised a number of key issues [20].

(4) The modern stage of international cooperation in combating climate change begins with the 2015 Paris Climate Agreement [1]. This Agreement involves the following steps: holding the increase in the global average temperature to well below 2 °C above pre-industrial levels and pursuing efforts to limit the temperature increase to 1.5 °C, recognizing that this would significantly reduce the risks and impacts of climate change; increasing the ability to adapt to the adverse impacts of climate change and foster climate resilience and low greenhouse gas emissions development, in a manner that does not threaten food production; and making finance flows consistent with a pathway toward low greenhouse gas emissions and climate-resilient development.

Meanwhile, despite the importance of this Agreement, it does not stipulate any specific sanctions or other consequences for the countries that fail to fulfill their obligations. This leads to a conclusion that this kind of consensus is extremely fragile. Even if only a few countries withdraw from the Agreement, this can have negative consequences, serving as an example for other countries, which can also cease performing their obligations.

One of the possible solutions to increase confidence between the countries participating in the 2015 Paris Agreement is to create a system of deposits within the framework of a special international fund for common property management: the parties to the Agreement will make an advance deposit to this special international fund understanding that all or part of their contribution will be canceled if they do not fulfill their obligations to reduce greenhouse gas emissions. In addition, we should note that the purpose of this scheme is not to eliminate specific cases of non-fulfillment of obligations, it is rather to stimulate an adequate degree of the initial confidence of countries in mutual promises, which can prevent breaches of the Paris Agreement [6].

Therefore, the need to promote international interaction in the area of combating global climate change is explained by the lack of a direct relationship between the place of greenhouse gas emissions and climate change [19]. To settle this issue, the international community has adopted a number of international documents, the most important of which at the present stage of climate cooperation is the 2015 Paris Climate Agreement. The slowdown of the increase in the average annual temperature, the reduction of greenhouse gas emissions and the preservation of human civilization in its current form depend on the implementation of its provisions. The search for guarantees for the implementation of the provisions of the Paris Agreement is still relevant at the moment, and the main issue is the creation of conditions for mutual confidence among the parties to the Agreement.

13.3.2 Development of National Legislation in the Area of Counteraction to Global Climate Change: Trends and Prospects

Two strategies for overcoming the effects of global climate change are considered commonly recognized in international science: mitigation and adaptation. Mitigation can be defined as anthropogenic intervention aimed at reducing the sources or emissions of greenhouse gases. On the contrary, adaptation is a regulator of natural and human systems in response to actual or expected climate change or its effects which mitigate the harm or provide new beneficial possibilities. Measures to mitigate the effects of climate change are often preventive and focus on the sources of climate change, while adaptation is created as a measure to respond to the effects of the already changed climate [17].

It appears that the climate response, both to anthropogenic impact and to measures to mitigate it, is characterized by its delay in relation to this impact. Within the framework of the national climate policy, this peculiarity predetermines an important role of timely adaptation to the climate change which is inevitable in the coming decades. At the moment, the main objectives of the Russian climate policy are to strengthen and develop the information and scientific basis for decision-making in the field of climate (in particular, to improve the scientific, technical, and technological potential of the Russian Federation), to elaborate and implement prompt and long-term measures to adapt to climate change and measures to mitigate anthropogenic impact on the climate, to ensure Russia's participation in initiatives of the international community in addressing questions associated with climate change and related issues. Adaptation to climate change is necessary for Russia to reduce losses and use the benefits associated with the observed and future climate change. Measures for the adaptation to climate change are stipulated by resolutions of Russian public authorities with consideration of international arrangements. The planning, organization, and implementation of measures for the adaptation to climate change, including anticipatory adaptation, are organized with consideration of industrial, regional, and local specific features as well as the long-term nature of these measures, their scope and depth of impact on various areas of the social life, economy, and state.

The following estimates are essential components in the development and planning of measures for the adaptation to climate change: vulnerability to the adverse effects of climate change and the risks of related losses; possibility to gain benefits associated with the favorable effects of climate change; cost-effectiveness, efficiency (including economic efficiency) and practical feasibility of the corresponding adaptation measures; adaptation potential with consideration of economic, social and other factors for the state, economic sectors, population and individual social groups.

To mitigate the effects of climate change, state authorities of Russia must focus a maximum of their efforts on the reduction of anthropogenic greenhouse gas emissions and the increase in their absorption by sinks and reservoirs. Measures are also required to improve energy efficiency in all economic sectors, develop the use of renewable and alternative energy sources, decrease market distortions, implement financial and fiscal policy measures that stimulate the reduction of anthropogenic greenhouse gas emissions, protect and enhance the quality of greenhouse gas sinks and reservoirs, including rational forestry, afforestation and reforestation on a sustainable basis.

Like any complex phenomenon, climate change within a particular country can have not only negative but also certain positive effects. For example, with respect to the Russian Federation, the negative effects of the expected climate change are as follows: increased health risks (higher morbidity and mortality) for some social groups of the population, increased frequency, intensity and duration of droughts in some regions, extreme precipitation, floods, overwatering of soil dangerous for agriculture in other regions, increased fire danger in woodlands, degradation of permafrost in the northern regions damaging buildings and communications, disruption of the ecological balance, displacement of some biological species by other ones, spread of infectious and parasitic diseases, increased energy costs for air conditioning in the summer season for a significant part of settlements. In turn, the possible positive effects of the expected climate change with which the significant potential of effective economic development is associated include reduction of energy consumption during the heating period, improvement of ice conditions and, accordingly, conditions for cargo transportation in the Arctic seas, easier access to the Arctic shelves and their development, improvement of the structure and expansion of crop production area as well as increase of the efficiency of animal husbandry (subject to compliance with a number of additional conditions and adoption of particular measures), increase of boreal forest productivity [5].

Despite certain theoretically possible effects of global climate change, there are obviously more negative effects, which required the adoption of the necessary measures at the level of national legislation. This work for the reduction of greenhouse gas emissions started in Russia when the Kyoto Protocol was still in effect.

Already then, it determined the procedure for the selection, approval, and monitoring of the progress of projects implemented in accordance with Article 6 of the Kyoto Protocol to the United Nations Framework Convention on Climate Change as well as receipt, transfer, and acquisition of greenhouse gas emission reduction units.

The bylaws adopted by the Government of the Russian Federation [4] stipulate that the projects implemented in accordance with Article 6 of the Kyoto Protocol to the United Nations Framework Convention on Climate Change are approved by the Ministry of Economic Development of the Russian Federation, and the limit of carbon units for the selection and approval of projects implemented in accordance with Article 6 of the Kyoto Protocol is 300 million units.

Afterward, when the Paris Climate Agreement had been already adopted, Russia provided for several stages of climate change adaptation. In particular, the first stage of adaptation planning involved the determination of a set of organizational, regulatory legal, methodical, informational, and scientific support of the implementation of the national plan for global climate change adaptation in Russia [7]. Accordingly, the Government of the Russian Federation was instructed to ensure greenhouse gas emission reduction up to 70% of the 1990 level by 2030 with consideration of the maximum absorption capacity of forests and other ecosystems and subject to sustainable and well-balanced socio-economic development of the Russian Federation, to develop the Strategy for Socio-Economic Development of the Russian Federation with low greenhouse gas emissions until 2050 with consideration of peculiar features of the economic industries as well as to ensure the creation of conditions for the implementation of measures to reduce and prevent greenhouse gas emissions and to increase their absorption [3].

Therefore, from the above review it follows that the issue of climate change has been at the center of attention of state authorities in the Russian Federation in recent years. A number of important political and legal acts have been adopted. They determine a long-term strategy of the state policy in ensuring climate change mitigation and adaptation. Some bylaws set a list of specific measures and obligations to be achieved in Russia as a result of actions of the authorities.

13.3.3 Current Issues of Legal Protection of Climate: Articulation of the Issue and Possible Solutions

Despite the international cooperation expanding every year as well as measures adopted at the level of individual states (including Russia), a number of issues in the field of climate protection and elimination of the consequences of greenhouse gas emissions are universal and they are still to be solved.

(1) Climate change and the resulting floods, droughts, and other emergencies inevitably cause harm to the life, health, and property of citizens. In Russia, there are still no claims for compensation for harm caused by climate change. However, this law enforcement practice has been already established in the United States. For example, in 2005, a tribe of Inuit living in the Arctic filed a petition with the Inter-American Commission on Human Rights (IACHR) that accused the United States of violating its human rights obligations. The petition detailed the effects of rising Arctic temperatures on the ability of the Inuit to enjoy a wide variety of human rights, including the rights to life (melting ice and permafrost make travel more dangerous), property (as permafrost melts, houses collapse and residents are forced to leave their traditional homes), and health (nutrition worsens as the animals on which they depend for sustenance decline in number). The petition connected the rising temperatures to increasing levels of greenhouse gases, and in particular to the failure by the United States to take effective steps to reduce its emissions [13]. The plaintiffs in Comer v. Murphy Oil USA filed suit against energy production companies, alleging that the defendants' greenhouse gas emissions contributed to climate change and the intensity of Hurricane Katrina. The plaintiffs sought monetary damages for property loss caused by Hurricane Katrina. In Connecticut v. American Electric Power, the plaintiffs filed suit against electric power corporations, claiming that the defendants' greenhouse gas emissions were contributing to climate change, and claiming that climate change harmed and continues to harm the plaintiffs' residences and property. The plaintiffs sought an injunction, which

would place a cap on the defendants' greenhouse gas emissions. However, all the mentioned cases were dismissed [12]. This review of the judicial practice of the United States can be continued; however, the main conclusion of its study is that rejecting compensation for harm caused by climate change, the courts considered that the causal connection between emissions of the particular companies and the climate disasters the plaintiffs had suffered was unproven. Therefore, the development of the theory of evidence in "climate cases" is the most promising and unsettled doctrinal legal issue.

(2) If the provisions of the 2015 Paris Climate Agreement are not implemented in the near future, global warming will become the main reason for the extinction of wild animals [11]. Climate change already has a significant influence on nature. The first signs of the impact of climate warming on fauna were found by scholars in many countries long ago. Despite the recognition of the hazard of these phenomena and the taken actions, the volumes of emissions continue growing and threaten the preservation of biodiversity, the normal functioning of the agricultural industry, the interests of the territorial integrity of individual countries, and the exercise of human rights [2]. Climate change is too fast, this is why many species of wild animals do not have time to adapt to it.

(3) By now, global climate change has caused the emergence of a new legal phenomenon—environmental migration, which results not from warfare or political persecution but from change in the human habitat. These changes can be due to droughts or floods caused by deterioration of the environment and climate change or by the increase in the level of the World Ocean as a consequence of melting ice, which entails the flooding of recently habitable islands [10]. The emergence of "climate migrants" requires a change in the available approaches and the development of a new international legal category; moreover, it is obvious that climate migration is a special case of environmental migration—a category that is not sufficiently developed in international law either [15].

(4) The occurring climate change causes a number of consequences which are not obvious; consequently, they are little discussed by experts and representatives of the political community. First, according to recent studies by criminologists, climate change and rising temperatures lead also to an increase in crimes (and not necessarily only because people have less access to drinking water) [16]. Second, there are scientific studies showing the interrelationship between the political system (authoritarian or democratic) and the climate (hot or cold) [8, 9]. Though all the conclusions made by these scholars require check and further research, they cannot be denied, which is indicative of the multi-faceted nature of the climate issue.

(5) There is now no mention of climate as an object of environmental relations in Russian environmental legislation (including the main environmental Federal Law "On Environmental Protection" of January 7, 2002). This means there is no consistency in the understanding of the issue of climate protection at the level of legislative acts. Undoubtedly, references to specific practical measures to mitigate or adapt to the effects of climate change in bylaws are important for

the fulfillment of the international obligations assumed by Russia; however, in order to increase the efficiency of statutory regulation of climate protection issues, it is necessary to supplement the above mentioned environmental law of the Russian Federation with a special chapter containing a list of state regulation measures, including a list of restrictive measures in certain areas of human activity (transport, construction, etc.) for the reduction of impact on climate.

With consideration of the fact that the principles of law are the core of the legal system (and each particular branch of law), determining not only its current condition but also performing the predictive function of the development of the relevant branch, it appears reasonable to supplement Article 3 of the Federal Law "On Environmental Protection" with a new environmental principle of "legal protection of climate", which now is still purely doctrinal.

13.4 Conclusion

The conducted study leads to a conclusion that climate change is of an objective nature and consists in the growth of average annual temperatures registered for several decades in many countries, including Russia. Undoubtedly, the causes of this phenomenon are complex; however, the growth of anthropogenic impact on nature, which manifests itself in the increase in greenhouse gas emissions, makes its contribution to the development of the dynamics of this process. This caused the development of international cooperation in mitigating the impact on climate and the elaboration of measures for the adaptation to climate change that has already occurred. To resolve these issues the 1992 United Nations Framework Convention on Climate Change was adopted as well as two international instruments developing and supplementing its provisions—the Kyoto Protocol and the Paris Climate Agreement. These international instruments stipulate a list of measures that must be taken by the countries joining these treaties, which will reduce anthropogenic impact on nature and slow the growth of average annual temperatures.

Each country adopted its own measures to fulfill these international obligations, and Russia is not an exception in this case. However, despite Russia's adoption of a range of bylaws dedicated to the specific questions of fulfilling the assumed international obligations, there is no systematic approach to the issue of climate protection at the level of federal laws, which requires correction. Some universal issues in the area of climate protection are still relevant; they are typical for all countries and are to be resolved, including the issues of climate refugees, compensation for harm caused by climate change and some other issues.

Acknowledgements The study was carried out with the financial support of the Russian Science Foundation (project No. 20-18-00314).

References

1. Agreement (2015). The Paris Agreement under the United Nations Framework Convention on Climate Change (Signed in New York on April 22, 2016)
2. Brinchuk, M.M.: Legal issues of climate change and the prevention of natural disasters in the agricultural context. Agrar. Land Law **11**, 78–85 (2009)
3. Decree (2020). On the Reduction of Greenhouse Gas Emissions (Signed by the President of the Russian Federation, No. 666 of November 4, 2020) (Russian)
4. Decree of the Government of the Russian Federation (2011). On Measures to Implement Article 6 of the Kyoto Protocol to the United Nations Framework Convention on Climate Change (with the Provision on the Implementation of Article 6 of the Kyoto Protocol to the United Nations Framework Convention on Climate Change) (Adopted by the Government of the Russian Federation, no. 780 of September 15, 2011) (Russian)
5. Directive of the President of the Russian Federation (2009). On the Climate Doctrine of the Russian Federation (Adopted by the President of the Russian Federation, no. 861-rp) (Russian)
6. Druzin, B.H.: A plan to strengthen the Paris climate agreement. Fordham Law Rev. Res Gestae **84**, 18–23 (2016)
7. Executive Order of the Government of the Russian Federation (2019). On the Approval of the National Action Plan of the First Stage of Climate Change Adaptation until 2022 (Adopted by the Government of the Russian Federation, no. 3183-r of December 25, 2019) (Russian)
8. Falileev, V.V.: Natural climate conditions as a determining factor in the formation of spiritual culture. Bull. Tomsk State Pedagog. Univ. **2**, 131–134 (2004)
9. Grinin, L.E.: Lecture: natural factor in the aspect of the theory of history. Philos. Soc. **2**(62), 168–198 (2011)
10. Inshakova, E.I., Inshakova, A.O.: Nanomaterials and nanotechnology: prospects for technological re-equipment in the power engineering industry. In: IOP Conference Series: Materials Science and Engineering: International Conference on Modern Trends in Manufacturing Technologies and Equipment 2019, ICMTME 2019, Sevastopol, 09–13 September 2019. Sevastopol: Institute of Physics Publishing, p. 033020 (2020). https://doi.org/10.1088/1757-899X/709/3/033020
11. Ivanova, S.V.: On the impact of climate change on wildlife: the view from Russia. Environ. Policy Law **48**(5), 322–330 (2018)
12. Jaffe, J.: The political question doctrine: an update in response to recent case law. Ecol. Law Quart. **38**, 1–39 (2011)
13. Knox, J.H.: Linking human rights and climate change at the United Nations. Harv. Environ. Law Rev. **33**, 477–498 (2009)
14. Matveeva, E.V.: Environmental policy of post-soviet states in solving the issue of global climate change. Bull. Perm Univ. Polit. Sci. **2**, 32–42 (2013)
15. Matytsin, D.E.: Quality management strategy in Industry 4.0 and building a cognitive economy based on production and technological factors in the Russian Federation and the EU. Law Econ. Psychol. **2**, 40–51 (2020)
16. Ovchinsky, V.S.: Global climate change and crimes. Russ. Law J. **3**, 106–116 (2008)
17. Parker-Flynn, J.E.: The intersection of mitigation and adaptation in climate law and policy. Univ. Calif. Davis **38**, 1–47 (2014)
18. Resolution of the Presidium of the Russian Academy of Sciences 2020. Climate Change and Issues of Adaptation to Them (Signed by the RAS, No. 133 of October 13, 2020) (Russian)
19. Sokolova, N.A.: Global climate change: prospects of international cooperation (outline of the strategy for the implementation of the Kyoto protocol). Lex Russ. **4**, 808–820 (2005)
20. Sokolova, N.A.: Evolution of the international legal cooperation in combating climate change. Curr. Issues Russ. Law **9**, 2057–2065 (2014)
21. Strategy (2017). Environmental Security Strategy of the Russian Federation until 2025 (Approved by Decree of the President No. 176 of April 19, 2017) (Russian)
22. The Kyoto Protocol (1997). (Adopted to the United Nations Framework Convention on Climate Change of December 11, 1997)

23. The United Nations Framework Convention (1992). On Climate Change (Adopted by the United Nations on May 9, 1992)
24. Zolotova, O.A.: Does Law Affect Climate Change? Russ. Law J. **6**, 119–124 (2010)

Chapter 14
Current Issues of Agriculture Digitalization in the Russian Federation

Agnessa O. Inshakova, Anatoliy Ya. Ryzhenkov, Viktor F. Pon'ka, and Davud A. Davudov

Abstract The Chapter states that the information digital technology of the twenty-first century conceals an enormous potential of economic growth due to the automation and optimization of production processes, improvement of monitoring and control that enhance the manageability of both individual production and the entire economy. The implementation of digital projects in agriculture becomes an increasingly common trend in the whole world, including Russia. Digitalization includes such phenomena and technologies as big data, the Internet of Things, robotics, sensors, 3D printing, system integration, artificial intelligence, blockchains, etc., improving the quantity and quality of agricultural products and the logistics of their sale as well as increasing the social support of rural residents, reducing and easing their workload due to the production robotization. A range of foreign countries have created the necessary legal framework and launched the process of introducing digital technologies in agriculture, which has made it possible to achieve considerable success on this way also with regard to a significant increase in labor productivity. Digital systems in these countries have become important tools for managing risks in agriculture, helped to assess and eliminate them, and reduced the risks of production, weather, market, or climate variability. The Ministry of Agriculture of the

A. O. Inshakova (✉)
Department of Civil and International Private Law, Volgograd State University, Volgograd, Russia
e-mail: ainshakova@list.ru; gimchp@volsu.ru

Department of Law, Volga Branch of Volgograd State University, Volzhsky, Russia

A. Ya. Ryzhenkov
Volgograd State University, Volgograd 400062, Russia
e-mail: 4077778@list.ru

V. F. Pon'ka
Russian Academy of Sciences, Honorary Worker of Higher Professional Education, Leninskiy avenue, d.14, corp.1, 117198, Moscow 119071, Russia
e-mail: ponka@pran.ru

D. A. Davudov
Department of Civil and Private International Law, Volgograd State University, 100, prosp. Universitetsky, Volgograd 400062, Russia
e-mail: dav0587@mail.ru

A. O. Inshakova and E. I. Inshakova (eds.), *New Technology for Inclusive and Sustainable Growth*, Smart Innovation, Systems and Technologies 288,
https://doi.org/10.1007/978-981-16-9808-8_14

Russian Federation actively works in this area, having developed the departmental project "Digital Agriculture". Its main goal is the digital transformation of agriculture by introducing digital technology and platform solutions to ensure a technological breakthrough in the agro-industrial complex and achieve the growth of productivity in "digital" agricultural enterprises. The implementation of the program of digital agriculture will lead to the achievement of the following results: automation of agricultural machinery, reduction of costs and demand for manual labor, remote satellite data and sensors will improve the accuracy and reduce the costs of the monitoring of crop growth and land and water quality. Tracking technologies and digital logistics services will optimize agricultural food supply chains and provide reliable information for consumers. However, to achieve the set goals, it is necessary to make amendments to the Federal Law "On Agriculture Development" specifying the incentives and benefits for developers and consumers of digital technology.

14.1 Materials

The legislative framework of the study consists of the Federal Law "On Agriculture Development", Executive Order of the Government of the Russian Federation No. 1632-r of July 28, 2017 on the approval of the Program "Digital Economy of the Russian Federation" as well as the Departmental Project "Digital Agriculture" implemented by the Ministry of Agriculture of the Russian Federation. The said regulatory legal acts laid down the legal foundation for digital transformations in Russia's agriculture.

Some issues and achievements in the area of agriculture digitalization in a number of foreign countries and in Russia have been considered in the works by [3, 7–9, 11–16] and many other researchers. Their works allowed the authors to formulate the key issues of agriculture digitalization on a global scale, to show the dynamics of the settlement of this issue in the Russian Federation as well as to give a number of suggestions for the development of Russian agrarian legislation.

14.2 Methods

Such general scientific methods as formal logical, dialectical, systemic structural methods, and the method of critical cognition were used in the course of the study. Synthesis, classification, and generalization techniques were applied in the interpretation of the results of the study. Specific scientific methods were used in the work as well: formal legal method, the principle of legal process assessment, the method of comparative analysis, etc.

14.3 Introduction

Humanity now shifts to a new technological order characterized by the development of both new technologies and computer programs used by businesses to produce products and provide services, create new digital communication channels, and rethink the approaches to the use of information.

Information has become the main resource of the digital economy. Affecting various industries of economy, including agriculture, digital technology changes the ways of social interaction and personal relations, this involves the convergence of fixed, mobile and broadcast networks, devices and objects connected to each other via the internet. Nowadays, Russia is at the stage of implementing the strategy for its digital economy development since economic industries, from agriculture to metallurgical industry, from medicine to housing and utilities, cannot do without automation as the first steps of digitalization. However, the growth of efficiency is almost always associated with high technology, materials, modern management, with the implementation of information and communication processes and with digital transformation in general. The transition to a digital economy involves comprehensive changes both at the level of the state and at the level of companies. Consequently, the digital transformation of the economy is inevitable and it is very important for Russia to be among the leaders in this case [3].

There are enormous possibilities for the modernization of the agro-industrial complex of the country. The adjustment of the areas of the state support of the industry, including new guarantees of ensuring the country's food security, could give a high-tech character to the industry and provide possibilities for the introduction of innovative technologies. However, the current digitalization of Russia's agriculture "is at a very low level: no experts who could make global forecasts for prices of agricultural products are involved; the system of logistics, delivery, storage and processing of products is not developed. Moreover, along with a mass of practical problems, there is no scientific approach to innovative modern agrotechnology and scientific methodology in general. It should be also noted that all this enormous system can develop in a stable and sustainable way if there is an effective legal framework, which should be worked out by the state thoroughly and thoughtfully" [16]. In these conditions, the legal support of the state help plays a considerable role in the digital economy development, which is especially important in the economic branches which are the most vulnerable from the perspective of national security (agriculture is among them). This requires the continuation of the enhancement of agrarian legislation, which defines the main areas of the state support of this sector of the economy, however, pays little attention to the legislative recognition of specific incentives and guarantees for producers of agricultural goods (in particular, small and medium business) that have decided to use digital technology in the course of agricultural production as well as for producers of this technology.

14.3.1 International Experience in Agriculture Digitalization Which is of Interest in Terms of the Russian Federation

In the most general terms, agriculture digitalization implies that the tasks of agricultural production management, both at the level of an individual enterprise and in the agrarian industry on the whole, will be fulfilled more efficiently in case of using the internet, blockchain technologies, drones, robots, etc. This will make it possible to forecast changes in weather conditions, improve weed and pest control, save electricity, monitor animal behavior, etc. Due to this, farmers and other producers of agricultural goods will be able to make more correct decisions optimizing both the production of agricultural goods and their sale. Digitalization can also contribute to the settlement of rural social problems, improving the access of rural residents to cultural, educational, and medical services.

To achieve these goals, it is necessary to resolve the issue of the interaction of multiple databases related to agriculture, including the information about features of seeds, climate and weather conditions, soil quality, management decisions made by farmers, the state of the stock of agricultural products and their market prices. The provision of the access to this set of data will allow certain farmers and trade organizations to make decisions increasing the productivity and profitability in agricultural production and at the target markets. Besides the increase in labor productivity, digital systems offer important tools for managing risks arising in agriculture due to production, weather, market, or climate variability, and these tools can help to assess and eliminate such risks. These improvements will make it possible to take preventive decisions for short-term protection of income and to work to improve soil and environmental quality management at the farm level in the long term [11].

Many countries now have mandatory farm animal traceability programs to track their movements, for example, in the United Kingdom. These systems generate significant amounts of data that are analyzed and modeled to provide early warning and rapid response, for example, in the event of an epizootic outbreak. In Switzerland, genetic and production data are routinely linked to movement databases to identify on-farm deaths and stillbirths that may indicate the emergence or re-emergence of disease (similar measures are considered in Ireland). Of course, these capital-intensive technologies promise to be lucrative and this has captured the attention of many companies, policymakers, and investors. For instance, in Canada the Finance Minister's 'Advisory Council for Economic Growth' published a major report in 2017 on how to grow Canadian middle-class incomes. One of their key recommendations was to invest in agro-food sector innovation.

Similarly, the UK Government's £4.7 billion 'Industrial Strategy Challenge Fund' includes artificial intelligence and data as one of four challenge areas, with specific schemes focusing on precision agriculture. As well, venture market investment in agricultural technologies has increased 80 percent annually since 2012. Similar investments are being made in the United States and across the EU. Future projections predict that the precision agriculture technologies market will exceed $10 billion globally by 2025 [13].

The growth of the market for robots and drones will play a big role in the processes of agriculture digitalization. In particular, experts estimate the market for drones in agriculture at $32.4 billion. Drones can be used to create three-dimensional maps for the analysis of soil and plant seeds, water agricultural crops, apply fertilizers, spray plants, and monitor crops.

Sensing systems and associated analytics can provide agricultural producers with better information to make more timely decisions with more predictable outcomes, while automating tasks using sensing technologies and machine learning can increase the reliability of agricultural production.

Rapid developments in the Internet of Things, cloud computing, robotics, and Artificial Intelligence are accelerating the transition to smart farming and the promotion of big data and precision agriculture to improve agri-food sustainability.

The expectation is that smart farming approaches will ultimately improve knowledge about an individual enterprise also via efficient sharing and learning from data from multiple enterprises. However, although this "fourth agricultural revolution" brings the promise of multiple gains, it also brings with it technical, social, economic, ethical, and other practical questions, with significant implications. Digital applications and platforms have the potential to dramatically change the way knowledge is processed, communicated, accessed and utilized. For farmers, digital applications will provide decision-making capabilities that were previously not possible, potentially leading to radical changes in farm management. As smart machines and sensor networks increase on farms and farm data grow in quantity and scope, farming processes will become increasingly data-driven and data-enabled. This raises critical questions about how digital agriculture will require new capabilities, support decision-making and interact with, and potentially disrupt, established modes of knowledge processing [6].

A number of other social risks of digitalization should be mentioned since technologies aimed at production automation and efficiency growth can deprive farmers and agricultural workers of their work, exclude or discriminate against those who are not digitally literate. This can have negative consequences for the demand for rural workforce and, therefore, have an impact on marginal groups, including migrants, in the context of the growing division of labor and capital in agriculture. Moreover, in scientific literature, there are already discussions about risks of cyberattacks and new ethical and philosophical issues of the relations of humans and animals that now live autonomously [8].

14.3.2 *Russian Legislation and Planning of the Introduction of Digital Technology in Agriculture: Trends and Prospects*

The National Program "Digital Economy of the Russian Federation" was approved in Russia according to Government Executive Order No. 1632-r of July 28, 2017 [4]. The main activities of this National Program are aimed at solving the following key tasks of the transformation of the Russian economy and social sector: formation of a new regulatory environment of relations between citizens, businesses, and the state emerging with the development of the digital economy; creation of a modern high-speed infrastructure for data storage, processing and transmission, sustainability and security of its functioning; formation of a system for training staff for the digital economy; support of the development of promising end-to-end digital technologies and projects for their implementation; improvement of the efficiency of public administration and state services through the introduction of digital technology and platform solutions.

Program documents taking into consideration the corresponding peculiarities were adopted at the level of individual economic sectors to achieve the said goals. With respect to agriculture, the country's official policy concerning the priorities in the agriculture digitalization development is presented in several executed program documents of the Ministry of Agriculture of the Russian Federation. Digital agriculture is the agriculture based on modern methods of production of agricultural products and foodstuffs with the use of digital technology (the Internet of Things, robotics, artificial intelligence, big data analysis, e-commerce, etc.) ensuring the growth of labor productivity and the reduction of production costs. Its development includes a number of stages.

The first stage consists in the creation and introduction of the national platform of digital public administration of agriculture "Digital Agriculture". This digital platform will be integrated with other sub-platforms for agriculture management at the regional and municipal levels, which will provide agricultural producers with the possibility to receive state support through the common unified national digital platform. This will require the appropriate work organization, first of all, development of the concept of digital agriculture, organization of the collection of the necessary information and interaction with other authorities and organizations.

To ensure further record, monitoring and analytics trough digital agriculture, it is necessary to work out in detail the regulations for agricultural land data transmission. Novosibirsk Region, Krasnodar Krai, and Altai Krai can be mentioned as an example of the involvement of regions in the implementation of this departmental project.

Well-established interaction between federal executive authorities will provide up-to-date data about agricultural land and the land used or allocated for agriculture as part of land of other categories. The collected information will help to solve a number of issues, in particular, to understand the need of creating comprehensive products for agricultural producers that optimally combine state support measures for commercial agroservices. The normal functioning of the platform requires the development and adoption of a range of regulations.

The second stage consists in the creation and introduction of the module "Agrosolutions" of the national platform of digital public administration of agriculture "Digital Agriculture" to increase the efficiency of activities of agricultural producers. The logical result of the module "Agrosolutions" is a double increase in labor productivity per worker as well as the reduction of costs of agricultural enterprises. The introduction of comprehensive digital solutions will ensure the productivity and efficiency of the use of this module. To appropriately realize the second stage, it is necessary to develop technical and content-related requirements of the module, taking into account the actual needs of agricultural producers.

The third stage is based on the creation of the system of continuous training of staff of agricultural enterprises with the purpose of formation of competencies in the area of the digital economy. The Competence Center "Digital Agriculture" with representative offices based on agrarian higher education institutions of the Ministry of Agriculture of the Russian Federation and other agricultural organizations will implement training and retraining programs for specialists of agricultural enterprises so that they could gain competencies in the digital economy.

The issue of training and retraining of staff working in agriculture became urgent long ago, since the introduction of new digital technologies rapidly becomes part of our life, however, their use often faces difficulties. The suggested industry-specific electronic educational environment "Land of Knowledge" will make it possible to acquire knowledge remotely in order to use digital technology in general and exchange experience among students. Therefore, the process of the implementation of the departmental project "Digital Agriculture" requires the compilation of a clear phased development plan which will present real mechanisms of its implementation and needs. The project also requires support from the state in the form of subsidies, the development of a regulatory framework and the creation of conditions for the training of specialists [1].

Hence it follows that the process of agriculture digitalization is a very complex system including a number of elements: the main areas and functions of the digital system, legal support of the agro-industrial complex in terms of digitalization and the management structure enabling the implementation of solutions. Like any other complex systems, digitalization of the agro-industrial complex and its efficiency depend on internal and external factors, which are both objective and subjective. The external factors influencing the transformation of the agro-industrial complex include, first of all, geographical features of the location of rural entities, which differ significantly from each other in their territory, density of the population, the financial component and legal support from the state and municipalities. The internal factors influencing the processes of the digitalization of the agro-industrial complex include the creation of the regulatory framework, which depends on the legal policy of the state at each particular stage and specific interests of authorities, the availability of staff in general and the organization of workplaces, management technologies being used, etc. [16].

There are now both certain achievements and particular difficulties in the course of the implementation of the agriculture digitalization management project. According to M.L. Vartanova (2018), "at the moment, there are enough resources for the collection of data from fields but the automated system for the processing of these data and the work with artificial intelligence has not been adjusted yet" [14]. It is still relevant and necessary to organize electronic mapping of agricultural land (formation of electronic datasheets of fileds), which will make it possible to identify uncultivated land and control the use of fertilizers as well as all operations of sowing agricultural machinery. In its turn, this will require from the industry to provide a new generation of drones and sensors, develop GLONASS and GPS systems assessing and managing crop production processes.

The implementation of the tasks set by the Ministry of Agriculture of the Russian Federation will be facilitated by both the digital technologies focused exclusively on the development of the agrarian sector (for example, the Unified Register of Farms, which is maintained by the Ministry of Agriculture of the Russian Federation) and the digital technologies which are of a more framework nature and are used also in other sectors of the economy (for example, the Unified State Register of Immovable Property includes information about the holders of rights to immovable property, restrictions of these rights, etc. also beyond rural areas).

We should agree with [12] that a range of digital technologies is already widely used in the agrarian sector, including closed water use systems for industrial fish farming and the establishment of pig breeding complexes and poultry farms with continuous chains of complete production and life cycles of products, from the creation of balanced animal feed rations to the sale of finished products, fully managed with the use of software, computers, and the internet [12]. The Internet of Things is used for free grazing of cattle. For example, loose housing for cows, when the animals can move freely, is used in the largest DolgovGroup Agroholding Company of Kaliningrad Region. A tracking chip is inserted into the earlobe of the cow, and the movement of the animal can be tracked from any part of the world, with the use of a computer, the necessary program and access to the internet. Financial costs are reduced several times in this case. Digital technology is used also in crop production (spraying and soil cultivation, fertilization, regulation of plant nutrition, and microclimate in greenhouses) and animal husbandry (maintenance, milking and feeding of animals, management of technological processes in poultry houses). New technologies manage the process of storage of vegetables, assess the economic efficiency of production, etc. [10].

One of the achievements of the robotization of agricultural machines is the creation of a robotized combine harvester manufactured by Rostselmash with artificial intelligence, which passed tests in the fileds of Belgorod Region in 2018, having harvested from 60 hectares of fields in an unmanned mode [2]. The "flying tractor" called "Skif", which is a cargo-type drone, was presented in Kazan. Today, it is the biggest multi-rotor drone in the world. Only two people are needed to manage the Skif, any technically competent person can learn this in a week. The maximum carrying capacity of the platform is 400 kg. If the mass of the cargo is 50 kg, the drone can fly up to 350 km. It is refueled with AI-95 gasoline and has an hourly flow rate of about

30 L. The maximum duration of the flight without refueling is eight hours. While the main area of use of this drone is agriculture, it is capable of round-the-clock treatment of fields. On average, the "tractor" can lift about 250 L of pesticides or fertilizers at a time, this will be enough for about 25 hectares of crops, i.e., the average rate is 10 L per hectare. The drone can spend half an hour on the processing of 25 hectares [15].

However, a number of questions remain open, in particular, the legal status of robots, the issues of compensation for harm caused by tractor robots or cows grazing freely and controlled with the use of remote automated technologies.

A separate question consists in the lack of a clear strategy for the creation of "digital villages" facilitating access of rural residents to the main benefits of civilization (education, medicine), the deficit of trained IT staff in rural areas and the lack of financial resources of producers of agricultural goods, which impede the introduction of digital technology [9]. The Federal Law "On Agriculture Development" could facilitate the resolution of these problems. It stipulates the informational support of the state agrarian policy as a state support measure; however, this area should be specified in detail [5]. It appears that the Law must be supplemented with a special article including specific measures for the state stimulation of agriculture digitalization. These can be benefits for taxation, credits, development of the informational sector of the support of agriculture, education, the system coordinating the work among agricultural authorities of all levels as well as with the public and producers of agricultural goods, measures for the planning of the development of the digitalization of the agrarian sector, etc. Also some administrative issues are to be settled, for example, the simplification of the procedure for obtaining permits for drone flights over the fields of farmers. Moreover, the development of digital technology can improve the procedure for mediated resolution of conflicts between agricultural producers as well as between them and authorities [7].

14.4 Conclusion

The information digital technology of the twenty-first century conceals an enormous potential of economic growth due to the automation and optimization of production processes, improvement of monitoring and control that enhance the manageability of both individual production and the entire economy. The implementation of digital projects in agriculture becomes an increasingly common trend in the whole world, including Russia. Digitalization includes such phenomena and technologies as big data, the Internet of Things, robotics, sensors, 3D printing, system integration, artificial intelligence, blockchains, etc., improving the quantity and quality of agricultural products and the logistics of their sale as well as increasing the social support of rural residents, reducing and easing their workload due to the production robotization.

A range of foreign countries have created the necessary legal framework and launched the process of introducing digital technologies in agriculture, which has made it possible to achieve considerable success on this way also with regard to a significant increase in labor productivity. Digital systems in these countries have become important tools for managing risks in agriculture, helped to assess and eliminate them, and reduced the risks of production, weather, market, or climate variability.

The Ministry of Agriculture of the Russian Federation actively works in this area, having developed the departmental project "Digital Agriculture". Its main goal is the digital transformation of agriculture by introducing digital technology and platform solutions to ensure a technological breakthrough in the agro-industrial complex and achieve the growth of productivity in "digital" agricultural enterprises. The implementation of the program of digital agriculture will lead to the achievement of the following results: automation of agricultural machinery, reduction of costs and demand for manual labor, remote satellite data and on-site sensors will improve the accuracy and reduce the costs of the monitoring of crop growth and land and water quality. Tracking technologies and digital logistics services will optimize agricultural food supply chains and provide reliable information for consumers. However, to achieve the set goals, it is necessary to make additions to the Federal Law "On Agriculture Development" specifying the incentives and benefits for developers and consumers of digital technology.

Acknowledgements The study was carried out with the financial support of the Russian Science Foundation (project No. 20-18-00314).

References

1. Departmental Project "Digital Agriculture": Official Issu. FGBNU "Rosinformagrotech", Moscow (2019)
2. Digitalization of Russia's Agricultural Production for the Period of 2018–2025. Research of the Cooperative Project "German-Russian Agricultural Policy Dialogue". Moscow, Berlin (2018)
3. Duplyakina, O.K., Miroshnichenko, M.A.: The necessary conditions for the development of the digital economy in Russia. In: Knowledge Economy in Russia: from Knowledge Generation and Innovation to Cognitive Industrialization. Proceedings of the 9th International Scientific and Practical Conference. Managing Editor V.V. Ermolenko, M.R. Zakaryan. Kuban State University, Krasnodar, pp. 225–232 (2017)
4. Executive Order: Program "Digital Economy of the Russian Federation" (approved by the Government of the Russian Federation, no. 1632-r of July 28, 2017) (Russian) (2017)
5. Federal Law: On Agriculture Development, (Adopted by the State Duma of the Russian Federation on December 22, 2006, no. 264-FZ) (Russian) (2006)
6. Ingram, J., Maye, D.: What Are the Implications of Digitalisation for Agricultural Knowledge? (2020). https://doi.org/10.3389/fsufs.2020.00066/full. Accessed February 1, 2021
7. Inshakova, A.O., Goncharov, A.I., Inshakova, E.I., Tumchuk, Y.A.: Digital technologies for alternative methods of resolving conflicts: the prospects of application in Russia and other BRICS countries. In: Inshakova, A.O., Bogoviz, A.V. (Ed.) Alternative Methods of Judging Economic Conflicts in the National Positive and Soft Law. A volume in the series Popkova, E.G. (Ed.) Advances in Research on Russian Business and Management. Charlotte, NC, USA, Information Age Publishing (2020)
8. Klerkx, L., Jakkub, E., Labarthec, P.: A review of social science on digital agriculture, smart farming and agriculture 4.0: new contributions and a future research agenda. NJAS Wageningen J. Life Sci. **90–91**, 1–16 (2019)

9. Kurbatova, S.M., Aisner, Yu, L.: Digitalization of agriculture as an area of the modern state policy of the Russian federation. Agrarian Land Law **4**, 102–104 (2019)
10. Matytsin, D.E.: Quality management strategy in industry 4.0 and building a cognitive economy based on production and technological factors in the Russian Federation and the EU. Law. Econ. Psychol. **2**, 40–51 (2020)
11. Phillips, P.W.B., Relf-Eckstein, J.A., Jobe, G., Wixted, B.: Configuring the new digital landscape in western Canadian agriculture. NJAS Wageningen J. Life Sci. **90–91**, 1–11 (2019)
12. Popova, O.V.: Issues of legal regulation of digitalization in the agro-industrial complex. Agrarian Land Law **9**, 96–102 (2018)
13. Rotz, S., Duncan, E., Small, M., Botschner, J., Dara, R., Mosby, I., Reed, M., Fraser, E.D.G.: The politics of digital agricultural technologies: a preliminary review. Sociol. Rural. **59–2**, 203–229 (2019)
14. Vartanova, M.L.: Increasing the intensification of agriculture through the introduction of modern technologies. Natural Humanitarian Stud. **21**(3), 13–17 (2018)
15. Voronin, B.A., Lorets, O.G., Mitin, A.N., Chupina, I.P., Voronina, Ya.V.: Issue of the digitalization of the Russian agriculture (overview of information materials). Agrarian Bull. Urals **2**, 46–52 (2019)
16. Yagofarova, I.D.: Theoretical and methodological approaches to the legal support of the transformation of the agro-industrial complex in terms of digitalization. State Power Local Govern. **11**, 11–15 (2019)

Chapter 15
Sustainable Development of Rural Areas in Russia: Legal and Technological Support

Denis E. Matytsin, Agnessa O. Inshakova, Maxim V. Sevostyanov, and Olesya P. Kazachenok

Abstract The chapter examines the legal category "sustainable development of rural areas" and gives its authors' definition. The authors suggest meaning by sustainable development of rural areas the regular economic activities on agricultural lands, providing growth of agricultural production, under condition that social rights of residents of rural settlements are properly guaranteed, as well as measures protecting the environment on agricultural land (including measures for farmland fertility) are taken in rural settlements as well. As a result of the study, it is noted that the transition of rural areas to sustainable development is hampered by the lack of an efficient system of interdepartmental interaction and coordination of specific issues related to the development of rural areas. In comparison with developed countries, government support for agriculture in Russia is significantly lower, what can be considered as one of the factors of a low profitability of agriculture. Poor support limits the possibilities of modernization and innovative development of the industry, negatively affects wages of workers in the industry and the tax base of local budgets. The authors study the problems and prospects for the development of agricultural tourism as one of

D. E. Matytsin (✉)
Insitute of Law, Volgograd State University, Volgograd, Russia
e-mail: Dmatytsin@ya.ru; gimchp@volsu.ru

Department of Civil Law and Procedure, Volga Branch of the International Law Institute, Volzhsky, Russia

A. O. Inshakova
Department of Civil and International Private Law, Volgograd State University, Volgograd, Russia
e-mail: ainshakova@list.ru; gimchp@volsu.ru

Department of Law, Volga Branch of Volgograd State University, Volzhsky, Russia

M. V. Sevostyanov
Volgograd State University, Volgograd, Russia
e-mail: sev_mv@mail.ru

O. P. Kazachenok
Department of Civil and Private International Law, Volgograd State University, 100, prosp. Universitetsky, Volgograd 400062, Russia
e-mail: o.kazachenok@yandex.ru

A. O. Inshakova and E. I. Inshakova (eds.), *New Technology for Inclusive and Sustainable Growth*, Smart Innovation, Systems and Technologies 288,
https://doi.org/10.1007/978-981-16-9808-8_15

the most important elements of transition of rural areas to sustainable development, identifying the reasons for its insufficiently rapid and effective development. This chapter also shows that the Concept of Sustainable Development of Rural Areas can be a tool to mitigate the severity of social, economic, and environmental problems in rural areas, taking into account the interests of citizens, business, and the government. A complex and multi-level system of criteria and indicators for sustainable development of rural areas is required for an efficient realization of this concept offered in this chapter.

15.1 Materials

The legal framework for the study was made up of international and national legal acts dedicated to various aspects of sustainable development. The chapter examines the UN declarations adopted at various conferences on sustainable development, UN resolutions, and decisions of EU bodies. Special attention is paid to the constitutions of various countries of the world, what reflect certain provisions of the concept of sustainable development, as well as the Constitution and legislation of the Russian Federation.

The authors analyze the provisions of the Urban Planning Code of the Russian Federation, federal laws "On environmental protection", "On development of agriculture", "On viticulture and winemaking in the Russian Federation", as well as some secondary legislation, for example, the Order of the Government of the Russian Federation of 02.02.2015 "On approval of the Strategy for sustainable development of rural areas of the Russian Federation for the period up to 2030" [4–6, 14, 16].

In the course of the study, the papers of Russian and foreign authors devoted to both general provisions on sustainable development and specific features of the concept of sustainable development of rural areas were used. In particular, the chapter deals with the papers of [1, 2, 9–12, 17–19] and a number of other authors devoted to certain aspects of sustainable development.

15.2 Methods

The study used general scientific methods (formal and logical, dialectical, systemic and structural, critical knowledge). When interpreting the research results, the methods of synthesis, classification, and generalization were used. The work also used such private scientific methods as formal legal, method of assessment of legal processes, method of comparative analysis, etc.

15.3 Introduction

Rural territories of the Russian Federation are the most important resource of the country, the value of which is rapidly growing in the context of deepening globalization, while the importance of natural and territorial resources in the development of the country increases. The development of rural areas today is extremely uneven, while rural areas have vast natural, demographic, economic, historical, and cultural potential, the rational use of which can ensure sustainable development, a decent level and quality of life for the rural population. The implementation of nationwide functions of rural areas is hampered by the structural crisis associated with the transformation of forms of ownership and the transfer of activities in all spheres of life to market principles. Currently, the share of people employed in the main sectors of the rural economy (agriculture, forestry, fish farming) has significantly decreased, but agriculture in most administrative regions continues to be an important sphere of employment of the rural population. At the same time, wages in agriculture over the past years were on average consistently two times lower than the average rate for the economy, what causes a number of deformations in the development of the social sphere.

In rural areas, the share of poor families is higher. While the share of the rural population of the total population is 26%, the share of families with per capita incomes below the subsistence level in rural areas is more than 40% of the total number of low-income families. If the prevailing trends in the development of rural areas, characterized by a low quality and standard of living, continue, the outflow of the most promising young people will continue, what is confirmed by sociological studies, according to which 50% of young people intend to leave the rural areas in which they permanently reside. The consequence of this is the degradation of human potential of rural areas, and when the labor force is attracted on a rotational basis, there is an increase in social tensions.

In such a situation, comprehensive diversification of the rural economy, support for farming and alternative forms of employment and self-employment are required. The support measures are expected to include the support for development of crafts and agricultural tourism, organization, and removal of administrative barriers to product marketing, access facilitation to natural resources, including land, material, financial and information resources, support for the activities of consulting centers and the development of infrastructure that allows the population to receive a decent income. The achievement of this task can be facilitated by the fact that Russia has a large land potential for agriculture. As of January 1, 2014, the area of agricultural land was 220.2 million hectares, including the area of arable land 121.5 million hectares. Despite such potential opportunities for the development of agriculture, in recent years there has been a tendency to reduce the area of agricultural land, which is caused by the unsatisfactory financial and economic situation of agricultural producers. In turn, this situation results in the rejection of the provided land, the transfer of land to other categories for building in suburban areas, the lack of measures for land reclamation and restoration of soil fertility. This situation is aggravated by environmental factors caused, on the one hand, by long-term extensive forms of agriculture, leading to soil depletion, and, on the other hand, by climate changes leading to an increase in risk of agricultural production [14].

These trends require further improvement of the concept of sustainable development of rural areas, taking into account the differentiated problems of rural residents and different levels of agricultural development in the constituent entities of the Russian Federation. Such a concept of sustainable development of rural areas should create conditions for ensuring a stable improvement in the quality and standard of living of the rural population based on the advantages of the rural lifestyle, which will develop the social and economic potential of rural areas and ensure that they perform national functions such as production, demographic, spatial, and communication in order to preserve historical and cultural foundations of the identity of the peoples of the country, to guarantee social control and the development of rural areas in Russia.

15.3.1 Main Provisions of the Concept of Sustainable Development of Rural Areas in International Law and Russian Legislation

The concept of sustainable development is one of the most discussed ideas of interaction between nature and society. It was first mentioned in 1987 in the report of the International Commission on Environment and Development, but it is finalized in the decisions of several UN Conferences on Environmental Protection [3]. In their decisions, "sustainable" means such a progressive development of the state and society, which ensures a balance of economic, environmental, and social needs of all private and public entities in the interest of the present and future generations [8]. It is thanks to the documents adopted at such UN conferences (starting with the decisions of the UN Conference on Environment and Development held in Rio de Janeiro in 1992) that the quality of the environment (along with the level of economic and social development) for the first time began to be considered as an essential element of sustainable development. The obligation of the state to preserve ecosystems and ecological processes that are vital for the biosphere of a country, to protect biological diversity, and to observe the principle of optimal sustainability when using biological resources and living ecosystems is now added to the economic and social obligations of the state which existed before after the recognition of the concept of sustainable development on a global scale.

These ideas were fully developed in the materials of the World Summit on Sustainable Development in Johannesburg in 2002.

The 2002 Johannesburg Declaration on Sustainable Development focused on poverty eradication, changing production, and consumption patterns, and the protection and sound use of natural resources as the main goals and needs of sustainable development. The decisions of the conference also highlighted the difference in opportunities for sustainable development for developed and developing countries.

The decisions of the UN Conference on Environment and Development "Rio + 20", which took place in 2012 in Rio de Janeiro, it was continued to analyze the issues of poverty eradication and measures to ensure sustainable development. At the same time, the term "green economy" was used for the first time in the decisions of the conference. Building a "green" economy in all countries of the world will mean ensuring the sound use of natural resources with the lowest losses for the environment, increasing the efficiency of their use and reducing the volume of generated waste.

Discussion of sustainable development strategies continued further, with special emphasis on the UN General Assembly Resolution of September 25, 2015 "Transforming our world: the 2030 Agenda for Sustainable Development" [15]. The 17 Sustainable Development Goals presented in this document were developed and supplemented in subsequent UN documents, for example, in the UN Sustainable Development Goals Report (2017). All these and many other documents contain a detailed set of criteria and indicators of sustainable development, and an algorithm for achievement of these goals through the efforts of the national authorities and the public.

These recommendations were accepted both by interstate associations (for example, in the European Union) and by nations in the world. Thus, on January 30, 2019, the European Commission presented a document entitled.

"Towards a Sustainable Europe by 2030", which summarizes the progress made in Europe and identifies the necessary priorities for moving forward: development of a circular economy, creation of a sustainable food system, ecologization of energy, mobility and man-made environment, and use of a range of instruments starting from education and digitization to finance and taxation which facilitate the transition to sustainable development. The document emphasizes the idea that without social sustainability there can be no sustainable development; therefore, it is fundamentally important to ensure social justice in the transition to sustainability, for the benefit of all, leaving no one behind.

At the national level, the provisions of the concept of sustainable development had different relevance in the national constitutions, as well as legislative acts of the countries of the world. The most explicit manifestation of the interest to sustainable development is the mention of sustainable development in the following constitutions: Article 17 of the 2010 Constitution of the Dominican Republic, Article 8 of the 1997 Constitution of Eritrea, and Art. 60 of the 2010 Constitution of Kenya, article 22 of the 2008 Constitution of the Republic of Maldives, article 81 of the 1976 Portuguese Constitution, article 33 of the 2003 Constitution of the State of Qatar, article 13 of the 1998 Constitution of the Republic of Sudan and a number of others (Environmental provisions of constitutions, 2012). The Constitution of the Russian Federation does not mention sustainable development; however, this legal category can be found in several federal laws, in particular, in the Civil Code of the Russian Federation (Articles 1, 30, 41, and a number of others), Art. 1, 3, and 36 of the Federal Law "On Environmental Protection", Article 36 of the Federal Law of December 27, 2019 No. 468-FZ "On Viticulture and Winemaking in the Russian Federation", as well as in several development strategies of the country (on the sustainable development of rural areas, sustainable development of tourism, development of indigenous peoples, etc.).

Federal Law of December 29, 2006 No. 264-FZ "On the Development of Agriculture" under the sustainable development of rural areas means their stable social and economic development, increase in agricultural production, increase in the efficiency of agriculture, achievement of full employment of the rural population and increase in their standard of living, sound use of land. Thus, the main emphasis here is made on social and economic aspects, while the environmental factor did not come to the attention of the legislator.

It follows from this description that Russian legislation lacks a balanced concept of transition to sustainable development of rural areas (although its specific aspects are undoubtedly reflected in laws and subordinate laws). International acts provide for the adoption (to track the implementation of the provisions of the idea of sustainable development) at the national level a number of special criteria and indicators showing the degree of compliance of legislation and the state of economic, environmental and social interests with the goals stated in the Concept. In the Russian Federation, the most developed are the criteria and indicators of sustainable development in forestry [1], as well as in the use and protection of wildlife [9]. For other sectors of public life (agriculture, transport, waste management, etc.), criteria and indicators exist only in scientific research.

Thus, the concept of sustainable development is directly connected with economic, environmental, and social spheres in their indissoluble unity, and also affects other spheres, including political, ideological, informational, cultural, and others [18]. It is necessary to support the suggested classification of measures to ensure sustainable development of agricultural territories by subject and object criteria. From the point of view of the subject, the taken measures are as seen follows. First, these are steps taken by government in form of competent authorities. Second, the measures are taken by agricultural producers themselves. Third, it is society as a whole which influences the process in the economy. From the point of view of the object, it is suggested to distinguish organizational, legal, economic, environmental, social, ethical, informational, and educational measures [11].

The mechanism identification for their implementation requires a study of narratives, both of the category "sustainable development of rural areas" and the criteria (indicators) of its development, including the definition of priority areas for sustainable development of rural areas in the Russian Federation.

15.3.2 Issue of Criteria and Indicators of Sustainable Development

The issue of the transition to sustainable development standards for rural areas during its discussion inevitably results in a discussion about its criteria and indicators. In recent years, a number of judgments on this issue have already been expressed in the agrarian science and law of the post-Soviet countries.

In particular, representatives of the Belarusian agrarian science and law suggest to measure the level of unemployment in rural areas, size of wages and general standard of living of rural residents, provision of rural residents with housing and infrastructure (kindergartens, hospitals, etc.), taking into account the gasification and supply of drinking water, roads, etc. [2]. Meanwhile, for all the value of this approach, the authors focus on the social parameters of life of rural residents, while the idea of sustainable development involves finding a balance of three groups of interests—economic, social, and environmental ones. It follows from this description that this approach needs to be supplemented by two more interest groups.

Russian scientists, in turn, often focus on economic interests at the expense of social and environmental ones. Thus, Ostrovsky, speaking about economic indicators of sustainable development of rural areas, adds to them international cooperation as a means of acceleration development of rural areas, changes in consumption characteristics, development of a financing mechanism for rural areas, introduction of environmentally friendly technologies in rural areas [13].

When analyzing environmental criteria and indicators, a number of scientists reasonably suggest analyzing the proportion of unused lands of agricultural organizations as environmental indicators of sustainability; soil fertility of arable land (average level of humus in the soils of arable land); total pesticide load per hectare of arable land; share of agricultural land used for organic production in the total land area; proportion of degraded drained land with peat soils in the total area of agricultural land; proportion of disturbed lands in the total area of lands of agricultural firms; and share of waste generated in agriculture in the total amount of waste [10].

Other authors, examining environmental criteria and indicators of sustainable development of rural areas, suggest criteria such as the state of soil fertility, level of emissions and discharges of harmful substances, and degree of use of pesticides and agrochemicals. Within the framework of social criteria, attention is paid to the employment of population, wages, vocational training of rural residents, intensity of use of rural labor resources, labor protection, development of health care in the countryside, and the preservation of the cultural landscape [7].

Thus, representatives of various scientific schools usually focus on only one group of interests (economic, environmental, and social) in the triad of sustainable development, while the urgent task is to build their balance, taking into account all three areas of sustainable development. Undoubtedly, the discussion about criteria and indicators should be continued, since they make it possible to assess not only the achievement of specific economic, environmental, or social goals but also make it possible to identify the extent to which the development of specific rural areas (for example, within the boundaries of a settlement or a municipal district) generally corresponds to the specified parameters of sustainable development set by federal or regional authorities, they allow assessing the effectiveness of the taken measures applied to achieve the specified indicators.

In our opinion, in the light of the existing international and foreign experience, the criteria for sustainable development of rural areas should look as follows. Economic criteria: technology level in the production of organic products of plant growing, animal husbandry and fish farming; competitiveness of agricultural products of a certain country, their demand in the world agricultural market; development of agricultural entrepreneurship in a particular region (municipality) or the country as a whole; development of markets for agricultural products, including cooperative forms; construction of roads and other infrastructure facilities in rural areas; dynamics of income growth of agricultural entrepreneurs, including the measures for state support of agriculture (loans, subsidies, etc.); use of renewable energy in rural areas (including biofuel production); number of facilities used for agricultural tourism; amount of unused land in the region (country).

Social criteria: population size (indicators of birth rate, mortality, population migration); provision of utilities; income of the rural population; dynamics of increase/decrease in the number of social facilities (kindergartens, schools, hospitals, cultural facilities) in the municipality (region); jobs growth in the region (municipality); growth of the construction of social housing in the municipality (region); number of dilapidated and dangerous housing, modern residential buildings built in their place; access of the population of rural areas to the means of communication (development of the Internet system); rural crime level; and development of the retail chain.

Environmental criteria: effectiveness of soil conservation measures; preservation of rural landscapes; growth of the system of protective afforestation; availability in the region (municipality) of program documents on adaptation to climate change; quality of drinking water within the boundaries of the municipality; measures for the protection of aquatic biological resources; presence / absence of unregistered waste dumps in the municipality; condition of hydraulic structures; and reclamation systems in the municipality.

15.3.3 Rural Tourism as One of Directions of Transition to Sustainable Development of Rural Areas

Agricultural tourism, which appeared in a number of European countries back in the 1970s of the last century, becomes more and more popular every year. Its appearance is a typical example (and proof) of the multi-purpose character of agricultural activities, which should not be limited to the cultivation of crops or livestock. The development of this direction of agricultural activity is the most characteristic manifestation of the dynamics of sustainable development of rural areas, since it allows balancing all three groups of interests, i.e., economic, environmental, and social ones.

In scientific papers, there are several approaches to understanding rural tourism, its goals and objectives. Thus, some authors believe that rural tourism is one of the branches of agriculture, within the framework of which rural residents are employed to make a profit organization when organizing rest and recreation. At the same time, two types of agricultural tourism are distinguished. The first one is carried out in small forms of agricultural business (for example, within the boundaries of personal subsidiary plots of citizens). The second one involves the opening of rural tourist villages, centers, and complexes [19]. Others write that rural tourism "is a purposeful travel to rural areas with relatively undisturbed ecosystems and ethnic and cultural complexes, providing a direct contribution to the solution of the issues of rural population and are subject to an adequate management regime based on sustainable development" [17].

The issue of rural tourism is connected with the issue of the boundaries of "rural areas". The fact is that the legislation does not provide a clear answer to the question: are rural areas the lands of rural settlements, or agricultural lands outside their borders? Is it advisable to set their boundaries, and if so, in what order? Without solving this issue, it is impossible to implement effectively the norms of urban planning, natural resources, environmental and other legislation regulating the participation of rural residents in planning the development of these territories and their resources [12]. When discussing this issue, it should be noted that the most optimal interpretation of the category "agricultural territory" is as follows. It is part of the territory of the Russian Federation located outside the boundaries of urban districts and cities of federal significance, whose residents are mainly engaged in agricultural production.

Taking into account this more precise definition, it is advisable to understand rural tourism as a type of activity carried out in rural areas aimed at providing tourists with a range of services for accommodation, food, recreation, and excursions within rural landscapes, with the possibility of voluntary participation of guests in certain traditional methods of farming (horse breeding, fishing, cooking, caring for farm animals, etc.).

The positive sides of rural tourism are economic development (profits of local firms, taxes, real estate construction, souvenirs production, food stuff production, etc.); social aspect (increase of employment of rural population, incomes' growth and living standards of rural population, demand for educational programs and master classes of organizers of agricultural tourism, decrease in the outflow of population to cities, preservation of the traditional peasant way of life); environmental aspect (implementation of programs for protection of land and other natural objects otherwise tourists will not come; production of environmentally friendly products and mitigation of the effects of chemicalization of agriculture; conditions for a healthy lifestyle for local residents).

One of the most characteristic features of agricultural tourism as a type of environmental and agricultural entrepreneurship is the lack of the need for large capital investments, since this type of recreation does not imply high-class hotels, swimming pools, expensive tourist infrastructure facilities (restaurants, casinos, etc.).

This activity allows entrepreneurs to show some flexibility. Rural tourism can act as the only type of activity for them, or be a source of additional income, when the entrepreneur is limited only to the accommodation of the tourist, and the tourist organizes excursions and recreation to his own will.

Meanwhile, the full development of this type of entrepreneurship is hindered by a number of objective and subjective reasons, including the following ones: political reason (reluctance of the authorities to fight corruption effectively, reduce the bureaucratic burden on business, pursue an effective tax and credit policy); economic reason (poverty of rural residents, lack of startup capital, as well as business skills); social reason (inactivity, alcoholism, low level of education and culture); legal reason (lack of a regulatory framework for the development of agricultural tourism, information and coordination of such activities, efficient judicial system that allows entrepreneurs to protect their rights).

In these conditions, the agrarian policy of the Russian government requires certain adjustments. Given that the support for rural tourism is not included in the list of government measures limited by the WTO agreements, measures of support for rural tourism at the federal and regional levels could contribute to the solution of three main issues in the agricultural sector: soil conservation, sustainable development of the rural economy, as well as the solution of social problems in the countryside.

15.4 Conclusion

The study showed that "sustainable development of rural areas" should be understood as a stable implementation of economic activities on agricultural land, growth of agricultural production, provided that the social rights of rural residents are properly guaranteed, as well as measures are taken to protect the environment on agricultural lands (including measures of preservation of the fertility of agricultural land) and in rural areas.

The transition of rural areas to sustainable development is hampered by the lack of an efficient system of interdepartmental interaction and coordination of individual issues connected with the development of rural areas.

Compared to developed countries, government support for agriculture in Russia is significantly lower, what is one of the factors of low profitability of agriculture, and the lack of support limits the possibilities of modernization and innovative development of the industry, negatively affects the level of remuneration of workers in the industry and the size of the tax base of local budgets. Development of a vital infrastructure for health care and education aims at economic efficiency at the expense of the population's access to these essential social services, what in fact leads to the violation of the constitutional rights of villagers to health care and education. The infrastructural development of rural areas in Russia, especially the road network and modern communications, is proceeding at a pace that does not allow in the near future overcoming the existing spatial and communication gap between the city and the countryside.

Departmental barriers which hinder the availability of development resources for the rural population remain. The reform of local self-government did not lead to the creation of the institution of rural self-government, which has sufficient funds to solve the problems of the rural population. Civil society institutions are poorly developed in rural areas. The loss of rural traditional culture is not recognized by society as a problem that deserves attention and there are no public resources helping to overcome it; as a result, unique monuments of the material and intangible heritage of the peoples of Russia, which serve as the basis of their identity, are destroyed and then disappear. Life in rural areas is not attractive for young people. The outflow of young people from rural areas to cities is a serious obstacle to the formation of a human resources base for rural development.

The lack of an effective system of environmental supervision, as well as the poverty of rural residents, entail poor use of modern agricultural technologies and fertilizers, increase in the number of abandoned and unused lands, and also lead to the appearance of illegal waste dumps.

The concept of sustainable development of rural areas can be a tool to mitigate the severity of social, economic, and environmental problems in rural areas, balancing the interests of citizens, business, and government. For an efficient realization of this idea, a complex and multi-level system of criteria and indicators for sustainable development of rural areas is required and they are suggested in this chapter.

Acknowledgements The study was carried out with the financial support of the Russian Science Foundation (project No. 20-18-00314).

References

1. Abanina, E.N.: Ideas of sustainable development in Russian forestry legislation: terminology, legal consolidation, implementation. Eurasian Law J. **10**, 206–209 (2016)
2. Blokhin, V.N.: Significance of a sustainable development strategy for rural areas. In: Bulletin of the Faculty of Business and Law: collection of scientific papers. Anniversary edition. Belarusian State Agricultural Academy, Gorki (2018)
3. Environmental provisions of constitutions: collection of papers. In: Vystorobets, E.A. (Ed). MIRmPOS Publication Moscow-Ufa (2012)
4. Federal Law: On Environmental Protection (Adopted by the State Duma on December 20, 2001, no. 7-FZ) (Russian) (2002)
5. Federal Law: On the Development of Agriculture (Adopted by the State Duma on December 22, 2006, no. 264-FZ) (Russian) (2006)
6. Federal Law: On Viticulture and Winemaking in the Russian Federation (Adopted by the State Duma on December 18, 2019, no. 468-FZ) (Russian) (2019)
7. Gurnovich, T.G., Bezlepko, A.S., Kuznetsov, R.A.: Main factors of sustainable and effective development of agricultural production. Bull. Adygea State University. Series 5: Econ. **4**, 124–130 (2010)
8. Imekova, M., Boltanova, E.: Prospects for sustainable development of Russia: The case of eco-innovations. IOP Conference Series: Earth and Environmental Science, 395, 012103 (2019)
9. Ivanova, S.V.: Sustainable use of wildlife indicators: formation and implementation issues in the Russian federation. Environ. Claims J. **30–3**, 237–250 (2018)
10. Kuzmich, I.P., Saskevich, V.V.: Legal regulation of environmental protection in agriculture of the Republic of Belarus. In: Anisimov, A.P. (Ed). Important Problems of the Theory of Environmental Law: Monograph. Yurlitinform Publ., Moscow (2019)
11. Kurman, T.V.: Theoretical and Methodological Foundations of Legal Support for Sustainable Development of Agricultural Production: Dissertation Abstract for the Degree of Doctor of Economic Sciences. National Law University named after Yaroslav the Wise, Kharkov (2019)
12. Minina, E.L.: Legal regulation of sustainable development of rural areas. J. Russian Law **12**, 31–37 (2009)

13. Ostrovsky, N.: Criteria for sustainable development: municipal aspects (2002). http://n-t.ru/tp/ns/kur.htm. Accessed 6 January 2021
14. Order of the Government of the Russian Federation: On Approval of the Strategy for Sustainable Development of Rural Areas of the Russian Federation for the Period up to 2030 (Adopted 02.02.2015 no. 151-r (as amended on 13.01.2017)) (Russian) (2015)
15. UN General Assembly Resolution: Transforming Our World: The 2030 Agenda for Sustainable Development (Adopted by the UN on September 25, 2015) (2015)
16. Urban Planning Code of the Russian Federation: (Adopted by the State Duma on December 22, 2003) (Russian) (2004)
17. Vorobchukov, S.A.: Competitiveness Increase of Agricultural Tourism in Russia: Dissertation Abstract for the Degree of Candidate of Economic Sciences. State University of Management, Moscow (2007)
18. Vyphanova, G.V.: Definition and legal support of the concept of sustainable development. Bulletin of the University named after O.E. Kutafin **1**, 64–93 (2016)
19. Zdorov, A.B.: Organizational and economic bases for the development of rural tourism: dissertation abstract for the degree of Doctor of Economic Sciences. All-Russian Research Institute for Organization of Production, Labor and Management in Agriculture, Moscow (2011)

Chapter 16
Integrated Water Resource Management in West African Basin Organizations—Toward Sustainable Economic Development

Ernest C. Peguita and Alexander M. Solntsev

Abstract To the hectic pace of creation of River Basin Organizations in West Africa in the 1960s and 1970s, it will follow a period of institutional lethargy of the organizations thus created. Meanwhile, on the international front, efforts to codify the customary rules forming the corpus juris of international water law will intensify, leading to new development in international water law. To maintain their relevance, the West African RBO had to adapt their legal framework, resulting in the successive adoption of various water charters. These charters, however, far from being mere actualization of the existing legal frameworks, will in many regards prove to be much more ambitious and comprehensive, especially on the issues of environmental protection, water resources management, recognition, and preservation of the rights of pastoral populations living on transhumance, but also on the delicate issue of the participation of water resource users in the decision-making process on the management of the resource, thus confirming the paradigm shift in international water law and the new trends in the development of the latter which, due to the complexity and dynamism characterizing an international watercourse, requires a more holistic approach with other branches of international law such as international environmental and human rights.

16.1 Introduction

Most basin organizations in West Africa were created in the 1960s and 1970s, in a context marked by the consideration of water as an economic resource and in some cases as a means of political integration of the newly independent States. The resulting legal framework will be influenced not only by internal factors (the need for development, joint management of water resources) but also by external factors (geostrategic) and the global paradigms of the time regarding water and environmental management.

E. C. Peguita · A. M. Solntsev (✉)
Peoples' Friendship University of Russia, (RUDN University), Moscow, Russia
e-mail: solntsev-am@rudn.ru

A. O. Inshakova and E. I. Inshakova (eds.), *New Technology for Inclusive and Sustainable Growth*, Smart Innovation, Systems and Technologies 288,
https://doi.org/10.1007/978-981-16-9808-8_16

On the international front, efforts to codify the rules and principles of international water law will result in the enshrinement of customary norms [2]. Numerous texts, such as the Helsinky Rule of 1996 [7], the UNECE Water Convention, the Rio Declaration, and the Watercourse Convention of 1997, were adopted, establishing fundamental principles of international water law as norms of international law [1].

To maintain their relevance vis-à-vis these major developments in international water law, but also to meet the new challenges posed by the management of water resources, from the early 2000s, the West African basin organizations will successively adopt water charters, new legal instruments to supplement the existing legal framework. These basin water charters, of which there are 4, one of them, notably the Volta Basin Water Charter, is still awaiting formal adoption and ratification, take advantage of the continental and international momentum generated by the adoption of various legal instruments aimed at equitable and sustainable use of water resources, such as the adoption in March 2000 of the African Water Vision 2025 by the Extraordinary Summit of the African Union [4, 15], the adoption of a subregional Action Plan for Integrated Water Resources Management in West Africa in Bamako on 16 December 2000.

16.2 Methodology

This study is based on a combination of general scientific methods and approaches, as well as special scientific methods of systematic and logical study. In other words, methods of deduction, induction, and dialectical analysis have been used. In addition, other special methods such as systemic and structural, but also historical-legal, and comparative methods have been used to investigate the subject, and to reach the following results.

16.3 Results

16.3.1 The Charter of the Waters of the Senegal River

Adopted on May 28, 2002, the Senegal River Water Charter is the first charter of its kind on the entire African continent. The document was elaborated with the expertise of the Société du Canal de Provence and completes the already rich legal framework of the OMVS. It is described by OMVS as "a legal instrument of international scope, innovative and forward-looking" combining law and operationality thanks to its principles, modalities, shared management mechanisms, and annexes relating to the modulation of optimal choices [10].

The charter redefines, with more precision, the guiding principles for the management of the Basin. Among other things, it determines the terms and conditions for approving new projects using the waters of the Basin, or likely to affect its quality, as well as the framework and modalities for the participation of water users in the decision-making process for the management of the Senegal River water resources.

Although the charter mentions "fair" uses of water in its article 8, it does not give priority to any particular use [9]. However, the charter specifies that in the event of shortage «of the resource, special attention will be paid to the supply of drinking water and domestic uses of water», an acknowledgement of the population rights to water.

The document is much more far-reaching on environmental issues. Indeed, in addition to the common principles of polluter pays [6], the charter imposes on the States in its title 4, the obligation to preserve the environment. This evident emphasis on ecological aspects is perceptible even in the elements to be taken into account in the distribution of water between uses [6].

Another aspect of the charter in contrast with the pre-existing legal framework in the Basin is the participation of water users in the decision-making process for the management of the Senegal River water resources. In spite of the initial reluctance of the States which saw in the direct participation of the users a delegation of their attributions likely to weaken their decision-making capacity, according to Guy Meublat, they were forced to do so by donors through the policy of financing conditionalities [3]. It is within this scope that the water charter proposes that the observer status of the standing water commission be granted to representatives of users, local authorities, non-governmental organizations, decentralized management committees, etc.

In short, the water charter has thus allowed a renovation of the legal framework of the Senegal River Basin while defining the strategic option for the development of the resource. In the same way, it now makes it possible to combine environmental protection and economic development of water resources in the spirit of the Rio declaration on sustainable development, while laying the foundations for the establishment in the basin of a true environmental participatory democracy.

16.3.2 The Niger Basin River Water Charter

The Niger Basin Water Charter was adopted at the 8th Summit of Heads of State and Government in Niamey on April 19, 2009. It represents a significant development of the legal framework for water management in the Niger Basin. Although broadly inspired by the charter of the waters of the Senegal River that preceded it, it is more ambitious than the latter on the subject of the environmental protection, which is the sole subject of an annex [5]. The desire to modernize the legal framework of the NBA is evident from the reading of the preamble where it refers to conventions such as the Helsinki Convention of 1992 and the New York Convention of 1997 [13].

The Charter thus includes a rather rich preamble which was elaborated in a participatory manner and its various provisions allow, within the framework of a clear objective and a relatively broad field of application, to guarantee the principles applicable to the management of international water resources, to renovate the institutional framework of the NBA, and to reinforce the obligations of the Member States in the concerted and rational management of the waters of the Niger Basin. To this end, the charter establishes 6 general principles to serve as a basis for the action of the States parties [13]:

· The Principle of equitable and reasonable utilization and participation [14], · the principle of non-damaging use (No-Harm rule), · the precautionary principle, · the principle of prevention, · the polluter-pays principle, and the collector-pays principle.

Furthermore, the question of priority between uses is addressed in the Charter, which instructs taking into account the priority given to basic human needs, which in substance restates the provisions of the 1997 New York Convention [13]. Furthermore, the thorny question of the proposed measures has been the subject of an initial regulation.

Once again, cooperation between the States remains a priority, thus, "The States Parties undertake to exchange information and to consult each other and, where appropriate, to negotiate on the possible effects of planned measures on the Niger Basin" [13].

With respect to dispute resolution mechanisms, just like the charter of the waters of the Senegal river, Chapter IX of the charter provides for two modes of settlement: ***Amicable settlement or non-jurisdictional*** means through the good offices of the Authority, mediation or conciliation or any other peaceful method of dispute settlement and ***jurisdictional settlement*** which consists in the referral to the AU conciliation commission in the first instance, before any referral to the ICJ, as illustrated by the example of the border dispute between Benin and the Republic of Niger [16].

16.3.3 The Lake Chad Basin Water Charter

Adopted in 2012 in a climate of heightened concern over the issue of water sharing in the basin and the environmental, economic, and social implications of global warming. This charter complements the Fort Lamy Convention, of which it serves as a protocol, as evidenced by the reference made in its preamble to the legal instruments signed on May 22, 1964 at Fort Lamy, the final provisions of which stipulate the following:

This Water Charter shall be ratified by the States Parties to the Agreement in accordance with the constitutional rules and procedures and the 1964 Convention and Statute and shall enter into force within thirty days of the deposit of two-thirds of the instruments ratified by Member States. It shall remain in force until the expiration of the Convention for the Development of the Lake Chad Basin and the Statute of the Lake Chad Basin Commission of May 22, 1964.

The objective of the Charter is to ensure the sustainable development of the waters of Lake Chad through an integrated, concerted, and equitable management of its surface water resources, the qualitative management of wetlands, and the management of groundwater.

The charter therefore brings up to date the cooperative framework of the different States of the Basin to reflect the numerous developments in international water law.

16.3.3.1 The Equitable and Rational Use of Water Resources and the Obligation to Do no Harm

The charter in its Chap. 2, Sect. 1 deals with the issue, emphasizing the quantitative aspect as stipulated in article 10 of the charter. The notion of equitable and reasonable utilization is thus framed by quantitative limits determined in Articles 11 and 12 and Appendices 1 and 2.

The charter also sets out a number of factors and criteria to be taken into account in determining what use is considered to be equitable and reasonable. A total of 18 factors and criteria are listed, ranging from hydraulic factors to environmental factors while taking into account certain legal and social aspects [9].

However, by not giving prevalence to any of its factors over others, the charter recognizes that a relative inequality exists between these factors in determining what is equitable and reasonable and leaves it to States Parties to determine their relevance by considering local circumstances and different water needs, while giving priority to basic human needs.

On the relationship between water uses, the charter also does not give priority to a specific use. However, in the event of conflict between uses, the satisfaction of the basic human needs of populations must be given priority.

Specificity of Lake Chad requiring [17], while being in line with the logic of sustainable development, the Charter remains pragmatic on the need not to sacrifice the water needs of current generations, depriving them of the right to enjoy water resources under the pretext of reserving water resources for future generations.

16.3.3.2 Cooperation and Integrated Water Resources Management

It is impossible to address integrated management without cooperation between the States sharing the water resource. That said, Article 1 of the Charter calls on the States riparian to the Lake to cooperate in order to maximize, in an equitable manner, the economic and social well-being of their populations without compromising the sustainability of vital ecosystems. This general obligation to cooperate (*Article 8*) is carried out within the Commission through "harmonization of their positions within the Commission in order to ensure coordinated participation in multilateral negotiations on the management of water resources and the environment." States must therefore work with the Commission to harmonize their different legal regimes relating to the use, protection, and preservation of water resources and their natural ecosystem.

Beyond the harmonization framework mentioned above, States shall also cooperate for the prevention, elimination, and compensation of damage that may result from the use of water resources by another State. This cooperation concerns in particular notification and consultation.

The charter, accordingly, requires prior notification to the Water Resources Advisory Committee and the Environment, Science, and Planning Advisory Committee of all planned measures by a State Party concerning the lake or associated watercourses that may cause significant adverse effects in another country in the basin [9].

The charter also imposes notification measures in the event of a natural disaster of anthropic origin affecting the basin, and any riparian state that causes significant damage to another state party in its use of the water resource is required by the Charter to enter immediately into consultation with the state affected by the damage to eliminate or mitigate the consequences of that damage as soon as possible. The two States must also, in consultation, define the modalities of compensation for the damage caused [9].

16.3.3.3 Environmental Protection and Fight Against Pollution

In terms of environmental protection and the fight against pollution, the charter is very ambitious, and through its provisions clearly demonstrates the willingness of States to make environmental issues one of the key objects of their cooperation. To this end, the charter requires the States, in the territories under their jurisdiction, to conduct environmental impact studies of any activity likely to have adverse effects on the environment [9]. In the event that these adverse effects could be felt by a State of the Basin, the State in its impact study should evaluate the external effects of this activity. In addition to the environmental impact study, the charter also contains provisions on environmental auditing [9] which should be conducted on a regular basis, and strategic environmental assessments for a critical analysis of the various policies, programs, and development plans for water resources and development in the basin.

For the institutional strengthening of the legal framework for environmental protection established by the charter, the states are enjoined to harmonize their various national legal instruments for environmental protection.

To fight against pollution, the States are invited to cooperate in the prevention and mitigation of pollution, by committing themselves, among others, to:

- To individually and collectively through the commission, control, and prevent actions likely to have a noticeable effect on the Lake and associated watercourses, on their health condition and physical and chemical composition, their biological characteristics and on the environment in general.
- prevent pollution at the source.
- Establishing methods and practices to prevent point and non-point pollution.
- Establishing lists of substances and concentrations prohibited, limited, or subject to research and testing in the waters of the Basin.

It is also envisaged that a system of authorization for the discharge of polluting substances into the waters of the lake will be issued to enable the application of pollution prevention standards, in particular those relating to the application of water quality standards and other measures set out in *Article 22* of the Charter. Thus, any polluting discharge into Lake Chad and into the surface or ground waters of its watershed is subject either to a discharge authorization or to a prior declaration in compliance with the prescribed pollution standards. The nomenclature of samples subject to authorization or declaration is determined by the LCBC. It is the same for any discharge of pollutant likely to endanger public health and safety or likely to constitute a threat to the biodiversity of the basin, and is subject to authorization.

To conclude, the water charter is a great step in the institutional strengthening of cooperation in the basin, it modernizes a 50-year-old institutional framework and enable the basin countries to face the various challenges related to the integrated management of water resources, while allowing the removal of legal barriers that may hinder cooperation in the basin.

16.3.4 The Volta Basin Water Charter

The Volta Water Charter is the last West African basin water charter to be established.

Still at the project stage, as it has yet to be formally adopted by the summit of heads of state and government of the Volta Basin Authority member countries [11], it is the result of a participatory drafting process which began in 2015 through the VBA capacity development support project: water charter and master plan [8] followed by the actual drafting of the Charter. The Water Charter was validated by the Committee of Experts of the six member states of the VBA in Lome in December 2018 and approved by the Council of Ministers of the Authority a few months later at its 7th ordinary session held in Accra, Ghana, on 10 May 2019 [11].

Elaborated following the previous charters mentioned above, it also completes and actualizes the existing legal framework of the Volta Basin. It should serve to strengthen cooperation between the States of the basin and to find a balance between the preservation of the good ecological status of the river and the economic development by satisfying the uses of the river (drinking water, irrigation, industry) and non-withdrawals (hydroelectricity, navigation).

16.3.4.1 An Instrument for the Concerted and Peaceful Management of Water Resources and for Reducing the Risk of Water-Related Conflicts

The Charter imposes a general obligation on States to cooperate, though already universally recognized in international conventions such as the New York Convention of 1997, the general obligation is recalled in the charter with a specification of the different areas or fields of cooperation. Thus, the cooperation extends not only to the issues of water management in the basin but also to the implementation of the charter.

A specificity of the charter is the recognition to pastoralists of transhumance in the basin and their right to exploit pastoral resources in accordance with community regulations and national legislation. This recognition aims to limit internal conflicts that are often between farmers and herders related to transhumance [12].

16.3.4.2 A Tool for Quantitative and Qualitative Management of Water Resources and Information Sharing

With increasing pressure on the waters of the basin due to population growth and the multiplicity of uses, the charter is presented as an instrument for regulating disputes over uses in order to guarantee the availability of water on the different sections. In its title devoted to the quantitative and qualitative management of water, the charter, while determining the various uses, recognizes the right of the State to share the waters or part of the waters of the basin subject to compliance with the provisions of the charter [11]. In the same vein, the fundamental principles of international water law are recalled. The charter also regulates offtake flows in order to ensure that upstream water usages do not interfere with the usages and proper functioning of downstream ecosystems.

In addition, the charter makes it mandatory to share hydrometric information for flood prevention but also for emergency management.

16.3.4.3 Protection of the basin's Ecosystems

Building on provisions and other international conventions for environmental protection, the Volta Basin Water Charter proposes a harmonization of national policies, strategies, and legal framework for environmental protection in the Volta Basin area. The prevention and control of pollution (environmental pollution in all its forms) is a conventional obligation of the States, which thus undertakes to work individually and collectively, to take various measures to protect the ecosystems of the basin. In addition, the States Parties also undertake to notify the Basin Authority of any changes in the physical, chemical, or biological characteristics of the Basin. It further develops provisions for the management of pollutants, measures to prevent and control aquatic invasive species and control the introduction of modified, enhanced and exogenous organisms With regard to areas with international status, the charter confirms its adherence to the international instruments committing States Parties to manage these areas in accordance with these international instruments, in particular the Convention on Wetlands of International Importance especially as Waterfowl Habitat, adopted in Ramsar on 2 February 1971; and the Convention concerning the Protection of the World Cultural and Natural Heritage, adopted in Paris on 16 November 1972 [11].

16.4 Conclusion

The wave of adoption of water charters or charter of waters by the various river basin management organizations in West Africa over the last two decades is more a result of the paradigm shift in integrated water management approaches than a simple overhaul of the legal and institutional framework. For if conventions such as the Rio declaration of 1992 and the New York convention of 1997 served as foundations and basis in the elaboration of these charters, they however go far beyond the principles established by these international instruments to provide more ambitious and elaborated standards. The complexity and natural dynamism of a transboundary watercourse forces a holistic management and regulation, a diversity of approaches and a multi-faceted consideration that invites a certain intricacy of different branches of international law: international environmental law, human rights, etc. As elaborate as they are, it is still very early to assess the implications of these charters; however, there is no ambiguity in the fact that they confirm the paradigm shift in international water law toward an increasingly green and humanized water law, a process in which West African basin organizations seem to have taken the forefront.

References

1. Muhammad, M.R.: Principles of international water law: creating effective transboundary water resources management. Int. J. Sustain. Soc. **1**(3) (2009). https://doi.org/10.1504/IJSSOC.2009.027620
2. Salman, M.A.: Salman. The Helsinki Rules, the UN Watercourses convention and the berlin rules: perspectives on international water law. Int. J. Water Resour. Dev. 23, 625–640 (2007). https://doi.org/10.1080/07900620701488562
3. Meublât G.: La gestion partagé des fleuves internationaux en Afrique. in Revue Tiers-Monde, tome 42, No. 166 (2001). https://doi.org/10.3406/tiers.2001.1513
4. Abashidze, A.Kh., Inshakova, A.O., Dementev, A.A.: New challenges of international law in the digital age. In: Popkova, E.G., Sergi, B.S. (Eds.) Modern Global Economic System: Evolutional Development vs. Revolutionary Leap, pp. 1125–1132. Springer Nature, Cham (2021)
5. Niger Basin Authority: Annex N°1 to the Niger Basin Water Charter relating to environmental protection of 2009. http://www.abn.ne/images/documents/Charte_Eau/annex1_protection_environment.pdf. Accessed 2 January 2021
6. ECOLEX: The Charter of the waters of the Senegal river of 2002 http://www.ecolex.org/server2neu.php/libcat/docs/TRE/Full/En/TRE-153511.pdf. Accessed 5 January 2021
7. Slavko, B.: the international law association helsinki rules, contribution to international water law. Brill Res. Perspect. Inte. Water Law (2019). https://doi.org/10.1163/23529369-12340012
8. Le Hub Rural: Launch of the study for the elaboration of the Volta Basin Water Charter - Le Hub Rural. http://www.hubrural.org/Demarrage-de-l-etude-de-l-elaboration-de-la-Charte-de-l-eau-du-Bassin-de-la.html. Accessed 7 January 2021
9. African Water Facility: The Lake Chad Basin Water Charter. https://www.africanwaterfacility.org/fileadmin/uploads/awf/Projects/MULTIN-LAKECHAD-Water-Charter.pdf (2011). Accessed 6 January 2021
10. Audrey, A.: Maîtrise en études internationales Maître ès arts (M.A.) - The Senegal River Basin : towards participatory management? Involvement of the population in the management of

resources of the basin and evolution of relations between riparians (2013). https://corpus.ulaval.ca/jspui/bitstream/20.500.11794/24477/1/30127.pdf. Accessed 2 January 2021
11. Volta Basin Authority: The Volta Basin Water Charter. http://abv.int/la-charte-de-leau-du-bassin-de-la-volta-the-volta-basin-water-charter/. Accessed 6 January 2021
12. Garané, A.: The International Legal Framework of the Volta Basin. UICN, Gland, Suisse (2009)
13. Niger Basin Authority: The Niger Basin River Water Charter. http://www.abn.ne/images/documents/Charte_Eau/niger_basin_water_charter_certified_copy.pdf (2009). Accessed 6 January 2021
14. United Nations (1997). Convention on the Law of the Non-Navigational Uses of International Watercourses of 1997. https://legal.un.org/ilc/texts/instruments/english/conventions/8_3_1997.pdf. Accessed 2 January 2021
15. African Development Bank Group: Africa water vision for 2025: Equitable and sustainable use of water for socioeconomic development, is a document published by the United Nations Economic Commission for Africa (ECA) in 2004 detailing the vision for an Africa where there is equitable and sustainable use and management of water resources for poverty reduction, socio-economic development, regional cooperation, and the environment. https://www.afdb.org/fileadmin/uploads/afdb/Documents/Generic-documents/african water vision 2025 to be sent to wwf5.pdf. Accessed 1 January 2021
16. International Court of Justice: Benin (2005). Niger Border Dispute Case, ICJ, 12 July 2005. https://www.icj-cij.org/en/case/125. Accessed 2 January 2021
17. Jacques, L., Géraud, M.: Le développement du lac Tchad / Development of Lake Chad: Situation actuelle et futurs possibles / Current Situation and Possible Outcomes [online]. Marseille: IRD Éditions (2014). https://doi.org/10.4000/books.irdeditions.11684

Chapter 17
Prospects for Russia's Accession to the 2013 Minamata Convention as an Important Step in the Transition to Mercury-Free Alternatives in Production Processes in Order to Implement the Concept of Sustainable Development

A. M. Solntsev and P. N. Yusifova

Abstract One of the challenges in implementing the concept of sustainable economic development is to reduce the use of mercury or mercury compounds in production processes. An effective international mechanism in this regard is the 2013 Minamata Convention, which entered into force in 2017, and to date, it has been ratified by 128 states. The Convention encourages states to move towards technically and economically feasible mercury-free alternatives. The Convention regulates the entire life cycle of mercury: supply, trade, use, emissions, waste, storage and management of waste and contaminated sites. Within the framework of this Convention, it is envisaged to promote measures to reduce the use of mercury obtained as a result of primary mining of mercury, to take measures to reduce emissions and releases of mercury into the environment; support for research and development on mercury-free catalysts and processes, and other measures. On September 24, 2014, Russia signed the Minamata Convention, within the framework of the article, and special attention is paid to the problems and prospects of Russia's ratification of this Convention.

17.1 Introduction

Mercury has been known to all mankind since ancient times. In one form or another, its application for practical purposes began several millennia ago. Since then, the concentration of mercury in the environment has increased several hundred times

A. M. Solntsev · P. N. Yusifova (✉)
Peoples' Friendship University of Russia, (RUDN University), Moscow, Russia
e-mail: 1042200218@rudn.university

A. M. Solntsev
e-mail: solntsev-am@rudn.ru

A. O. Inshakova and E. I. Inshakova (eds.), *New Technology for Inclusive and Sustainable Growth*, Smart Innovation, Systems and Technologies 288,
https://doi.org/10.1007/978-981-16-9808-8_17

over natural levels. An active release of mercury-containing products began with the beginning of the industrial revolution, as the extraction of mercury increased sharply.

Mercury has special chemical properties and acts as a global environmental pollutant. Mercury emissions know no national or continental boundaries: it is capable of being transported over long distances, which means that, being released on one continent, it can be deposited on others and in general anywhere in the world.

Mercury occupies a special place among heavy metals due to the fact that it is highly toxic for all forms of life, including humans, and even in small quantities it can cause serious health problems [1].

Sources of mercury in the environment are of both natural and man-made origin. "About a third of all mercury in the environment is of natural origin, and about two-thirds is associated with man-induced activities" [2].

The problem of environmental pollution with mercury and its negative impact on the human body became especially acute in the second half of the twentieth century. Among the most serious incidents that shocked the world community are the following events: 1956 in Japan and 1971 in Iraq, where some of the most massive mercury poisoning occurred.

Awareness of the dangers and harmfulness of mercury and its compounds on human health and the environment was the main reason for starting the process of creating an international convention on mercury. The world community understood the need to consolidate the efforts of all states for the benefit of solving this problem. On October 10, 2013, the Minamata Convention on Mercury was signed under the auspices of the United Nations in the Japanese city of Kumamoto [3]. The Convention got its name from the events that took place in 1956 in the city of Minamata, Japan. Activities of a Japanese company that used mercury compounds in its production and dumped waste from the enterprise containing inorganic mercury and methylmercury into Minamata Bay, resulted in a massive mercury poisoning of residents of the city, who subsequently died. The Convention entered into force on August 16, 2017, after it was ratified by 50 countries; at the moment 128 countries have signed it and 127 have ratified it. The main goal of the convention is "to protect the human health and the environment from anthropogenic emissions and releases of mercury and mercury compounds" [3]. The process of ratification of the Minamata Convention in the Russian Federation is quite complicated. In July 2014, Russia signed the Minamata Convention but has not yet ratified it.

17.2 Methodology

Despite the fact that mercury is a global environmental pollutant and at the same time has a harmful effect on human health, today at the domestic level this problem is not given due attention by the state. The methodological basis of the research includes the use of a philosophical and ideological dialectical approach; philosophical method—the method of formal logic (analysis, synthesis, induction, deduction);

general scientific method—systemic, private scientific method; and comparative legal method.

17.3 Results

In Russia, even before the decision on the need to develop the Minamata Convention was made, some issues related to the regulation of the handling of mercury were regulated at the legislative level. So in 1983, GOST was approved, dedicated to the occupational safety system when working with mercury [4]. According to this standard, specific safety requirements were established in the process of using mercury in work and its application in various industries. The peculiarity of this standard was that it did not apply to activities associated with mercury-containing compounds. In 1998, Federal Law No. 89 "On Production and Consumption Waste" [5] established five classes of waste, depending on the degree of their negative impact on the environment. Mercury and its compounds were assigned to the first class—extremely hazardous waste, which, under normal conditions, is either very difficult to process or not subject to processing at all, while activities aimed at transportation, disposal, decontamination of this class of waste are subject to mandatory licensing. In 1999, the Federal Law No. 52 "On the Sanitary and Epidemiological Welfare of the Population" [6] was adopted, according to which the maximum permissible concentrations (MPC) of chemicals in soils, water resources and the atmosphere were established (the provisions of the Federal Law apply to mercury and its compounds).

Particular attention to the issue of mercury pollution of the environment was paid in Russia in 2006 by the Security Council at a meeting of the Interdepartmental Commission on Environmental Safety following the results obtained in 2005 within the framework of the project "Assessment of the release of mercury into the environment from the territory of the Russian Federation" [7]. According to the results presented in this document, the main sources of mercury pollution of the environment in the Russian Federation are: industrial and household waste, mercury-containing products to be disposed of, mercury deposits, mercury and its compounds used in various industries, combustion of coal, gas, wood, oil and gas production, cement, phosphate production. The results obtained served as a reason for a more detailed study of this problem and the introduction of certain changes in the current legislation. Thus, changes were made to the technical regulations concerning the maximum level of mercury in food, additives, in school supplies, in perfumery and cosmetic products, in textiles. According to the new requirements, if the maximum permissible level of mercury in these products is exceeded, then the import/export of these goods from the territory of the Russian Federation is prohibited. All of these changes were made before the Convention entered into force. Starting from August 16, 2017, a number of measures were also taken to restrict the use of mercury lamps and fluorescent lamps, which are not produced in Russia.

From 2013 to 2017, Russia, together with UNEP, implemented a pilot project to create a register of mercury pollution of the environment in the Russian Federation

[8]. On the basis of this, an inventory was taken of the sources of mercury release into the environment and the volumes of such releases in 2012. According to the inventory data, mercury emissions into the atmospheric air amounted to about 97.8 tons, and according to the data provided by Rosstat −2.993 tons. The main reason for the differences between these two numbers is that in the Russian Federation the accounting system is based on hygienic rationing, where substances that do not exceed the MPC for mercury are not taken into account. At the same time, the MPC indicators for mercury in substances differ depending on the ingress medium (water, air, etc.), which leads to an underestimation of mercury releases from enterprise owners. To date, there are no accurate and reliable data on the sources of mercury entering the environment and the volume of these releases on the territory of the Russian Federation [9]. On the information portal for assessing the release of mercury into the environment in the Russian Federation [10], in the section of the inventory of mercury pollution, no information is provided, all columns are empty.

Particular attention should be paid to the protocol on heavy metals to the 1979 Convention on Long-Range Transboundary Air Pollution [11], according to which each country shall maintain emission inventories of heavy metals (including mercury) and submit them to the Centre on Emission Inventories and Projections. Since the Russian Federation has not ratified this protocol, it does not submit this information. All this testifies to the fact that "a reliable inventory of the sources of mercury entering the environment of the Russian Federation is the basis for the implementation of targeted actions to reduce the use of mercury in the country and the negative consequences of its impact on humans and the environment" [12].

Article 13 of the Minamata Convention on Mercury, which establishes a financial mechanism to support developing-country parties and parties with economies in transition in implementing their obligations under the Convention, and that the mechanism includes the Global Environment Facility Trust Fund and a specific international programme to support capacity-building and technical assistance [13].

Today, the provisions of the Minamata Convention mostly affect such industries in the Russian Federation as mechanical engineering, metallurgy, chemical-engineering complex, consumer goods industry, timber industry, pharmaceutical and medical industry. Upon ratification of the Minamata Convention, Russia will have to undertake a certain range of obligations aimed at reducing the use of mercury-containing products, limiting the use of mercury in industrial and production processes, maintaining an inventory and controlling the release of mercury [14]. Despite restrictions in national legislation, a standard has been established for the permissible level of mercury in various substances, requirements for the control and monitoring of mercury and its compounds in the environment have been established. All these indicators and requirements differ from those enshrined in the Convention, due to which Russian industries will be seriously exposed to certain changes they are not ready for right now.

17.4 Conclusion

The analysis of the current legislation of the Russian Federation, and economic and legal provisions of the Minamata Convention showed that many Russian industries will have to make certain changes in their production activities, as a result of the ratification of the Mercury Convention. In this connection, it seems necessary to develop a national action plan aimed at the implementation of the Convention:

- Introduce green technologies into the production process, which will prevent mercury from entering the environment;
- Improve the system for handling mercury-containing waste;
- Improve state environmental control and supervision over the activities of enterprises;
- Change the system of state monitoring and accounting of mercury in the natural environment.

All these actions will allow the Russian Federation to soon ratify the Minamata Convention and assume obligations for its implementation.

References

1. Inshakova, A.O., Frolova, E.E., Galkina, M.V., Rusakova, E.P.: Development of social market economy under the influence of noneconomic factors: modeling and regulation. Int. J. Sociol. Soc. Policy (2020). https://doi.org/10.1108/IJSSP-03-2020-0067
2. Aimakhan, B.B.: Problems associated with mercury pollution of natural environments. Scientific Pract. Res. **5–4**(28), 5–10 (2020)
3. Minamata Convention on Mercury (Kumamoto, October 10, 2013). http://www.mercuryconvention.org/Portals/11/documents/Booklets/Minamata%20Convention%20on%20Mercury_booklet_Russian.pdf (2013). Accessed 23 March 2021
4. GOST 12.3.031-83 Occupational Safety Standards System (OSSS). Work with mercury. Safety Requirements (October 10, 1983). http://docs.cntd.ru/document/1200008484 (1983). Accessed March 2021
5. Federal Law No. 89-FZ June 24, 1998 On production and consumption wastes (June 24, 1998). http://www.consultant.ru/document/cons_doc_LAW_19109/ (1998). Accessed 24 March 2021
6. Federal Law No. 52-FZ of March 30, 1999 On the Sanitary and Epidemiological Welfare of the Population (30 March 1999). http://www.consultant.ru/document/cons_doc_LAW_22481/ (1999). Accessed 23 March 2021
7. Assessment of the release of mercury into the environment from the territory of the Russian Federation (2005). https://www2.mst.dk/udgiv/publications/2005/87-7614-541-7/html/indhold_rus.htm. Accessed 23 March 2021
8. Pilot project to create a register of mercury pollution of the environment in the Russian Federation. https://docplayer.ru/56639052-Rezultaty-vypolneniya-pilotnogo-proekta-po-sozdaniyu-kadastra-rtutnyh-zagryazneniy-okruzhayushchey-sredy-v-rossiyskoy-federacii.html (2013). Accessed 23 March 2021
9. Matytsin, D.E.: Digital technologies of the bank of Russia for regulating investment relations. In: Inshakova, A.O., Frolova, E.E. (eds.) Smart Technologies for the Digitisation of Industry: Entrepreneurial Environment, pp. 227–240. Springer, Singapore (2022)

10. Information portal for assessing the release of mercury into the environment in the Russian Federation (2013). http://www.mercury.nii-atmosphere.ru/. Accessed 23 March 2021
11. Protocol to the 1979 Convention on Long-Range Transboundary Air Pollution on Heavy Metals (Aarhus, 24 June 1998). https://www.un.org/ru/documents/decl_conv/conventions/heavy.shtml (1998). Accessed 24 March 2021
12. Romanov, A.V., Ignatieva, Yu.S., Morozova, I.A., Speranskaya, O.A., Tsitser, O.Yu.: Mercury pollution in Russia: problems and recommendations, 97, Liga, Kolomna (2016)
13. Matytsin, D.E.: Internet-investing as a remote algorithm of the retail investment financing. In: Popkova, E.G., Sergi, B.S. (eds.) Modern Global Economic System: Evolutional Development Vs. Revolutionary Leap. Lecture Notes in Networks and Systems, pp. 1850–1857. Springer, Nature Switzerland (2021)
14. Abashidze, A.K., Inshakova, A.O., Solntsev, A.M., Gugunskiy, D.A.: Socio-economic inequality as a barrier on the path of sustainable development: "institutional traps" and perspectives of overcoming them. Int. J. Sociol. Soc. Policy **41**(1–2), 62–75 (2021)

Chapter 18
Space Resource Utilization as Sustainable Economic Growth Perspectives: Legal Aspects

Irina Chernykh and Denis Gugunskiy

Abstract Off-Earth mining has become a more and more popular topic for discussion at the national and international levels these days. It is predicted that such utilization could start in the nearest possible time and become a remunerable new type of economic area, aiming at inclusive growth and subsequent economic one. It establishes a safer and more secure working environment as it is stated in SDG 8 "Decent Work and Economic Growth." Also, it is very important to bear in mind that the space sector allows non-governmental sectors (business) to benefit from space. However, such utilization is required to be well regulated including ensuring its long-term sustainability. Thus, the aim of this research is to comprehensively analyze and compare national legislation and policy of States in the sphere of space resource utilization and single the commonalities out.

18.1 Introduction

Humanity has started using space resources, when people launched the first satellite into outer space and become using the outer space as a resource from a broad point of view. Despite this fact, the use, exploitation and utilization of space resources are being discussed actively, firstly, since 2015 when the United States adopted its national space law granting the right of U.S. citizens to space resources. Secondly, the interest in space resources has increased when humanity understood that some of Earth's natural resources are in the process of completion, i.e. non-renewable natural resources (oil, nonferrous metal ores, marble, etc.) [25]. The last problem has been highlighted in the context of 17 interrelated Sustainable Development Goals (hereinafter SDG) adopted in 2015 in the final document of the United Nations Development Summit for the post-2015 period "Transforming our world: the 2030

I. Chernykh (✉) · D. Gugunskiy
RUDN University, Moscow, Russia
e-mail: chernykh-ia@rudn.ru

D. Gugunskiy
e-mail: gugunskiy-da@rudn.ru

A. O. Inshakova and E. I. Inshakova (eds.), *New Technology for Inclusive and Sustainable Growth*, Smart Innovation, Systems and Technologies 288,
https://doi.org/10.1007/978-981-16-9808-8_18

Agenda for Sustainable Development" [17]. Two of these goals are closely related to resource utilization including space resources, for example, SDG 8 "Decent Work and Economic Growth" and SDG 12 "Ensure sustainable consumption and production patterns". The goals enshrine that it is required to improve resource efficiency in consumption and production (target 8.4) and to achieve the sustainable management and efficient use of natural resources (target 12.2). Fulfillment of these targets and goals deals with the complex plan of action basing on the concepts of inclusive economic growth and sustainable development [29], as well as the sustainability of space activities.

The mentioned concepts cover a wide range of circumstances and include different doctrines. Firstly, inclusive growth does not have a universally recognized definition [22, 27, 32]. One of the definitions says inclusive growth is "economic growth that results in a wider access to sustainable socio-economic opportunities for a broader number of people [not only to poor people], regions or countries, while protecting the vulnerable, all being done in an environment of fairness, equal justice, and political plurality" [22]. Such a growth should be based on the estimation of the following factors: indicators of inequality, poverty and development, labor markets, governance and climate [27]. Secondly, inclusive economic growth connects with sustainable development, which appeared as a reaction to the negative impact of economic growth and globalization on the environment and the Earth's ecosystem (climate change, water scarcity, etc.) [31]. Thirdly, sustainable development in the framework of space activities is considered to provide its sustainability, when future generations will be able to explore, use and benefit from it at equal bases. It has been established that space activities help to achieve the 2030 Agenda for Sustainable Development [19]. Moreover, this correlation has been recognized in the following documents: the Sendai Framework for Disaster Risk Reduction 2015–2030 [16], the Paris Agreement of 2015 [14], the General Assembly Resolution "Fiftieth anniversary of the first United Nations Conference on the Exploration and Peaceful Uses of Outer Space: space as a driver of sustainable development" of 2018 [18], as well as Declaration on International Cooperation in the Exploration and Use of Outer Space for the Benefit and in the Interest of All States, Taking into Particular Account the Needs of Developing Countries of 1996 [15] and the Guidelines for the Long-term Sustainability of Outer Space Activities of the Committee on the Peaceful Uses of Outer Space of 2019.

Consequently, the research on the use, exploitation and utilization of space resources from the legal point of view and for the benefits and increase of inclusive economic growth is considered to be advisable and relevant.

18.2 Methodology

The adoption of national legislation and policies of States in the field of the exploration and use of outer space is based on the sectoral principle of international space law, which enshrines the international obligation of States to apply an authorization procedure and to constantly supervise any national space activity. Certain issues of commercial exploration of outer space and its resources began to appear in national legislation and policies of individual States, for example, the United States, starting in the 1980s. At the same time, regulation of research, using and development of space resources are associated with the adoption of the U.S. Commercial Space Launch Competitiveness Act of 2015.

Thus, to study the approaches of States to the regulation of activities in the field of exploration, exploitation and utilization of space resources, the method of comparative law was applied, as well as the legal and technical method to develop and identify examples of best practices of States. In addition, general scientific methods of cognition were used, such as analysis and synthesis, generalization, dialectical, systemic and structural methods.

18.3 Results

18.3.1 The U.S. National Law and Policy on the Exploration, Exploitation and Utilization of Space Resources

The United States is considered to be one of the first States in which national space legislation was adopted. In 1958, it has already had the National Aeronautics and Space Act [9]. The commercial use of space in the United States was later enshrined in legislation such as the Commercial Space Launch Act of 1984 [5] and the Land Remote Sensing Commercialization Act of 1984 [6], which were subsequently amended. In 2002, the Commercial Reusable In-Space Transportation Act was passed to support private space companies in the development of commercial reusable space transportation systems [1]. The National Aeronautics and Space Administration Authorization Act of 2008 set the U.S. space industry to expand the presence of humans and robots in the solar system, including exploration and use of the Moon, nearby Earth asteroids, Lagrange points, and ultimately Mars and its satellites [7].

It should be noted that in 2010, within the framework of the Code of Laws of the United States, a new section 51 "National and Commercial Space Programs" was introduced [2], which codifies practically all legislative acts adopted on this topic.

In 2015, under President Barack Obama, the U.S. Commercial Space Launch Competitiveness Act (hereinafter the U.S. Act of 2015) was passed [13]. Thus, the United States became the first state to consolidate its position at the national level regarding the status of space resources and the right of U.S. citizens to these resources. Under this Act, any U.S. citizen wishing to recover asteroid resources or other space resources on a commercial basis has the right "to possess, own, transport, use, and sell" it in accordance with applicable law, including the international obligations of the United States (Sec. 51,303). However, these provisions do not mean the proclamation of sovereignty, exclusive rights, jurisdiction or ownership of any celestial body.

Under the U.S. Act of 2015, a space resource is defined as "*an abiotic resource in situ in outer space that includes water and minerals*" (Sec. *51,301*). Separately, the definition of an asteroid resource is given, meaning "*a space resource found on or within a single asteroid*."

Also, this Law amends other interrelated regulatory legal acts in the field of insurance, licensing and liability. In general, it is aimed at supporting the U.S. national economy.

In the period from 2017 to 2020, several directives were signed in the field of exploration and use of outer space, including Space Policy Directive-2 "Streamlining Regulations on Commercial Use of Space" of 2018 [11]. It amended the licensing process of private space activities.

In May 2019, the National Aeronautics and Space Administration (hereinafter NASA) announced the launch of a new lunar space program "Artemis", which pledged NASA to develop an innovative solar system exploration program to begin with the return of Americans to the Moon. Among the partner states of this program, Australia, Canada and Japan [20], as well as the European Space Agency (hereinafter ESA) and large private companies such as Lockheed Martin and Boeing, Deep Space Systems, Blue Origin, SpaceX and others were named [12].

About a year later, in the development of this space program, on April 6, 2020, Executive Order on Encouraging International Support for the Recovery and Use of Space Resources was signed [3]. This Order reflects the official position of the United States on key issues governed by international space law, including the international legal regime for the utilization and use of the natural resources of the Moon and other celestial bodies. The Order states that it is U.S. policy to encourage international support for the government and commercial extraction and use of resources in space "*consistent with applicable law.*"

The Executive Order of 2020 instructs the U.S. Secretary of State to negotiate with other states on the adoption of joint statements or conclude bilateral and multilateral arrangements on "safe and sustainable operations for the public and private recovery and use of space resources" and report the results to the U.S. President in 6 months. After that, during the year, including at the 71st International Astronautical Congress, held in October 2020, such states as Luxembourg, Australia, Canada, Japan, Italy, Great Britain and the United Arab Emirates signed such a "new international agreement"—the Artemis Accords, which was later joined by Ukraine and Brazil.

The Artemis Accords consists of 13 articles and, in fact, is presented in the form of principles of cooperation in the civil exploration and use of the Moon, Mars, comets and asteroids for peaceful purposes. It was developed by the United States to implement the lunar space program Artemis announced in May 2019. The Artemis Accords states: "the extraction and utilization of space resources, including any recovery from the surface or subsurface of the Moon, Mars, comets, or asteroids, should be executed <...> in support of safe and sustainable space activities."

Thus, the U.S. national policy and legislation in the field of exploration, exploitation and utilization of space resources (1) is based on the existing system of normative legal acts, which since 2010 have been codified in the Code of Laws of the United States; (2) is closely related to the U.S. national space program; (3) involves the active participation of private business in space activities and the attraction of international partners.

18.3.2 National Space Legislation of the Artemis Accords' Participants

In accordance with the earlier analysis of the US national legislation and policy in the field of exploration, exploitation and utilization of space resources, including the commercialization of space activities, the US legislation has formed a clear regulatory legal framework in this regard, which does not contradict the sectoral principles of international space law. Do the US partner States, involving in the framework of individual projects, including the new lunar program "Artemis", have the appropriate mechanisms and tools?

18.3.3 The Legislation of Luxembourg on Space Activities and Space Resources

Luxembourg is one of the top U.S. partners in the lunar program "Artemis". Also, this state became the first European and second state in the world which drafted national law on the use of outer space, as well as through the exploitation of the celestial bodies' resources. In 2017, the Law on the Exploration and Use of Space Resources (hereinafter the Luxembourg Law of 2017) was adopted, the second legal act on space activities in Luxembourg after the Law of 1991 on Electronic Media [30]. It consists of 18 articles. In 2018, Luxembourg established the Luxembourg Space Agency. At the end of 2020, Luxembourg adopted the Law on Space Activities, which entered into force on January 1, 2021 (hereinafter the Luxembourg Law of 2020) [8]. It consists of 18 articles too.

It is important that the Luxembourg Law of 2020 does not apply to the exploration and use of space resources' activities regulated by the Luxembourg Law of 2017.

Space activities in Luxembourg have to be authorized and supervised. The Ministry of the Economy is responsible for it. Luxemburg has a registry of space objects. The insurance contracts under which space objects registered by Luxembourg have the tax exemption. Operators of space objects have a possibility to use tax credit (for investments).

The Luxembourg Law of 2017 does not apply to satellite communications, orbital positions or the use of frequency band (Art. 2 (4)). It has no description of space resources but in accordance with Art. 1 it is said that "space resources are capable of being owned." The authorization of such activity shall be granted only to a certain company type (public company limited by shares, a corporate partnership limited by shares, a private limited liability company), which has been registered under Luxembourg law or to the European Company which has a registered office in Luxembourg.

Nowadays, Luxembourg continues to build infrastructure and legal base on the space resources' use, exploitation and utilization. At the end of 2020, it was announced about creation of the European Space Resources Innovation Centre (hereinafter the ESRIC) in collaboration with ESA in Luxembourg. The ESRIC will gather different directions of the space resources' usage, conduct expertise for scientific, technical, business and economic aspects as well as set up a transfer between space and non-space industries [26]. Luxembourg cooperates with U.S. companies such as Planetary Resources or Deep Space Industries.

18.3.4 The UAE 2019 Federal Law No. 12 on the Regulation of the Space Sector and Its Policy

The United Arab Emirates is a party to three major treaties governing space activities—the Outer Space Treaty of 1967, the Rescue Agreement of 1968, and the Liability Convention of 1972 and the Registration Convention of 1975. Like many states, the Emirates has not ratified the Moon Agreement of 1979.

In 2006, Mohammed Bin Rashid Space Centre (MBRSC) was established in the UAE. The center builds and operates Earth observation satellites, offering data visualization and analysis services to customers around the world. The center has already launched several satellites,[1] and space object KhalifaSat was developed by a 100% team of engineers from Emirates.

The center is responsible for developing a mission to Mars which is designed to orbit Mars and study the dynamics in the Martian atmosphere on a global scale, and on both diurnal and seasonal timescales, as well as for the Astronaut Training Program, which launched the first Emirati astronaut to arrive on the ISS on a scientific mission on September 25, 2019.

The main objectives of the center include developing the UAE space science sector and training new generations of Emirati engineers, experts and researchers; training qualified leaders in the UAE space industry, implementing sustainable training programs and creating a stimulating work environment that encourages innovation; raising awareness of the UAE population about the importance of science, innovation and research; upgrading scientific facilities and infrastructure to support and develop scientific and technical tools and systems; launch of large scientific projects and missions, implement effective management and use of project results; establish small and medium-sized scientific and technological enterprises in the private

[1] Including DubaiSat-1, DubaiSat-2, Nayif-1.

sector through knowledge transfer programs and incubators; develop and implement intelligent systems and applications; support the development of policies and laws governing the scientific sector, research and advanced technologies; launch strategic initiatives.

In 2014, the UAE established the first national space agency in the region, which is responsible for organizing, regulating and supporting the national space sector. It is responsible for expanding and strengthening the Emirates' international position in space fields, for enacting policies and laws for the space sector, licensing in the space industry, establishing international partnerships for the development of the space sector and promoting the transfer of knowledge in space technologies. It represents the country at space conferences and international forums. The UAE Space Agency is also responsible for monitoring business and activities in the country's space sector.

In 2016, the UAE National Space Policy was adopted, which aims to create a strong and sustainable space sector that supports and protects national interests and vital industries, promotes economic diversification and growth, enhances competencies, develops scientific and technological capabilities, strengthens the culture of innovation and strengthens the status and role of the UAE at the regional and global levels.

And in March 2019, the UAE Government adopted the National Space Strategy until 2030, which sets out a common framework for the UAE space industry and public and private sector activities for the period up to 2030 [10].

The increased activity in the exploration and use of outer space has given a new impetus to the Emirates to prepare a national law regulating this sector. The Federal Law on the Regulation of the Space Sector was signed on December 19, 2019 [4], and on February 24, 2020, the details of the new law were announced during a seminar on the space law of the United Arab Emirates, organized by the Space Agency and held in Abu Dhabi [33].

The Law regulates the mechanism for issuing permits for space activities; registration of space objects and vehicles; rules for liability and insurance of space activities; rules for the elimination of space disasters and risk prevention; the transition period for the current rules of operators; provisions governing the construction of objects on other planets, as well as the use of space resources and the development of measures to reduce the formation of space debris.

The new law aims to protect the interests of the UAE by striking a balance between economic and commercial requirements, encouraging innovation, complying with safety and environmental requirements, and encouraging investment and promoting private sector participation in the space industry. It is important to note that the law was adopted at a time of transformation of the global space sector, and opens up new opportunities for startups, as well as small and medium-sized enterprises.

The law strengthens the UAE's position in the international space sector, where they are now among several countries that have such national laws, especially because it covers relatively new activities that are not yet regulated by other laws around the world, as well as other future activities for which the UAE intends to develop appropriate infrastructure in the near future. When developing this law, 20 relevant treaties and agreements were taken into account, and its elements were compared with 18 other national space laws of countries such as the United States, Russia, France, Germany, Korea, Hong Kong, Brazil and Kazakhstan.

The law also addresses advanced and modern concepts that attract international attention and are constantly on the agenda of the UN Committee on Outer Space, including space launches, manned flight operations, space tourism and related activities, educational and scientific activities, the construction and use of artificial objects in space and on other planets, the ownership and use of space resources, as well as other commercial activities such as mining and space logistics services, in addition to the mechanism for dealing with space debris, meteorites and space risks.

The Director of the UN OOSA, Simonetta Di Pippo, said in 2020 that in connection with the expansion of work with the UAE in space, an additional office of the Office in the UAE will be opened in the near future, whose activities will focus on space exploration and strengthen the UAE's position as a regional space center.

As mentioned earlier, by analogy with the laws adopted in the United States and Luxembourg, the United Arab Emirates pays attention to resources in its law. The law defines space resources as any non-living resources present in outer space, including minerals and water.

Article 4 of the Law states that the document regulates space activities, which include, inter alia, activities for the exploration or extraction of space resources and activities for the exploitation and use of space resources for scientific, commercial or other purposes. Also, in accordance with Artice 18, it was noted that the conditions and controls relating to Permits for the Exploration, Exploitation and Use of Space Resources, including their acquisition, purchase, sale, trade, transportation, storage and any space activities aimed at providing logistics services in this regard, are determined by a decision of the Council of Ministers or by a person authorized by it.

It is important to note that such a complex and sensitive issue as the extraction of space resources certainly requires discussion of the international legal regime governing this activity, and for this activity to be carried out properly, on a practical basis and under international law, all interested parties, including both public and private entities, should establish close cooperation with each other.

In July 2020, the Space Agency launched the "Pioneers of Arab Space" program aimed at accumulating Arab experience in the field of space science and technology. The program provides young Arab researchers, scientists, inventors and creative talents with the skills and knowledge they need to build a career in the growing space sector.

The UAE's space sector has provided 1,500 jobs in 57 space-related organizations, five space research and development centers, and three universities offering advanced degrees. In addition, the commercial space sector in the UAE includes the world's seventh largest satellite operator by revenue, Al Yah satellite communications company and Thuraya Telecommunications Company.

Another indication of the active participation of the United Arab Emirates in space activities is the holding of the 72nd International Astronautical Congress on October 25–29, 2021, in Dubai. It is the first Arab country to host the Congress since its founding in 1950.

During the congress, the host organization, the Mohammed bin Rashid Space Center, which has been a member of the International Astronautical Organization since 2012, will demonstrate the work of the Hope spacecraft, launched as part of the Emirates Mars Mission program—the first Arab spacecraft to explore Mars. Hope was the first of three space missions sent to Mars, with missions also launched by the national space agencies of China (Tianwen-1) and the United States (Mars 2020). All three are expected to arrive on Mars in February 2021. The space probes will study daily and seasonal weather cycles, weather events in the lower atmosphere, such as dust storms, and how the weather changes in different regions of the planet. It will also try to find out why it is losing hydrogen and oxygen to space, and other possible causes of drastic climate changes. The United Arab Emirates has also launched a program that plans to build a scientific settlement on Mars by 2117, for which The Hope spacecraft will now conduct exploration.

The UAE's space industry obviously benefits the national economy, as well as various industries, including the environmental sector, the transport sector, the insurance sector and others. In addition, in September 2020, the launch of the Lunar Program 2024 was announced.

18.3.5 National Space Legislation of Australia, Canada, Japan, Italy, the UK, Ukraine and Brazil on Space Activities and Space Resources

In comparison with Luxembourg and UAE such States as Australia, Canada, Japan, Italy, the UK, Ukraine and Brazil do not have special national legislation on the exploitation and utilization of space resources. At the same time, these States have different legislation on space activities.

The following States have the space legislation on the licensing, authorization and supervision: the UK (Outer Space Act of 1986), Australia (Space Activities Act of 1998 and Space Activities Regulations of 2001), Canada (Canadian Space Agency Act of 1990), Japan (Basic Space Law No. 43 of 2008, Act concerning Launch and Control of Satellites No. 76 of 2008, Regulation for Enforcement of the Act concerning the Launch and Control of Satellites № 50 of 2009, etc.), Ukraine (Ordinance of the Supreme Soviet of Ukraine on Space Activity Law of Ukraine of 1996, On Amendments to Certain Laws of Ukraine on the State Regulation of Space Activities of 2019) and Brazil (Resolution No. 182 of 2020 and different regulations).

Italy has no specific legislation, but it has Italian Space Agency.

At the same time, some tendencies could be underlined in this context. The Brazilian government is interested in the consolidation of the rules in the space activities' area; it is required by the active industry growth and academia. Thus, "Brazilian space law may still be regarded as a work in progress" [21].

Japan is drafting specialized space resource legislation. It has already had Remote Sensing Space Systems acts and regulations, but at the end of 2020 "the ruling Liberal

Democratic Party approved a bill to allow private businesses to own mineral and other samples collected outside Earth" [28]. In this context, Japan can become the fourth State orienting on off-Earth mining after the U.S., Luxembourg and the UAE.

18.4 Conclusions

Although the development of space activities is not one of the SDGs and is only mentioned in passing in the 2030 Agenda for Sustainable Development, it is becoming clear that space activities are one of the most important support tools for the achievement of each Sustainable Development Goal [24].

It is important that new players and new cooperatives can adequately distribute the benefits received and direct them to the development not only of their states, but also of all mankind, in accordance with SDG 17 "Partnership for Sustainable Development" [23].

One of the more significant findings to emerge from this study is that national space law is developing in favor of space resources' use, exploitation and utilization. More and more States announce plans on enacting new space legislation or its consolidation. Taken together, the results of the research show that States have space legislation regulating licensing, supervising and authorization of space activities.

Unfortunately, there is still no unified basis for the use, exploitation and utilization of space resources due to the absence of detailed international legal regime and the small amount of national space legislation in this context. For instance, only U.S. space legislation has the definition of "space resource". Thus, the present study lays the groundwork for future research in this direction.

Acknowledgements The article is published in the framework of the grants of the President of the Russian Federation for state support of young Russian scientists—candidates of science (Competition—MK-2021) project "Problems of the exploration and utilization (including exploitation) of space resources: international legal and comparative legal analysis" MK-2118.2021.2. This paper has been supported by the RUDN University Strategic Academic Leadership Program.

References

1. Commercial Reusable In-Space Transportation Act dated October 23, 2002, Public Law 107–248. The uscode.house.gov. https://uscode.house.gov/statviewer.htm?volume=116&page=1573. Accessed 17 March 2021
2. Enactment of Title 51—National and Commercial Space Programs dated December 18, 2010, Public Law 111–314. The govinfo.gov. https://www.govinfo.gov/content/pkg/PLAW-111publ314/pdf/PLAW-111publ314.pdf. Accessed 17 March 2021
3. Executive Order on Encouraging International Support for the Recovery and Use of Space Resources. Issued April 6, 2020. The federalregister.gov. https://www.federalregister.gov/documents/2020/04/10/2020-07800/encouraging-international-support-for-the-recovery-and-use-of-space-resources. Accessed 17 March 2021. Accessed 18 March 2021

4. Federal Law No. (12) of 2019. Issued on 19/12/2019. Corresponding to 22 Rabi' Al-Akhar 1441H. On the regulation of the space sector. The moj.gov.ae. https://www.moj.gov.ae/assets/2020/Federal%20Law%20No%2012%20of%202019%20on%20THE%20REGULATION%20OF%20THE%20SPACE%20SECTOR.pdf.aspx. Accessed 1 May 2021
5. H.R.3942—Commercial Space Launch Act dated October 30, 1984, Public Law No: 98–575. The Congress.gov. https://www.congress.gov/bill/98th-congress/house-bill/3942. Accessed 17 March 2021
6. H.R.4836—Land Remote Sensing Commercialization Act of 1984. The Congress.gov. https://www.congress.gov/bill/98th-congress/house-bill/4836. Accessed 17 March 2021
7. H.R.6063—National Aeronautics and Space Administration Authorization Act dated October 15, 2008. Public Law No: 110–422. The Congress.gov. https://www.congress.gov/bill/110th-congress/house-bill/6063. Accessed 17 March 2021
8. Loi du 15 décembre 2020 sur les activités spatiales. Legilux.public.lu. http://data.legilux.public.lu/eli/etat/leg/loi/2020/12/15/a1086/jo. Accessed 17 March 2021
9. National Aeronautics and Space Act of 1958, Public Law No. 85–568, 72 Stat., 426. The history.nasa.gov. https://history.nasa.gov/spaceact.html. Accessed 17 March 2021
10. National Space Strategy 2030. The space.gov. https://space.gov.ae/Documents/PublicationPDFFiles/2030-National-Strategy-Summary-EN.pdf. Accessed 1 May 2021
11. Space Policy Directive-2 "Streamlining Regulations on Commercial Use of Space" dated May 24, 2018. The whitehouse.gov. https://www.whitehouse.gov/presidential-actions/space-policy-directive-2-streamlining-regulations-commercial-use-space/. Accessed 17 March 2021
12. The Artemis Accords. 2020. NASA. The nasa.gov. https://www.nasa.gov/specials/artemis-accords/img/Artemis-Accords-signed-13Oct2020.pdf. Accessed 17 March 2021
13. The U.S. Commercial Space Launch Competitiveness Act dated November 25, 2015. Public Law 114–90. The Congress.gov. https://www.congress.gov/bill/114th-congress/house-bill/2262/text. Accessed 17 March 2021
14. UN Doc. (2015). Report of the Conference of the Parties on its twenty-first session, held in Paris from 30 November to 13 December 2015 (dated January 29, 2016), FCCC/CP/2015/10/Add.1. https://undocs.org/FCCC/CP/2015/10/Add.1. Accessed 17 March 2021
15. UN General Assembly Resolution A/RES/51/122 of 4 February 1997. https://undocs.org/A/RES/51/122. Accessed 17 March 2021
16. UN General Assembly Resolution A/RES/69/283 of 3 June 2015. https://undocs.org/A/RES/69/283. Accessed 17 March 2021
17. UN General Assembly Resolution A/RES/70/1 of 25 September 2015. https://www.un.org/ga/search/view_doc.asp?symbol=A/RES/70/1&Lang=E. Accessed 17 March 2021
18. UN General Assembly Resolution UN General Assembly Resolution of 26 October 2018. https://undocs.org/en/A/RES/73/6. Accessed 17 March 2021
19. Abashidze, A., Chernykh, I.: The role of international space law in the achievement of the sustainable development goals. Public Int. Private Int. Law **1**(112), 12–19 (2020)
20. Christensen, I.A., Johnson, C.D.: Putting the White House executive order on space resources in an international context. The Space Review (2020). https://www.thespacereview.com/article/3932/1. Accessed 27 April 2021
21. de Bittencourt Neto, O., Freire e Almeida, D.: Brazilian Space Law. Oxford Research Encyclopedia of Planetary Science. https://oxfordre.com/planetaryscience/view/https://doi.org/10.1093/acrefore/9780190647926.001.0001/acrefore-9780190647926-e-213. (2021). Accessed 5 April 2021
22. Dooley, M., Kharas, H.: Friday How inclusive is growth? The Brookings Institution (2019). https://www.brookings.edu/blog/future-development/2019/11/22/how-inclusive-is-growth/. Accessed 22 April 2021
23. Gugunskiy, D., Emelianova, N.: Space activities and sustainable development goals. In: Popkova, E, Sergi, B. (Eds.) Springer book series Lecture Notes in Networks and Systems "Modern Global Economic System: Evolutional Development vs. Revolutionary Leap.", vol. 198, pp. 702–708. Springer Nature, Cham (2021)

24. Inshakova, A.O., Frolov, D.P., Davydova, M.L., Marushchak, I.V.: Institutional factors of evolution and strategic development of general purpose technologies. Espacios **39**(1) (2018)
25. Inshakova, A.O., Goncharov, A.I., Inshakova E.I.: Innovative technologies of oil production: tasks of legal regulation of management and taxation. In: Inshakov, O.V., Inshakova, A.O., Popkova, E.G. (Eds) Energy Sector: A Systemic Analysis of Economy, Foreign Trade and Legal Regulations, pp. 79–94. Springer Science + Business Media, Cham (2019)
26. Jamasmie, C.: Luxembourg to set up Europe space mining centre (2020). Mining.com, https://www.mining.com/luxembourg-to-create-space-resources-centre/#:~:text=Luxembourg's%20administration%20has%20also%20set,then%20dated%20back%20to%201967 Accessed 18 April 2021
27. Koob, S.: Inclusive economic growth: are we talking about the same thing? The Forum OECD (2019). https://www.oecd-forum.org/. Accessed 23 April 2021
28. Kyodo, J.: Japanese bill to allow ownership of samples from outside Earth. The Japantimes (2020). https://www.japantimes.co.jp/news/2020/11/06/national/science-health/japan-bill-space-samples/ Accessed 6 April 2021
29. Matytsin, D.E.: Internet-investing as a remote algorithm of the retail investment financing. In: Popkova, E, Sergi, B. (eds.) Springer book series Lecture Notes in Networks and Systems "Modern Global Economic System: Evolutional Development Vs. Revolutionary Leap," pp. 1850–1857. Springer Nature, Cham (2021)
30. Official web site of the Luxembourg Space Agency. The space-agency.public.lu. https://space-agency.public.lu/en/agency/legal-framework.html. Accessed 1 April 2021
31. Reichart, J.E., Werhane, P.H.: Sustainable Development and Economic Growth. In: Westra, L., Lemons, J. (Eds) Perspectives on Ecological Integrity. Environmental Science and Technology Library, vol. 5, pp. 254–264. Springer, Dordrecht (1995)
32. Stott, J.: What is inclusive growth and why does it matter? The Joseph Rowntree Foundation (2017). https://www.jrf.org.uk/blog/what-inclusive-growth-and-why-does-it-matter Accessed 12 April 2021
33. UAE Space Law Details Announced to Facilitate Space Sector Development (2020). Spacewatch.global. https://spacewatch.global/2020/02/uae-space-law-details-announced-to-facilitate-space-sector-development/. Accessed 1 May 2021

Chapter 19
Prerequisites and Conceptual Directions for the Development of Legal Regulation of "Green Banking" in the Legislation of the Russian Federation

Tatyana V. Kokoreva

Abstract This publication will explore the concept of "green finance" and its main elements, including "green banking". The main attention will be paid to the grounds and procedure for financing environmental measures from the state and local budgets (in the Russian Federation and other countries), as well as the use of public–private partnerships for financing certain environmental measures. The existing mechanisms of financing preventive measures (for example, accidents at oil pipelines) by big businesses are studied. The economic interest of the world's leading banks in environmental lending activities, as well as economic, environmental, social and other benefits of the state and society as a result of the activities under study, are substantiated. The role of legal regulation of "green banking" functioning in the conditions of neo-industrial development is established, and preconditions and conceptual directions of its formation and fixation in the current legislation of the Russian Federation are revealed.

19.1 Introduction

Environmental protection requires significant financial expenditures, accumulation and proper spending of funds for the implementation of environmental protection measures, appropriate environmental programs and resource-saving programs. Therefore, there is an obvious functional correlation between the rational use of natural resources, environmental protection and the solution of problems of financing in this area.

There is a gradual growth of interest in green financing tools, including green banking, as a way to implement sustainable development in world practice. Moreover, in the context of active processes in the field of digitalization of the economy, information products and classic banking tools (lending, investment) are merging [9, 16]. Nevertheless, many financial institutions are only beginning to familiarize

T. V. Kokoreva (✉)
Volgograd State University, Volgograd, Russia
e-mail: t.korotkova@teplovolgograd.ru; gimchp@volsu.ru

A. O. Inshakova and E. I. Inshakova (eds.), *New Technology for Inclusive and Sustainable Growth*, Smart Innovation, Systems and Technologies 288,
https://doi.org/10.1007/978-981-16-9808-8_19

themselves with the opportunities of sustainable development in the field of environmental management and existing environmental tools. In this connection, the development of unified principles, standards and rules for "green finance" and environmental management in banks at the national and international levels, and the reflection of the studied clusters in specific areas of law (civil law, banking law and environmental law) are of critical importance for the development of this area.

19.2 Methodology

The conducted analytical work is based on general scientific and private scientific methods of knowledge. The general dialectical method allowed us to investigate the dialectical relationship between the mechanisms of "green banking" in the concept of sustainable development of the economy with the possibility of application of products of digitalization and fintech.

The structural–functional method was also used to characterize the principles of "green banking" aimed at the formation of a sustainable system of financing environmental measures.

Among the private-scientific methods to disclose the content of the legal regulation of "green banking" in the legislation of Russia, the formal-legal method was applied. The comparative legal method is taken as a basis for revealing the general tendencies of dialectical interaction at the financing of nature protection measures in separate countries and revealing preconditions of reforming of civil and banking legislation of Russia.

19.3 Results

One of the most common instruments of "green finance" is climate or environmental/green bonds. When issuing these securities, in addition to such characteristics as coupon yield, price and circulation period, the environmental assessment of the project is also taken into account. The first issuer of this type of bond was the International Bank for Reconstruction and Development in 2007. Subsequently, issuers of environmental bonds were commercial and non-commercial companies, states and international organizations.

To unify and standardize the issue of climate bonds, in 2009 the Climate Bond Initiative was established. While initially the issue of climate bonds was aimed at financing projects related to climate change, subsequently this list was expanded to include a variety of environmental projects.

The green bond market is growing rapidly. Among the areas of green bond financing are transport, agriculture and forestry, construction and industry, waste control and recycling, energy, water resources and multi-sector. Green bonds are considered a fairly low-risk financial instrument, which is in demand among institutional investors such as banks, pension funds and investment companies [12]. The advantages of green bonds compared to conventional debt instruments contribute to their active implementation. This is especially evident in the countries of Northern Europe, which in 2019 held 7% of the global and 19% of the European green bond market. Among the most popular areas of investment are projects aimed at the development of alternative energy sources, low-carbon transport and energy efficiency [13].

There is also another source of financing environmental measures, proven by the experience of various countries. This is the mobilization of future government revenues to solve urgent current problems by issuing government loans and forming a market of government securities (environmental bonds). This, based on market mechanisms, method of financing environmental activities can be effective in times of financial crisis. Environmental loans are a way to attract funds of legal entities and individuals in terms of voluntariness, urgency, repayment and payment for additional financing of environmental needs of more current opportunities of the state and enterprises. Issue of bonds of the state ecological loans could be carried out by the specially authorized banks, but for this purpose it is necessary to accept the corresponding standard-legal acts regulating their emission activity [1].

Abroad, in order to improve the conditions of investment of ecological directions of activity, ecological banks are functioning. The main sources of finances in Ecobanks are budgetary funds, which are consolidated in ecological accounts, innovative resources, attraction of funds of enterprises, institutions, organizations, population, investments of states and citizens.

Banks accumulating conventionally called "pollution rights" represent a development of the previous approach. Firms that save "pollution rights" can invest them in a special bank for future use or sell them. The bank becomes an intermediary that holds a stock of rights for those who sell and buy them. The banks keep records, ensuring that the rights are redeemed and not reused. Banks can also provide issue credits, that is, temporary rights to increase emissions, to polluting businesses. As the market for pollution rights expands, there is a need for intermediary organizations like emission rights exchanges (pollution rights exchanges).

Another utilitarian tool for green finance is green investment accounts and green investment portfolios, fot example, the possible introduction in Russia of individual investment accounts (hereinafter IIAs) of the third type. Since 2015, Russians may open a type 1 IIA and receive tax deductions, as well as a type 2 IIA and be exempt from personal income tax, provided that the funds are held in the account for at least three years. Despite the fact that this type of savings has not yet become widespread, the Central Bank of Russia is studying the possibility of introducing a third type of IIM with a minimum investment period of 10 years. The funds of these funds can be allocated to infrastructure and green projects, the payback period of which is quite long [15].

If we consider the corporate sector, in the financial world there is more than one investment platform specializing in sustainable projects [11]. For example, the American platform EarthFolio offers investors to form portfolios with social, managerial or environmental focus [14]. Investment corporation OpenInvest also works in the field of responsible investing. To work on the electronic platform OpenInvest, the investor only needs to open a brokerage account, credit the funds and specify the direction of investment (environmental protection, social equality, etc.). The formation of the investment portfolio and its management is engaged in financial advisors through technical capabilities offered by the platform [6].

Among the tools of "green finance", of course, it is necessary to highlight green lending. For example, a green mortgage is the lending of housing, which meets certain environmental standards. This kind of loan can be issued for the construction/purchase of new housing, classified as energy-efficient buildings. The borrower can raise funds for the renovation of a residential structure in order to improve the environmental class of his/her house/apartment. The energy-efficient mortgages initiative (EEMI) [8] promotes the development of this type of financing in Europe.

"EEMI is a global market initiative aimed at mobilizing capital markets and best practices for sustainable development in the financial sector, as well as supporting the goals of the European Green Deal and the Renovation Wave Strategy. EEMI focuses on building a new ecosystem, which involves the collaboration of all stakeholders and a macro-prudential framework for the financing of energy-efficient buildings and the implementation of energy-efficient housing renovation solutions.

As part of this initiative, an action plan (EeMAP) has been developed, which is necessary for the development of the green mortgage market. Often the borrowing costs of green lending are lower compared to traditional lending instruments, as these types of loans are considered less risky for banks for at least two reasons. First, greener and more energy-efficient housing reduces utility costs, and thus increases the ability of borrowers to make loan payments on time. Second, such housing is more attractive and, accordingly, its value is usually higher than that of similar structures without green features. In the case of the sale of energy-efficient houses/apartments, the bank will be able to get a larger financial return [7].

"Green lending" is not limited to the sphere of housing construction. Bank loans designed to finance sustainable projects are also used in agriculture, industry and services [10]. As for the incentives for the introduction of green lending, according to the results of a survey of the banking community of developing countries and countries with economies in transition, conducted by responsAbility Investments AG, among the most frequent reasons can be identified the interest of banks' customers themselves in such financing. An important motivation is the requirement of foreign creditors to adhere to the principles of green finance. Finally, the growing importance of responsible financial activities and the formation of a corresponding image as an environmentally responsible bank also affect the interest of financial institutions in the development of "green" direction. At the same time, risk management and the need to expand the product line, based on the study, are not strong motivators for the development of green lending [4].

Finally, green financial instruments include special bank cards. For example, the Russian Post Bank offers to issue a "Green World" bank card to a savings account. By paying with the card for every 4,000 rubles spent, the bank finances the planting of one tree in one of the country's national parks. At the same time, the client receives an electronic certificate indicating the location of the planted trees [2].

The tools of green banking can also include products in the field of digitalization, such as the Internet and online banking. Remote banking reduces paperwork, which reduces the consumption of paper during transactions. Customers also visit bank branches less frequently, which is reflected in the frequency of use of personal and public transportation, and consequently reduces the carbon footprint.

Recently, there has been a penetration of ESG initiatives in the fintech sector. Some banks and financial institutions offer specific applications and tools to help investors operate in the sustainable finance market. For example, IMP + ACT Alliance (IMP + ACT Alliance) has launched a special IMP + ACT Classification System (ICS) application that allows asset managers to independently generate classification and ESG management reports to implement financial risk mitigation strategies and contribute to the UN's Sustainable Development Goals. Thanks to this innovative development, it is expected to establish cooperation between various financial market participants in the field of strengthening control over social, environmental and managerial risks. The City of London Corporation, Deutsche Bank, Bridges Insights and other companies have acted as strategic partners to promote this product. This application may be of interest both to investors who want to make their portfolio more sustainable and to investors who want to increase the return on their investment portfolio by investing resources in companies that implement more efficient projects with an environmental and/or social component [3].

Another organization in the field of sustainable fintech is the Danish company Matter, which together with pension funds and other asset management institutions offers to distribute funds into sustainable investments. In 2019, the fintech company entered into an agreement with the AP Pension fund. Funds from this fund are allocated to ESG lines of business. The portfolio is formed based on the innovative expertise of the Matter resource database, and policyholders can use a special app to see how their savings affect environmental sustainability factors (green energy production, reduced carbon emissions, etc.). Also with the help of the Matter app, customers can view information about the types of insurance they have taken out, the profitability of savings and areas of investment of funds [5].

19.4 Conclusion

As a result of the study, we came to the conclusion that in modern conditions of climate transformation and exhaustible resources of the planet, as well as the reorientation of the world economy to resource-saving and environmental technologies, "green banking" is one of the ways to preserve the environment, combat climate change and ensure real economic growth through investment in eco-technology in the twenty-first century.

The formation of a sustainable system of financing environmental measures requires the creation of a harmonized financial and credit mechanism for regulating the use of natural resources, including the following components:

Financing of environmental programs and environmental protection measures from budgets of various levels;

A developed system of environmental funds, innovation environmental funds and enterprise environmental funds;

- a system of environmental banks;
- attraction of funds from environmental insurance funds;
- Use of own funds of enterprises for environmental needs;
- a system of preferential environmental investment loans.

The system of environmental banking should create conditions that interest both clients of the credit institution (environmental users) and creditors/investors (banks). For users of natural resources, this can be preferential lending (at low interest rates), priority lending for environmental purposes. It is possible to interest banks in concessional lending only if they are fully compensated for the funds they have spent. Such compensation can be done by granting tax benefits by reducing the tax base of the bank's income, reducing the tax rate or exempting the bank from certain types of payments.

An important aspect of the greening of credit policy is the introduction of a bank interest rate rating depending on the environmental reliability of natural resource users, since a high level of environmental hazard of an enterprise reduces the degree of guarantee of the repayment of the bank loan. The greening of credit policy should be based on the principle of "credit neutrality," that is, the higher interest rate for lending to environmentally dangerous enterprises should be compensated by preferential lending to environmentally reliable ones. However, a necessary condition for the introduction of such a rating in our country is a general recovery of the economy, because high inflation inevitably leads to high interest rates.

At present, there are two main directions of activity of environmental banks: responsible financing of projects, limits lending to projects with high energy consumption and high pollution, provides environmental and social expertise of projects, provides support for projects in energy conservation and development of alternative and renewable energy, provides priority support for environmental infrastructure in wastewater treatment and reduction of air emissions of harmful substances and.

However, despite the positive changes that have occurred in the process of financing the environmental sphere, only a small proportion of enterprises and organizations have used bank services for the implementation of measures to prevent environmental pollution. This indicates that the financial support of the environmental sphere is insufficient not only due to the unsatisfactory volume of credit resources in the environmental sphere, but also due to the lack of prolongation of "green banking" in the domestic legislation (branches of civil law, banking law, insurance law, etc.).

Acknowledgements The research was supported financially by Russian Science Foundation within the scientific project No. 20-18-00314 "Transformation of social relations in the conditions of Industry 4.0: legal prevention".

References

1. Bik, S. et al.: Green finance: the agenda for Russia. In: Expert Council on the market of long-term investments at the Bank of Russia. Working Group on Responsible Financing (ESG—finance), including "green" financing. Diagnostic note. Moscow. Oct., 63 p. (2018)
2. Brown, L., Larsen, J., Roney, M., Adams, E.: The Great Transition: Shifting from Fossil Fuels to Solar and Wind Energy, p. 178. London, WW Norton and Co, New York (2015)
3. Buchner, B., Trabacchi, C., Mazza, F., Abramskiehn, D., Wang, D.: Global Landscape of Climate Finance 2015: A Climate Policy Initiative Report (2015). http://climatepolicyinitiative.org/wp-content/uploads/2015/11/Global-Landscape-of-Climate-Finance-2015.pdf
4. Chernovol, K.: We have developed the principles of responsible financing, and the BRICS banks have signed! In: Russian Center for Competence and Analysis of OECD Standards. Access mode: https://oecd-russia.org/news/my-razrabotali-printsipy-otvetstvennogo-finansirovaniya-a-banki-briks-podpisali.html
5. Climate Change: Implications for Investors and Financial Institutions. Key Findings from the Intergovernmental Panel on Climate Change (IPCC) Fifth Assessment Report (AR5). Cambridge (UK), University of Cambridge, Institutional Investors Group on Climate Change and UNEP Finance Initiative. 16 p. (2014)
6. Formation of the green bond market in Russia taking into account the best international practice. Results of the survey of members of the Russian Council on the primary capital market. J. National Financial Organization. Access mode: https://new.nfa.ru/upload/iblock/0af/0afb9635754162eebb058f7a87200449.pdf
7. Frolova, E.E.: Green bonds in the Nordic countries: features of development and legal regulation. In: Journal of the Belarusian State University. Right. No. 2 ((2020)). Access mode: https://elib.bsu.by/bitstream/123456789/252886/1/43-48.pdf
8. Increasing the scale of "green" investments and financing in Belarus. In: EaPGreen. Access mode: http://www.green-economies-eap.org/ru/resources/Belarus_Scaling%20up%20Finance_Russian_FINAL.pdf
9. Inshakova, A.O.: Law as the basis of infrastructural support for the digital economy and Internet of Things technology. J. Legal Concept = The Legal Paradigm **18**(3), 6–11 (2019). https://doi.org/10.15688/lc.jvolsu.2019.3.1
10. Inshakova, A.O.: Law and information technology transformations of public relations in the context of industry 4.0. J. Legal Concept = Legal Paradigm **18**(4), 6–17 (2019). https://doi.org/10.15688/lc.jvolsu.2019.4.1
11. Matytsin, D.E.: Blockchain Turnover of Securities Placed for Financing Foreign Trade Transactions by the Business Entities under the Jurisdiction of the EEU and BRICS. In: The Transformation of Social Relationships in Industry 4.0: Economic Security and Legal Prevention. Cham: Information Age Publishing Inc. (2021)
12. Porfiriev, B.N., Vladimirova, I.L.: The role of "green construction" in the formation of a socially oriented economy in the world and in Russia. J. Problems Theory Pract. Manag. **6**, 8–19 (2015)
13. Porfiriev, B.N., Roginko, S.A.: Renewable energy: problems and prospects in the world and in Russia. Prospects for the development of renewable energy: challenges for Russia. Moscow. INP RAS, pp. 9–31 (2015)
14. Sustainable Development Sector. Moscow Stock Exchange. Access mode. https://www.moex.com/s3019
15. The project of MTBank and Nordic Environment Finance Corporation. Aleinikov & Partners. Access mode. https://argument.by/projects/detail.php?ID=290
16. The Economic and Legal Implications of Industry 4.0 for Russian Business and Management. In: Inshakova, A.O., Frolova, E.E. (eds.). Cham: Information Age Publishing Inc. (2020)

Part III
New Procedural and Technological Standards of Legal Proceedings in the Context of Digital Development in the Legislation of Russia and Foreign Countries

Chapter 20
Procedural Standards for Civil Proceedings in China's Internet Courts

Ekaterina P. Rusakova and Evgenia E. Frolova

Abstract The research purpose of the given article is the identification of the procedural features of conducting civil proceedings in the Internet courts of China. The creation of new methods for solving commercial disputes related to the Internet, primordially on e-commerce has increased the number of transactions concluded and, consequently, the volume of sales of goods and services, the net profit of which amounts to billions of US dollars. The emergence of new judicial instances required the consolidation of new procedural basis, and acts of the Supreme People's Court of the People's Republic of China plays an active role in this process. Despite the reform of the procedural legislation, the process of dispute resolution in Internet courts has not been fully regulated by the norms of the Civil Procedural Code and the Law on Administrative Proceedings. Actions performed by participants in an online court session have the same legal consequences as in the normal procedure of legal proceedings, since the civil procedural form of protection of rights must comply with mandatory legal features.

20.1 Introduction

The civil procedure legislation of China is in the process of constant reform, modern trends in dispute resolution have required the introduction of new procedural norms in the legislation [1]. The main legislative act, regulating civil proceedings, is the Civil Procedure Code of the People's Republic of China of 1991. From the time it was accepted, the legal provisions of the code have constantly changed, and the

E. P. Rusakova (✉) · E. E. Frolova
Peoples' Friendship University of Russia, (RUDN University), Moscow, Russia
e-mail: rusakova_ep@pfur.ru

E. E. Frolova
e-mail: frolova_ee@rudn.ru

E. E. Frolova
The Institute of legislation and comparative law under the government of the Russian Federation, B. Kharitonievsky Lane., bldg. 22-24/1A, 1BV, 107078 Moscow, Russia

A. O. Inshakova and E. I. Inshakova (eds.), *New Technology for Inclusive and Sustainable Growth*, Smart Innovation, Systems and Technologies 288,
https://doi.org/10.1007/978-981-16-9808-8_20

last version of the code was adopted in 2017, but many procedural issues related to Internet courts, have not been resolved [2]. In 2018, the Supreme People's Court published a number of acts that were supposed to fill this gap, one of which is the Provision of the Supreme People's Court on Certain legal matters of consideration of cases by Internet courts. The increased number of Internet-related disputes and the e-commerce law adoption in 2018 played a significant role in the given process. The active promotion of digital technologies in the judicial sphere creates the necessary conditions for the development of special procedural rules in the digital civil process [3].

Each of China's three Internet courts has adopted rules of procedure for conducting judicial proceedings, so the Hangzhou Internet Court in 2017 published the "Norms of Online Judicial Proceedings"; "Judicial Procedure of the Hangzhou Internet Court's Judicial Platform" 2019; "Guangzhou Internet Court Rules on Certainx Issues of Online Litigation" 2019; and Beijing Internet Court Electronic Litigation Standard 2020.

20.2 Methodology

The method of comparative analysis was utilized in this research to compare the regulation of legal proceedings concerning the Internet courts in China. The causal and system analysis methods made it possible to identify particular and broad features of the dispute resolution process. Based on the results of the analysis, an assessment of the existing regulation of digital legal proceedings is provided.

20.3 Results

20.3.1 Hangzhou Internet Court

The first Internet court in China has started to deal with disputes entirely in digital format, but the possibility of considering cases in the usual way also exists. For online proceedings, the acts "Online Trial Norms of Hangzhou Internet Court" of 2017 (Online Trial Norms of Hangzhou Internet Court) and "Judicial Procedure on the court platform of the Hangzhou Internet Court" of 2019, regulating this process, were adopted.

It is highly important to notice the preservation of the stages of the civil process. When the case is being prepared for trial, it is the obligation of the registrar to familiarize the parties to the process with the procedural rights and obligations, discipline, and fundamental principles, as well as to ensure that the addresses are confirmed and that the pleadings and evidence are uploaded to the court platform [1]. A special place at this stage is occupied by the authentication process of participants, which ends with the presentation of their identity cards online or by other means. The absence of a party in the process may be regarded by the court as a rejection of the claim, unless a valid reason for the inability to join the meeting is proved. Additionally, the claim may be considered by the court if the parties are absent, if it considers it to be appropriate. The electronic court platform has the technical capabilities to manage the participants' entry into the process, as well as the video recording of the sessions.

The presiding judge or judge announces the name of the case, it looks like this: plaintiff (X) vs. defendant (X) (the basis of the claim), the court applies (X) proceedings to consider this case according with the legislation. Before the court session starts, the parties must perform a number of procedural actions online: confirm that they did not challenge the judge or the court secretary and are also notified of their procedural rights and obligations.

The task of the judge is to check the compliance of the claims stated by the plaintiff on the court platform and in the hearing of the court. The stage of the judicial investigation clarifies the actual side of the legal dispute, where there can be also clarified, changed, or supplemented the subject and the basis of the claim, as well as the evidence in the case, is examined.

A cross-examination is then conducted in which new evidence, if any, may be presented. The interrogation begins by the defendant, and continues by the judge, who asks questions to the parties. If a large number of new facts are revealed during the meeting, they can be uploaded to the platform to simplify the proceedings, where each party can get acquainted with them online [4].

After making sure that all the circumstances in the case are investigated, the judge announces the finalization of the legal investigation and the commencement of the judicial debate, which usually takes place in one stage. In addition, in some cases, within the framework of the ongoing parts of the trial, only disputable circumstances can be investigated. When the court debate ends, the parties announce their final speeches.

Within the framework of legal proceedings, the judge is obliged to invite the parties referring to the legal dispute to the judicial mediation procedure, which shall be done under the direction of the court. With the consent of the parties, the presiding judge or the judge of the first instance begins online mediation, all case materials, court practice in similar cases, and other information are transferred to the mediator. In case of successful completion of the procedure, the mediation agreement concluded with the consent of the parties must be confirmed on the judicial platform.

If the parties have no dispute about the facts and the law, or the circumstances of the case can not affect the outcome of the case, the judge makes a decision, taking into account the credit history of the party. After announcing the decision, the judge announces the end of the trial.

After the judge announces the end of the trial, the parties have the right to review the transcript of the court session and, if necessary, ask the registrar to make changes or corrections by showing the image in real-time. After that, each of the parties confirms the change, and the secretary presses the "end the trial" button, and everyone leaves the meeting [5].

20.3.2 Guangzhou Internet Court

On October 1, 2019, the Guangzhou Internet Court adopted the rules for conducting online court sessions in order to standardize them. In many respects, the provisions of this regulation repeat the provisions of the Hangzhou Internet Court, however, there are differences. This consists in establishing the exact time frame for the commission of certain procedural actions. So, during the pre-trial stage of the proceedings, the applicants in the legal process are required to test the operation of the court platform in test mode three days before the date of the court hearing to make sure that the network is stable and uninterrupted, the video image is clear and the sound transmission is smooth. After testing, the parties must complete a test form in accordance with the guidelines of the judicial platform. In addition, the participants of the trial must connect to the court platform 15 min before the scheduled time of the court session. Despite holding the court session online, when the court enters the courtroom, all participants in the process must stand up, and their faces must be displayed throughout the court session.

During the trial, the persons participating in the case speak and ask questions with the permission of the presiding or sole judge and may not interrupt the speeches of other participants in the trial. There are requirements for the speech of participants, they must speak moderately loudly and not rush, as well as not use insults. The judge is responsible for compliance with all procedural requirements, as he manages the entire process.

20.3.3 Beijing Internet Court

On February 21, 2020, The trial Standard of Electronic Litigation in Beijing Internet Court was adopted.

A characteristic feature is a clearer regulation of the process and the consolidation of the basic principle of conducting hearings in this court, based on the online mode. Offline hearings are an exception to the rule when there is a reasonable need, for example: on-site identification, verification of originals, and providence of physical examination. However, in any case, the party must submit an application for this.

A lot of attention is being paid to identify the people involved in this case, so each participant must be logged in to the system, for this purpose, an account of this person is created. Participants are prohibited from sharing their username and password with anyone, as any action taken under a specific username and password is considered a proper action by that person.

Before the start of the court session, the judge identifies all participants in the process by comparing documents and certificates, as well as biometric identification.

The standard prescribes sanctions for violating the procedure for conducting legal proceedings by participants, in the flesh to criminal liability, if the actions will be traced to the corpus delicti.

In addition, the principle of transparency of the trial is emphasized, namely, representatives of the public and the media can be present during the court session, excluding cases considered in a closed court session. The only restriction is the prohibition of broadcasting the court session live, without obtaining permission.

The Judicial Committee may give a broad interpretation of the provisions contained in this standard, if they are not clear.

20.4 Conclusion

The adoption by the Supreme People's Court of acts on the provision of registration services for court cases involving foreign persons online through the judicial Internet platform in 2021 opened up the opportunity to apply for protection of their rights to foreign persons in the courts of China online [4]. Currently, online case registration is available in many Chinese courts, but most case registration platforms provide services only in Chinese and for individuals who have been identified based on the Chinese identity verification system. But from February 3, 2021, a foreign person only needs to register on China's mobile social network WeChat, which offers its users an application to obtain judicial protection based on China Mobile MiniCourt, in order to go to court. This opportunity brought not only foreign citizens closer to the jurisdiction and competence of the PRC but also mainland citizens and Taiwanese citizens living abroad, to file lawsuits in court.

The analysis of the acts of Internet courts regulating the dispute resolution procedure allowed us to draw a number of conclusions:

- The provisions enshrined in the internal acts of the courts must comply with the procedural regulation of the PRC, and the clarifications of the Supreme People's Court of the People's Republic of China;
- comprehensive regulation of the whole course of the legal process and the commission of individual procedural actions;
- fixing the ethical and behavioral foundations in the acts of Internet courts;
- consolidation of legal principles of legal proceedings.

Statistics indicate that the established dispute resolution mechanism in such areas as e-commerce and the protection of rights on the Internet increases the attractiveness of China's jurisdiction for doing business in this country, while guaranteed by the state [6].

Acknowledgements This work was given financial support by the Grant of the President of the Russian Federation No.NSH-2668-2020.6 "National-Cultural and Digital Trends in the Socio-Economic, Political and Legal Development of the Russian Federation in the 21st Century".

References

1. Rusakova, E.P., Inshakova, A.O.: Industrial and manufacturing engineering in digital legal proceedings in the Asia-Pacific region: a new level of quality based on data, blockchain and ai. Int. J. Quality Res. **15**(1), 273–289 (2021)
2. Rusakova, E.P., Frolova, E.E., Arzumanova, L.L.: Challenges of the judicial systems of the Russian federation and people's republic of China in the Era of the pandemic. In: Popkova, E.G., Sergi, B.S. (Eds). Modern Global Economic System: Evolutional Development vs. Revolutionary Leap. ISC 2019. Lecture Notes in Networks and Systems, vol. 198, pp. 1541–1549. Springer, Cham (2021). https://doi.org/10.1007/978-3-030-69415-9_170
3. Rusakova, E.P., Frolova, E.E.: Procedural aspects of proof in China's internet courts: opportunities for receiving BRICS jurisdiction. In: Popkova, E.G., Sergi, B.S. (Eds). Modern Global Economic System: Evolutional Development vs. Revolutionary Leap. ISC 2019. Lecture Notes in Networks and Systems, vol. 198, pp. 1598–1605 (2021)
4. Dudin, M.N., Frolova, E.E., Protopopova, O.V., Mamedov, A.A., Odintsov, S.V.: Study of innovative technologies in the energy industry: nontraditional and renewable energy sources. Entrepreneurship Sustain. Issues **6**(4), 1704–1713 (2019)
5. Rusakova, E.P., Inshakova, A.O., Frolova, E.E.: Legal Regulation of internet courts in China. In: Popkova, E.G., Sergi, B.S. (Eds) Modern Global Economic System: Evolutional Development vs. Revolutionary Leap. ISC 2019. Lecture Notes in Networks and Systems, vol. 198, pp. 1515–1521. Springer, Cham (2021). https://doi.org/10.1007/978-3-030-69415-9_167
6. Gaivoronskaya, Y.V., Miroshnichenko, O.I., Mamychev, A.Yu.: The immodest charm of digitalization. Legal Concept **18**(4), 40–47 (2019)

Chapter 21
On Some Aspects of Case Management in Electronic Courts of Indonesia

Irina A. Gronic

Abstract The chapter is devoted to "case management" in the electronic courts of the Republic of Indonesia concerning the implementation of some reforms in 2018 in the country on the provision of effective and efficient services for the delivery of cases in the electronic court. Electronic case management is the orderly use of an electronic system for conducting electronic office work, obtaining court documents, including requests, copies, duplicates, and verdicts, delivery, and storage of documents. Thus, the electronic management system applies to civil, religious, military, and state administrative cases, a significant event radically different from the pre-reform period. The whole process is carried out based on digital standards of information technology service and management, provided for by the Supreme Court's Chief Justice. The president of the court also acts as a controlling entity. He oversees the electronic litigation process and services. The active introduction of information technologies in Indonesia predetermined the country's economic growth since it is a prerequisite for this process.

21.1 Introduction

One of the positive trends in litigation declared in the Bureaucratic Reform 2015–2019, carried out by the Supreme Court and the judicial authorities of the Republic of Indonesia, along with improving the quality of public services, increasing the efficiency of the system for checking the files of cassation, supervisory authorities, modernizing the processing of assistance when delegating calls and notifications, an electronic record-keeping system (e-litigation) was developed and implemented [1].

Besides, the Program "Judicial Modernization for 2010–2035," adopted by the Supreme Court of the Republic of Indonesia, as one of the priority areas, provides for judicial renewal through the use of information technology. According to this Program, the role of information technology in the ongoing judicial reform is in

I. A. Gronic (✉)
Peoples' Friendship University of Russia (RUDN University), Moscow, Russia
e-mail: gronik-i@rudn.ru

A. O. Inshakova and E. I. Inshakova (eds.), *New Technology for Inclusive and Sustainable Growth*, Smart Innovation, Systems and Technologies 288,
https://doi.org/10.1007/978-981-16-9808-8_21

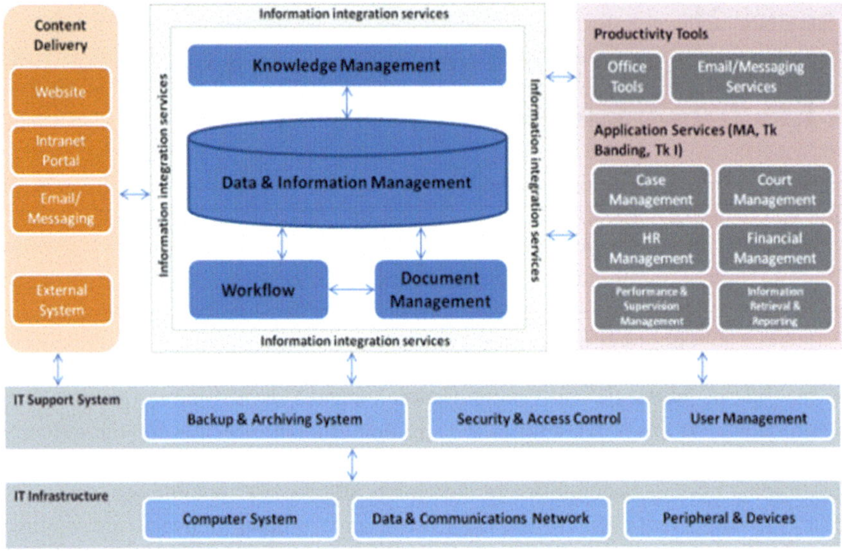

Fig. 21.1 Information integration services [1][1]

the formation of an electronic record and the maximum optimization of improving the efficiency of the judicial system. However, the information technology initiatives undertaken did not immediately produce satisfactory results for the country's judiciary.

The objectives of introducing information technology into the litigation were as follows:

a. Improving the quality of work by providing access to the necessary information, including court decisions, legal journals, and others;
b. Improving the judicial administration system, including access to online judicial activities;
c. Establishing the efficiency of the working process in the judicial system by reducing the use of human resources, replacing them with digital and electronic processes;
d. Formation of results-oriented activities, using information technology as a tool for monitoring and control over the implementation of assigned tasks;
e. Creating an electronic learning environment through the use of "distance learning" [1].

In the Program "Judicial Modernization for 2010–2035," a project for implementing and using information systems, consisting of several modules, is clearly illustrated (Fig. 21.1). The main module of the electronic platform is located in the center (knowledge management). Its functional significance is as follows:

[1] The figure is taken from Cetak Biru MA 2010–2035. Cetak biru pembaruan peradilan. Available at: https://mahkamahagung.go.id/media/198, pp. 66.

- Regulation of access to data services,
- Data management (workflow management),
- Pre-processing knowledge management,
- Workflow and document management.

Interaction with the user takes place through the use of an electronic application (platform).

Besides, the project proposes a case management module and a court management module (Fig. 21.1).

The functions of the module (court management) are to organize and distribute the flow of court cases, provide access to the module to judges and court clerks, and change the status of a court decision. This function was implemented using an electronic document management application. The module (court management) is a courtroom in which witnesses and parties to the case can be present. Besides, this module implements the management of the trial, records the court session, automatically transcribes, and there is video conferencing for witnesses' interrogation.

The IT support, identified in the 2010–2035 Forensic Modernization Program, was expected to include an e-Learning module to promote education and training, such as e-learning modules for judges and judicial officials.

In 2019, the Supreme Court of the Republic of Indonesia issued Decree No. 1 "On electronic review of cases and court proceedings," according to which the implementation of online electronic court proceedings is carried out using the e-Litigation application. This application is one of the elements of the electronic court and is used in cases of civil, religious, military, and government administration [2]. It is important to note that Indonesia has taken a huge step forward over the past three years by working through and improving on past mistakes. Before the reforms in the Republic in 2018, the electronic court system was actively used only for registering cases or public administration. In 2021, electronic litigation became available in the courts of the first instance for appeal and cassation proceedings. According to para 5 of Art. 1 of the Decree of the Supreme Court of the Republic of Indonesia No. 3 of 2018, "On the management of cases in the electronic court" [3], electronic management of the case is understood as a list of processes carried out to receive claims, requests, responses, copies, duplicate documents, management, filing and storing documents in civil, religious affairs, and in the military and public administration affairs using the electronic judicial system.

21.2 Methodology

The method of analysis, thanks to which it was possible to comprehend the methodological aspects and theoretical and conceptual approaches to the concepts of "case management," "electronic court," "electronic court proceedings," to identify and concretize some of their features. Based on the analysis of empirical material, the author reveals the essence and property of the institution of case management, its relationship with information technology [4]. The logical cognition of the material under study served the author to reproduce the events taking place related to judicial reform. We also engage the dialectical method in analyzing the ratio of the general and the particular in providing electronic consideration of cases in the Republic of Indonesia during the formation and amendments. The use of these methods made it possible to more clearly study the stated research topic, generate the necessary material for further work on its improvement, and also reveal the most effective technical and legal methods that regulate the management of a case in an electronic court, taking into account changing trends.

21.3 Results

The idea of the reform was to create maximum conditions for resolving the cases in the electronic court. For the implementation of the described Program, the Indonesian authorities began to introduce information technology in stages.

The goal of the first stage was to optimize existing investments in information technology, data integration, and prepare regulations and change the way of working by bringing them in line with the requirements of the digital age.

The purpose of the second stage was to create a unified information system for the judicial system, allowing the use of information to maintain legal integrity and opening up opportunities to expand access to judicial services.

The third stage aims to integrate the judicial process within the framework of the integration justice.

In the years following the judicial reform in Indonesia, the quality of the services provided has improved significantly by the achievements of the reforms, including through the implementation of the stated in the Roadmap for 2015–2019 provisions in the form of the following aspects:

1. Maintain and improve available resources.
2. Continue making the necessary changes.
3. Identify existing problems and find solutions to fix them.
4. Ensure the internalization of bureaucratic reform in the Supreme Court.

The four steps are strategic because their implementation gives hope to strengthen the results obtained from the implementation of judicial reform, which will allow the idea of a national bureaucratic reform to be realized by 2025 [1].

Thus, by implementing the goals of the judicial reform aimed at increasing public confidence in the expected state of the electronic judicial management system in 2025, all information systems of the Supreme Court will be integrated.

21.3.1 E-Court Application

The Republic of Indonesia is currently actively using the E-Court application. This application consists of several modules that make it easy and efficient to use the system.

The module for electronic registration of materials (E-Filing) will help register a lawsuit or submission and upload electronic documents for downloading and uploading documents in the context of duplication, copying, and verdict, management, filing, and storage of documents in civil, religious affairs, military, and state administration.

The electronic payment module (E-Payment) can be used to pay court fees set through the e-SKUM application. Registered users should pay close attention to the amount of the advance payment of the court fee, the billing account number (virtual account), the due date for the advance payment of the court fee that the system has determined, and understand and agree that any errors, delays, and other additional costs arising from the difference between the bank used by the Registered User and the official court account in which the claim is filed are at the expense of the Registered User.

Failure to make a payment to the billing account within a predetermined time will result in the expiration of the payment number, and the registered user must receive a new payment number during the same registration via electronic payment in electronic court.

Registered users are required to make payments following the invoice value in the payment number received during the registration of the case.

The application also contains the Electronic Notification Module (E-Pbt) and the Electronic Calling Module (E-Pgl).

Thus, the e-court application can be used to register, payment, and deliver documents related to court cases.

Registered users are warned that they are solely responsible for actions performed on their behalf.

Another topical and priority area of the modern Republic of Indonesia is the security of electronic systems. Therefore, registered users are warned about the prohibition to perform any actions that jeopardize the security and stability of the E-Court application and all data stored in it.

For illegal actions aimed at using the electronic application for other purposes, the following sanctions are provided:

- sanction in the form of a warning;
- sanctions in the form of temporary deprivation of access rights;
- sanctions in the form of irrevocable deprivation of access rights following the weight and impact of the violations found on the integrity of the E-Court application.

The application of the sanctions does not preclude the possibility of filing a civil claim for compensation or criminal prosecution of registered users if the Supreme

Court of the Republic of Indonesia considers that there has been an illegal act or a criminal act.

Besides, in 2019, several important agreements were signed between the National Cyber and Cryptographic Agency (BSSN) and the Supreme Court of Indonesia, including an agreement on cooperation in the security of electronic systems. This collaboration aims to "implement corporate governance within the framework of integrated regulation, governance, and control," as well as the administration of electronic government systems. Also, to promote and avoid actions that could undermine the court's authority as a whole and initiate by the Chairman of the Supreme Court of the Republic of Indonesia, modern electronic justice, E-Litigation, was implemented.

The electronic management of court cases is carried out based on the digital information technology service and management standards provided for by the decision of the Chief Justice of the Supreme Court (Art. 23 of the Supreme Court of the Republic of Indonesia Decree on the management of the case in the electronic court of 2018) [5].

The next step in implementing the "Judicial Modernization 2010–2035" Program was the declaration of creating an electronic learning environment through the use of distance learning [1] and creating an e-Learning module that promotes education and training. The materialization of this idea was embodied in February 2021 by the opening of the virtual application E-RIS—"electronic research information system" on the topic "The use of information technology in legal research to achieve legal unity and consistency of judges' decisions" [1]. The E-RIS App is an internal application of the electronic platform of the Supreme Court, E-Court, which will help judges find relevant information when considering cases. The E-RIS electronic application is a multifunctional module that allows performing the following actions:

- Keeping minutes of discussion of topical issues of current legislation in electronic court management and new draft laws submitted for discussion. This function contains a summary of an explanatory memorandum on the use of information technology in case management in an electronic court, and bills on combating corruption, the draft law on the criminal procedure code of the Republic of Indonesia, and others.
- Brief description of materials of judicial acts. This feature contains concise material on selected judgments that have gone through the peer-review process (practitioners/scholars).
- Function "expert opinion" on the interpretation or application of the rule of law.
- Function "statement of decisions of the Constitutional Court."
- Research results. This function allows getting acquainted with the results of scientific and practical research on relevant topics.

The introduction of the E-RIS virtual application into the work of the electronic platform of the Supreme Court, E-Court, is an expression of the active position of the Republic of Indonesia in the field of improving the work of the electronic court when considering cases.

21.4 Conclusion

The Republic of Indonesia seeks to improve legislation in the regulation of e-government, digital economy, information security, and the application and use of the electronic case management system in electronic court in civil, religious, military, and state administrative cases. The study showed that the use of information technology in the process of legal proceedings in the Republic of Indonesia [6] has a significant impact on implementing the orderly conduct of cases in electronic court, which is effective and efficient. It also contributes to implementing simple, fast, inexpensive, and transparent litigation and the reliable operation of the electronic court system as a whole. Over the years, there has been a noticeable improvement in the quality of the electronic resources of the judiciary to improve the service to the Indonesian population.

Acknowledgements The reported study was funded by RFBR, project number 20-011-00276 A.

References

1. Cetak Biru MA 2010–2035: Cetak biru pembaruan peradilan. https://mahkamahagung.go.id/media/198. Accessed 25 Feb 2021
2. Matytsin, D.E.: Features of the legal infrastructure of the turnover of investment objects in the Russian Federation. In: Inshakova, A.O., Frolova, E.E. (Eds.) Smart Technologies for the Digitisation of Industry: Entrepreneurial Environment. Smart Innovation, Systems and Technologies, vol. 254. Springer, Singapore (2022). https://doi.org/10.1007/978-981-16-4621-8_20
3. Ketua Mahkamah agung Republik Indonesia: Peraturan mahkamah agung Republik Indonesia Nomor 3 tahun 2018 Tentang Administrasi perkara di pengadilan Secara elektronik (2018). https://ecourt.mahkamahagung.go.id/perma_03_2018.pdf. Accessed 4 Feb 2021
4. Inshakova, A.O., Frolova, E.E., Rusakova, E.P., Kovalev, S.I.: The model of distribution of human and machine labor at intellectual production in industry 4.0. Journal of Intellectual Capital (2020). Vol. ahead-of-print No. ahead-of-print. ISSN: 1469-1930, https://doi.org/10.1108/JIC-11-2019-0257
5. Official page of the Supreme Court of the Republic of Indonesia. https://mahkamahagung.go.id/id/berita/4527/ketua-mahkamah-agung-luncurkan-aplikasi-peluncuran-aplikasi-e-ris-electronic-research-information-system-sistem-informasi-riset-elektronik-secara-virtual. Accessed 19 Feb 2021
6. Rusakova, E.P., Inshakova, A.O.: Industrial and manufacturing engineering in digital legal proceedings in the Asia-Pacific region: a new level of quality based on data, blockchain and AI. J. Qual. Res. 15(1), 273–290 (2021). ISSN: 1800-7473. https://doi.org/10.24874/ijqr15.01-16

Chapter 22
Applicable Law to International Commercial Arbitration in Panama (Digital Aspects)

Ekaterina P. Rusakova and Edgar Josue Young Dominguez

Abstract International arbitration proceedings give rise to a variety of choice-of-law issues, particularly when you add the increasing implementation of digital technologies. The success of the economic collaboration between the counterparties of various jurisdictions depends on the choice-of-law by the parties in dispute. This article focuses on how to properly choose the applicable law in arbitrations for international commercial disputes in the Republic of Panama, adding the particularities of the increasing use of digital technologies. The author explores the nuances and specificities when parties have expressly chosen an applicable law to the dispute and the legal reasoning and theories behind properly selecting the law applicable when the choice-of-law by the parties is absent, considering different approaches such as conflict-of-law rules, the closest connection test, the cumulative method, international conventions, and transnational principles of laws.

22.1 Introduction

International commercial arbitration as a special mechanism of solving commercial disputes at the international level has peculiar characteristic features [1, p. 91], one of these characteristics is that parties have the capacity and availability to choose the laws and/or rules that shall be applied to solve a current or future dispute. The economic growth of a country demonstrates remarkable changing in legal behavior on enumerated country's private parties' activities as well [2, p. 166]. The introduction of new methods of using means of production, the improvement of technology leads to the active development of the economy. The increasing degree of complexity and even the sophistication of financial markets and commercial have led to the need to apply more sophisticated dispute resolution methods that may come up in all types of disputes including cross-border banking and financial transactions. The above-mentioned modernization of commercial disputes resolution entails many problems

E. P. Rusakova (✉) · E. J. Y. Dominguez
RUDN University, Moscow, Russia
e-mail: rusakova-ep@rudn.ru; ermakova_ep@pfur.ru

A. O. Inshakova and E. I. Inshakova (eds.), *New Technology for Inclusive and Sustainable Growth*, Smart Innovation, Systems and Technologies 288,
https://doi.org/10.1007/978-981-16-9808-8_22

of legal nature [3, p. 754]. Additionally, digital technologies are currently affecting the settlement of dispute by arbitration, not only when it comes to consent to arbitrate the actual arbitration proceedings but also the selection of the legislation to the merits or substantial aspects. Furthermore, arbitrators may encounter circumstances when the parties did not agree or did not expressly select the legislation, or set of rules to be applicable. In this situation, it is of utmost importance to correctly determine the applicable norms to the dispute. In this regard, it is the arbitrators' responsibility to properly decide on the applicable law and the legal grounds of this determination such as transnational principles of law, the closest connection test, international conventions, the national arbitration act, and the arbitration rules from the corresponding arbitration institution.

The party autonomy principle intrinsically affects and has enormous and different impacts on the legislations and rules to be used in an arbitration. This principle affects the law governing the substance or mertis of the dispute, the procedural law governing the arbitration, the law applicable to the arbitration clause, and also the further recognition and enforcement of the arbitral award. The present article is devoted to properly choosing the mechanisms to rightfully determine the material law applicable to the merits of the case, particularly for international commercial disputes in the Republic of Panama.

Correctly determining the law governing the substance part of the dispute is significantly important, since this is the legal act or rules governing the agreement that is the basis for the controversy. Consequently, this substantive law shall determine the legal rights and obligations of the parties involved, including the types and calculation of damages, the limitation of defenses, and the substantive remedies available to the parties.

22.2 Methodology

The present article applies a comparative method analysis of different legal frameworks, arbitration legislations, conflict-of-law rules, arbitration rules, judicial decisions, and arbitral awards from several jurisdictions which in turn allows us to understand the approach accepted in the Republic of Panama.

This research by utilizing a formal-legal methodology observes from a legal point of view, the nature of the governing law applicable related to the legal disputes of international commercial disputes solved by arbitration in the Republic of Panama. Particularly, when the parties have chosen such set of rules or in contrast when there is an absence of choice by the parties. Additionally, this article uses the normative legal method to study the legal norms and regulation in the Republic of Panama of the material law that ought to be applied to the merits or substance part of international arbitrations, including national legal acts, international private law code, and internal arbitration center regulations.

The research's purpose consists of identifying the correct approach to pick the governing law of the merits in arbitrations of international commercial cases in

Panama. This purpose is achieved by analyzing the different approaches applied by arbitrators worldwide, by examining standards used in national arbitration laws and arbitration rules of well-renown institutions, and by studying the reasoning in case law when arbitrators have determined the material law of a specific case.

22.3 Results

22.3.1 Expressed Choice by the Parties—Digital Aspects

Panama's national arbitration legislation recognizes the party autonomy principle to freely indicate the legislation, acts, and arbitration rules applicable to a future or current dispute. Panama's law for national and international commercial arbitration (Law No. 131 of 2013) expressly stipulates under article 56 "the arbitral tribunal shall decide the dispute in accordance with the rules of law chosen by the parties as applicable to the merits of the dispute." In doing so, arbitrators shall consider "any indication of the law or legal system of a particular state shall be understood to refer, unless otherwise stated, to the substantive law of that state and not to its conflict of laws rules."

Indeed, Panama's legislation follows the approach of most jurisdiction considering that "any designation" of the law or legal system of a given State shall be construed as "directly referring to the substantive law of that State and not to its conflict of law regulation." However, this designation may be affected by digital aspects. Parties can consent the use of a material law by digital means. This may include, the use of electronic communication or data messages, as long as the information contained therein is accessible for consultation afterwards.

Article 5 of the law 131 of 2013 specifies the meaning of "electronic communication" and "data message." This regulation defines electronic communication as "any communication that the parties make by means of data messages"; and data message as "information generated, sent, received or archived by electronic, magnetic, optical or similar means, such as electronic data exchange, e-mail, Telegram, telex or fax, among others."

Digital aspects are also present when it comes to communications and notifications to the parties [4]. Art. 8 of the National Arbitration law states that notifications or communications made by telex, fax, or other means of electronic, telematics, or other similar types of telecommunication, which allow the sending and receipt of writings and documents stating their sending and receipt and which have been designated by the interested party, shall be valid.

22.3.2 Determination of the Applicable Material Law in Absence of Choice by the Parties

In circumstances where the parties to a dispute have not agreed about the substantive law governing the merits of dispute, the task of properly selecting the regulations governing the main contract, genesis of the dispute, is typically an obligation of the arbitral tribunal. Consequently, arbitrator may refer to national and international elements. In doing so, different arbitral tribunals may have different approaches to solve this issue. Some might refer to international private law, conflict norms of the legal place or seat of the arbitration, others may consider different criteria such as the cumulative method, the closest connection test, non-national legal systems, and international conventions; while others may select the conflict rules which they considered to be appropriate. Additionally, arbitrators must consider exceptions to the application of conflict rules, e.g., directly applicable laws, overriding mandatory rules, and/or public order policies [5].

Panamas' national legal system provides the arbitral tribunal, the authority and power to select the legislation that governs the substance or merits of the dispute, stating under article 56 of the law 131 of 2013 that if the parties do not indicate the applicable law, the arbitral tribunal shall apply the applicable legal norms to the dispute.

22.3.3 Conflict-of-Law Norms of the Seat of the Arbitration

An approach that was frequently used by arbitral tribunals in the past was to select the conflict-of-law norms of place or arbitral seat to indirectly determine the law applicable to the legal dispute. This point of view required arbitrators to refer to the choice-of-law norms concerning the seat or place of the arbitration to resolve the dispute.

Supporters of this theory considered that parties, whether individuals or legal entities, who have chosen a place for arbitration have consequently agreed that this law should be implicitly applicable to both procedural and substantive matters. Little reference was done to which conflict of norms dictated this, or were to be applied by the arbitral tribunal. Rather, the parties' selection of the seat was deemed to be an indirect choice of the seat's substantive law [6, p. 300].

This perspective, however, has been abandoned by most national laws and arbitration rules. It is regarded as an antiquated criterion and there is currently a minimum number of arbitrators that follow this approach.

22.3.4 Cumulative Method of Choice-of-Law Norms

The cumulative method is an approach that is frequently utilized by international arbitral tribunals. This method consists of simultaneously considering all of the choice-of-law norms of all legal systems with which the dispute in question is connected. If all of these diverse choice-of-law norms point to the same substantive law, the arbitrators apply this law regarding the merits of the case. The idea behind this method consists of demonstrating a "false conflict" to show that all analyses would lead to the same applicable law [7, p. 921]. Arbitral Tribunals normally make special efforts to show that the substantive solution found for the dispute is either one pointed out by the international private law legal systems of the national jurisdictions reasonably connected with the dispute (false "conflit de systems") or by a generally accepted conflict-of-laws rule.. On a practical level, the cumulative approach also provides some insulation against a challenge for failure to apply the proper conflict-of-laws or substantive rules [8, p. 191].

Arbitrators sometimes apply the conflicts rules of each of the states with a connection to the dispute. As a practical matter, this "cumulative" approach virtually always concludes that all potentially relevant conflicts rules select the same law. Alternatively, some awards apply a variation of this analysis that considers the application of all potentially applicable national (or other) substantive laws [6, p. 301].

This methodology has been criticized by some scholars as Yves Derains, when stating that "the method known as cumulative application of system of conflict laws of states involved in a dispute has no sense unless the approach results in convergence" [9, p. 529].

22.3.5 Closest Connection Test

Some national arbitration laws and arbitration rules demand that the arbitrators need to apply the legislation of the state which has the "closest connection" to the parties' dispute. Indeed, some arbitration legislation prescribes a "closest connection" standard for tribunals seated on national territory; where such legislation is applicable, tribunals typically have applied the closest connection standard. Indeed, even where no such statutory rule applies, some awards have applied a "closest connection" choice-of-law rule. This approach draws support, in regards to the choice-of-law analysis applicable to contracts in the international legal sphere, from the connected points of view of the Rome I Regulation on the Law Applicable to Contractual Obligations and its predecessor, Rome Convention, both adopting a "closest connection" standard, and the Restatement (Second) Conflict of Laws adopting a "most significant relationship" standard [6, p. 301].

This approach has been accepted by German Code of Civil Procedure ("the arbitral tribunal is to apply the laws of that State to which the subject matter of the proceeding has the closest ties"); Swiss Arbitration Law ("the arbitral tribunal shall decide... in the absence of a choice-of-law, by applying the rules of law with which the dispute has the closest connection"); and Italian Civil Procedure Code ("... if the parties are silent, the law with which the relationship has its closest connection shall apply").

Even when the tribunal that is deciding the arbitration case is not required to use the closest connection standard, some tribunals have considered that this test is a transnational principle of private international law and have gone on to apply it as the most appropriate conflicts rule to determine the substantive law. This method appeals to the notion that the law will be tailor-made for the particular contractual circumstance. However, in this modern and digital era, there can be too many relevant factors connected to the dispute [7, p. 921].

The criteria to understand which law has the "closest connection" can be quite complicated. Arbitrators may consider different factors: the place where the contract was concluded by the parites (lex loci contractus), the law of the place of performance of the contract (lex loci solutionis), the place of business or habitual residence of the parties, and the jurisdiction for future enforcement. Nevertheless, this approach might lead to ambiguity in international arbitrations and uncertainty for the parties, especially taking into account electronic commerce and the fact that more than one legislation may be applicable to one case.

Panama's national arbitration law does not expressly accept the closest connection test as a standard to determine the substantive law to a dispute. Nonetheless, this standard is recognized under Article 160 of the International Private Law Code, stating that "in case of silence of the contracting parties as to the applicable law, it refers to the determination made by the judge of the forum of the specific State with which the contract in question has the closest proximity or closeness based on its objective and subjective elements."

22.3.6 Choice-of-Law Norms that the Arbitral Tribunal Considers to Be "Appropriate"

As seen from above, this approach is currently being accepted by leading international organizations and arbitration centers around the world. This mechanism gives arbitrators the power and faculty to select the legislation to be applicable to the merits of the dispute based on what they considered to be more "appropriate."

Under this standard, arbitrators may have the freedom to select the most appropriate conflict-of-law norms that will ultimately determine the substantive law to solve the dispute. This freedom, however, should not be understood to permit unfettered discretion. On the contrary, the arbitrators remain obligated to select the conflicts rules that are "appropriate" in light of the procedural law to the arbitration and the arbitration agreement; this is a selection with right answers and wrong answers, and not a purely discretionary matter. For example, an arbitrator cannot select the conflicts rules of his home jurisdiction, if it has no connection to the dispute, merely because it is familiar [6, p. 300].

From the reading of Art. 56 (2) of the law 131 of 2013, it is clear Panama's national arbitration law also accepts this standard when stating: "If the parties do not indicate the applicable law, the arbitral tribunal shall apply the rules of law it deems **appropriate**." However, if we further analyze this article one may observe that it departs the Model Law considering that the article does not expressly stipulate a reference to conflict-of-law norms to determine the applicable law.

22.3.7 Direct Application of Substantive Law

Another standard recognized by arbitration laws when determining the material law applicable to the arbitration is to "directly" chose the substantive law. This standard allows arbitrators to directly apply the material law that shall be applicable to the substance of the legal dispute. In doing so, arbitrators may not necessarily refer to conflict-of-law norms, but they will have the possibility to directly select the substantive law.

After examining Panama's national arbitration law, it is clear that Panamanian legislation primordially follows this standard stating under Art. 56 (2) that "If the parties do not indicate the applicable law, **the arbitral tribunal shall apply the rules of law** it deems appropriate." In this regard, this provision does not expressly stipulate that arbitrators shall apply the "conflict-of-law rules" they deem appropriate, rather it only refers to the use of "rules of law" ("normas jurídicas"). Therefore, it is arguable, whether arbitrators shall mandatorily refer to conflict-of-law norms to figure out the applicable law or whether they may directly choose the substantive law to solve a concrete case. As mentioned previously, this constitutes a departure from the original wording of the UNCITRAL Model Law (Art. 28). In the author's view, this deviation is clearly intentional. This is precisely to not restrain arbitrators to a choice-of-law system and to grant arbitrators not only power to directly choose the material legislation applicable to the substance of the arbitration case but also to open the possibility for arbitrators to select non-national rules to govern the dispute.

This approach has been accepted by leading arbitration institutions such as the International Chamber of Commerce (ICC). Under article 21(1) 2017, the ICC Rules [10] provide that "[t]he parties shall be free to agree upon the rules of law to be applied by the Arbitral Tribunal to the merits of the dispute. In the absence of any such agreement, the Arbitral Tribunal shall apply the rules of law which it determines to be appropriate."

Recognition and acceptance of such approach shall ultimately depend on the legal provisions of the national arbitration regulation applicable to international commercial arbitrations. Jurisdictions such as France, Switzerland, the Netherlands, India, and Canada are in favor of the point of view that arbitrators are able to select on their own the rules they consider to be appropriate to solve the dispute. Article 1511 of the French Code of Civil Procedure authorizes arbitrators an international arbitration to "directly" apply the substantive law that they consider appropriate. Indian Arbitration and Conciliation Act, Art. 28(1)(b)(iii) ("apply the rules of law it considers to be appropriate given the circumstances surrounding the dispute") and Netherlands Code of Civil Procedure, Art. 1054(2) ("in accordance with the rules of law it considers appropriate"). Nonetheless, other influential jurisdictions such as England, Germany, and Japan have declined to accept this approach when determining the law applicable to the material issues between the parties.

22.3.8 Transnational Principles of Law (Application of Non-national Legal Systems)

Another method that international arbitrators may use to solve the issue of applicable law to the substance or merits of the arbitration dispute is applying transnational principles of law, which constitute non-national legal systems, not attached to a particular state or choice-of-law system. Such non-national legal systems may include reference to lex mercatoria, the UNIDROIT Principles of International Commercial contract (the UNIDROIT Principles), the Principles of European Contract law ("PECL"), or the "TransLex-Principles.

Whatever the case is, either choice-of-law or direct application of a substantive law; arbitrators must observe the stipulations of the agreement, and follow commercial usages. When it comes to international arbitrations in Panama, specifically, arbitrators must follow the UNIDROIT Principles as indicated in Art. 56 of the law 131 of 31 December 2013, stating that "... In all cases, the arbitral tribunal shall decide in accordance with the provisions of the contract and shall take into account the commercial uses applicable to the matter. International arbitrations shall also take into account the Principles on International Commercial Contracts of the International Institute for the Unification of Private Law (UNIDROIT)."

Moreover, Art. 79 of the Code of Private International law of Panama complements this norm by stating that "Parties may use the principles of international commercial contracts regulated by the International Institute for the Unification of Private Law, known as **UNIDROIT, as a supplementary rule** to the applicable law or as mean of interpretation by the judge or **arbitrator**, in the contracts or relations of international trade law."

22.3.9 International Approach

An alternative approach that arbitral tribunal have recognized to escape the peculiarities of national laws is to consider "international conflict of law norms." This method considers conflict-of-law norms that are not necessarily connected to any national conflict rules.

The reasoning behind this method was explained in ICC Case No. 7071 [10] in the following way: "… much to be said in favour of adopting generally accepted principles of international conflict of laws. The fact that the dispute arises out of dealings between one government and an instrumentality of another government gives them a unique international flavour. Hence, the parties could reasonably have contemplated that arbitrators would apply generally accepted international conflicts-of-law rules in arriving at the applicable law by which their dispute would be resolved. In the circumstances of the present arbitration, which is truly international in character, the Arbitral Tribunal is of the opinion that it should adopt generally accepted international conflict of laws rules."

Nonetheless, our current global scenario cannot offer yet a single set of organized international conflict-of-law rules that might be applicable universally and particularly to international arbitration cases. Arbitrators may refer to international conventions applicable to an arbitration case (e.g., the United Nations convention on Contracts for the International Sale of goods) or regional systems (e.g., Rome and Brussels regulations). The reality is that different legal systems regulate conflict-of-law rules differently and the same issue may vary from jurisdiction to jurisdiction.

22.4 Conclusion

Different approaches exist when determining the substantive law ought to be applicable to the merits of a case. Even when parties have expressly chosen the law governing their legal relation, the issue may turn more complex, when parties have decided to do so by the use of technological or digital means. Furthermore, when parties do not expressly choose a governing law that is to be applicable to the commercial dispute, it is the arbitral tribunal's responsibility to correctly determine such law. To do so, arbitrators may follow guidelines and standards established recognized and accepted by the national arbitration legislation and the arbitration rules of arbitration centers or institutions.

These methods include the use of conflict-of-law norms of the seat of the arbitration, the cumulative method of choice-of-law norms, the closest connection test, the application of choice-of-law provisions that the arbitral tribunal considers "appropriate," the direct application of material laws, and the application of non-national or transnational [11].

Out of all of these methods, Panamanian legislation follows a combination of approaches. If the parties involved in the arbitration process did not indicate the applicable law, the arbitral tribunal that is deciding the case shall apply the rules of law it deems appropriate. In this sense, one may suggest that the Republic of Panama accepts the "appropriate" method, however, there is a distinct peculiarity. When arbitrators choose the law they deem to be appropriate, it is not mandatory to refer to or consider conflict-of-law rules, rather they are able to choose the rules of law. This allows arbitrators to directly apply the law to be applicable to the merits of the dispute (including substantive law and non-national laws). After examining the Panamanian legislation, one must have reached the conclusion that the Republic of Panama primordially follows the direct application of rules of laws, without being restricted to the application of conflict-of-law norms.

In all cases, the arbitrators shall decide in accordance with the provisions of the contract and shall take into account the commercial uses applicable to the matter. When it comes to international arbitrations specifically, arbitrators must follow the UNIDROIT Principles, as stated in the national arbitration law and in the private international code of Panama.

Acknowledgements This work was financially supported by the Grant of the President of the Russian Federation No. NSH-2668-2020.6 "National-Cultural and Digital Trends in the Socio-Economic, Political and Legal Development of the Russian Federation in the 21st Century."

References

1. Rusakova, E.P., Gronic, I.A.: International Commercial Arbitration in African Countries: Development Prospects. In: Popkova, E., Sergi B., Haabazoka, L., Ragulina, J. (eds.) Supporting Inclusive Growth and Sustainable Development in Africa, vol. I. Palgrave Macmillan, Cham (2020). https://doi.org/10.1007/978-3-030-41979-0_7
2. Rusakova, E.P., Frolova, E.E., Kupchina, E.V., Ocaqli, U.: Recognition and enforcement of foreign arbitral awards in the Federative Republic of Brazil. In: 7th International Conference on Education and Social Sciences. Abstracts & Proceedings, pp. 166–170 (2020)
3. Rusakova, E.P., Frolova, E.E., Zankovsky, S.S., Kupchina, E.V.: Problems of implementation of leadership in dispute resolution of the BRICS countries (on the examples of the Russian Federation, China, India). In: 6th International Conference on Education, Social Sciences and Humanities, pp. 754–759 (2019)
4. Matytsin, D.E.: Neoindustrial tools for the circulation of uncertified securities: digital technologies for the implementation and protection of the rights of investors and issuers. Legal Concept = Legal paradigm **19**, 73–83 (2020)
5. Rusakova, E.P., Inshakova, A.O.: Industrial and manufacturing engineering in digital legal proceedings in the Asia-Pacific region: a new level of quality based on data, blockchain and AI. Int. J. Qual. Res. **15**(1), 273–289 (2021)
6. Born, G.: International Arbitration: law and practice; Alphen aan den Rijn. Kluwer Law International, The Netherlands (2016)
7. Jones, D.: Choosing the Law or Rules of Law to govern the Substantive Rights of the Parties. A discussion of voie directe and voie indirecte. Singapore Acad. Law J. **26**(Special Ed), 911–941 (2014)
8. Grigera, N.: Horacio. Choice-of-Law Problems in International Commercial Arbitration, 289 Recueil des Cours **9**, 191 (2001)
9. Derains, Y.: Jurisprudence of International Commercial Arbitrators Concerning the Determination of the Proper Law of the Contract. Int. Bus. Law J. **514** at 529 (1996)
10. Award in ICC Case No. 7071. In Naón, G. (eds.) Choice-of-Law Problems in International Commercial Arbitration, 289 Recueil des Cours 9, 236 n. 249 (2001)
11. Inshakova, A.O., Marchukov, I.P.: Applicable Law as an essential condition for a foreign trade contract for the supply of energy resources. Lect. Notes Netw. Syst. **44**, 53–64 (2019)

Chapter 23
Robotization of Civil Proceedings: Reality or Future

Ekaterina P. Rusakova

Abstract The consistent introduction of artificial intelligence technologies in Russia and abroad creates the necessary prerequisites for changing the legal and ethical standards for this phenomenon. The judicial form of legal protection was also changed in a number of countries, due to the replacement of judges with computers in human form. For China, this experience was successful, as main and pioneer participants in the simplest cases were absolutely not against the replacement. A number of countries have already adopted special laws and legal acts regulating the integration of robots in judicial processes. In almost all countries of the world, there is an irreversible process of replacing the basic legal framework. In the course of the study, the positive and negative features of this phenomenon will be figured out and because of the fact that many of the basic provisions governing the legal regulation of artificial intelligence technology were borrowed from various acts of an economic nature, since in many respects this area is autonomous and does not involve performing any additional actions other than pressing a button on a computer.

23.1 Introduction

The idea of transforming jurisprudence into an automated process is not new, the creation of special programs that allow you to design various legal documents has been used for a long time. The next step was the creation of robots that can replace a person in the process of legal advice. So, in 2017, a robot named Xiaofa "Xiaofa" was put into operation in the Beijing Internet court, which speaks in a child's voice and whose main function is to give explanations on complex legal terms in ordinary language. This robot can answer more than 40,000 judicial and 30,000 legal questions, which has significantly accelerated the process of applying to the court.

The Xiaofa robot developed by Professor Wang and his team develops according to 4 stages: in the first generation, the knowledge database is created manually with

E. P. Rusakova (✉)
Peoples' Friendship University of Russia (RUDN University), Moscow, Russia
e-mail: rusakova_ep@pfur.ru

A. O. Inshakova and E. I. Inshakova (eds.), *New Technology for Inclusive and Sustainable Growth*, Smart Innovation, Systems and Technologies 288,
https://doi.org/10.1007/978-981-16-9808-8_23

a very limited content coverage and provides users with answers using keywords; in the second generation, the knowledge database is built semi-automatically and covers certain areas. It can answer certain user questions; in the third generation, the knowledge database is built semi-automatically and covers more areas. It can answer complex user questions; in the fourth generation, the knowledge database can automatically expand its data beyond the database and can engage in a dialog with users through emotional expressions. Currently, the Xiaofa robot is being developed between the second and third generations [1].

In October 2016, China's first artificial intelligence robot, Faxiaotao, was introduced in Hangzhou, Zhejiang Province, which helps people analyze the best way to resolve a dispute, as well as help them choose lawyers, by analyzing and searching for the necessary information on the Internet [2].

China currently has more than 100 robots in courts across the country, because the process of creating intelligent justice is more efficient, so some of the robots already have specialization, for example, in business, law, or specific disputes [3].

The ubiquity of the process of robotization often raises a lot of questions about how justified such an introduction to law is.

The next step was the creation of a robot judge, as a result, on June 27, 2019, the Beijing Internet Court announced the opening of the "Online smart Litigation Service Center". The virtual judge demonstrated the mobile micro-court of the Beijing Internet Court. A "robotic" trial is more appropriately referred to by ordinary citizens as an "intelligent trial".

The Vice-President of the court reported on the plans to build an intelligent online justice system based on an integrated full-featured online judicial platform that provides a complete list of procedural actions in electronic form.

The mobile micro-court allows the parties themselves to manage the trial in real-time, as well as to turn to video mediation.

However, the main advantage of such legal proceedings is the transformation of the main procedural tools into an online format, starting with the download and storage of evidence and ending with judicial services, which is how the judicial form of legal protection is considered in modern society. Communication between all participants in the process is carried out in real-time and does not stop due to the use of modern means of communication, as well as providing the parties with more effective and convenient services: interactive questions and answers, obtaining the expert opinion of Weitao and intelligent legal proceedings.

23.2 Methodology

The study used comparative analysis to compare the judicial process regulation in China's Internet courts, as well as cause-effect and framework analysis to define common and unique features of the dispute resolution process. Based on the results of the analysis, an assessment of the existing regulation of digital legal proceedings is provided.

23.3 Results

Information reform of judicial proceedings. Representatives of the Chinese Academy of Social Sciences' Institute of Law stated in their report on the informatization of the Chinese judicial system in 2019 and their expectations for 2020 that 2019 was the first year when intellectual courts moved from the stage of creation to full growth, through standardization, systematization, and informatization [4].

It should be noted that there are a variety of issues that must be addressed in order for this method to be implemented successfully, the majority of which are technical in nature. People's courts at all levels effectively implement the digital agenda of the country and actively integrate the idea of creating intelligent courts and all judicial proceedings [5].

According to figures from the People's Republic of China's Supreme People's Court, the number of online applications at all levels of the judicial system increased dramatically, reaching 99%; 67% of courts used online evidence exchange; 33% of cases were broadcast live, and all these indicate the widespread informatization of legal proceedings. The ongoing changes in the informatization of the Chinese judicial system are taking place simultaneously with its reform, which corrects both processes in the announced programs (2019–2023).

Chinese President Xi Jinping stressed that it is not about the informatization of a particular area, but the complex informatization of the judicial system, enforcement proceedings, and its management.

Another breakthrough step was the creation of a special "Intelligent cloud Platform of the People's Court", which allowed the unification of multi-cloud resources, equipped with artificial intelligence technologies: text recognition, voice interaction, and machine translation.

In addition, the Shanghai High People's Court, together with the High People's Courts of Jiangsu Province, Zhejiang Province, and Anhui Province, created the "Yangtze River Delta Smart Court Information and Data Exchange Platform" to jointly cooperate, provide judicial services, and enforce the law. Initially, the court platform included the ability to pay court costs, online registration of the case, intellectual research of the case, service of documents, communication with judges, and submission of documents. In addition, a mediation platform was created, which brought together representatives of various industries and organizations, which made this procedure more effective.

In 2019, thanks to artificial intelligence technologies, courts at all levels accelerated the judicial process, due to the emergence of smart judicial proceedings, smart case management, smart enforcement proceedings, and smart administration.

The introduction of modern innovative technologies has radically changed the approach to legal proceedings and made it more efficient and secure.

23.3.1 A Virtual Robot Judge is the Basis of Intelligent Legal Proceedings

Intelligent court proceedings consist of three components: a manual of judges with artificial intelligence; a mobile micro-court; and an account on the court platform. The virtual judge who conducts the proceedings is a collaboration of two technologies: text-to-voice transmission and image synthesis, which is responsible for making its actions emotional and transmitting the original text. The prototype of the virtual judge was the real judge Liu Shuhan, who heads the Second Judicial Chamber of the Beijing Internet Court. He said that since the court works 24 h a day, the judicial services should work in the same mode.

Having identified common problems in the filing of documents, mediation procedure, legal advice, and technical operations, 20,000 keywords were identified for which the parties may have questions and answers were prepared based on the guidelines of the trial [6].

One of the advantages of a mobile micro-court is the provision of integrated legal proceedings "at any time and from any place", which shows its cross-border nature. So, the parties located in different places, one of which was abroad, and the other in another province, were able to apply to the court for mediation and by signing the mediation protocol, quickly and simply, and most importantly, at a distance, resolved their dispute.

Thus, using the Weeshat program on a mobile phone, the parties get access to legal proceedings, mediation, as well as to all procedural actions related to the consideration of the case in the mobile micro-court. The "My Case" function of the application allows you to view the details of the case, present evidence, and contact all participants in the process. Moreover, this service also simplifies the procedure for familiarizing the parties to the dispute with judicial acts, since the judge can hand over the documents on the case to the parties, make a transcript of the case, and also act as a mediator. The parties and the court create a group on WeChat and can use it to communicate with each other, submit, and create documents.

The online trial is recorded on video and the parties can view it in real-time, as well as after its completion.

In addition, it allows the public to interact with the virtual judge and ask him difficult questions and get answers to them.

However, the main purpose of digital platforms is the ability to carry out entrepreneurial activities, namely, online retail, whose revenues in China amount to multibillion-dollar figures. Receiving judicial services with a single click of a key has a powerful social response, which not only popularizes the mechanism for implementing electronic commerce on such platforms but also the judicial form of protection of law as universal and guaranteed by the state. The platform provides the following services: intellectual court; interactive questions and answers that allow you to demonstrate the professionalism of judges in the field of consumer protection; special cases that describe various cases that were considered in simple language; and professional legal issues [7].

It should be noted that the collaboration of four parties plays an important role in the integration of intellectual courts: government agencies, judges, IT firms, and entrepreneurs, each of whom plays an important role in this process. However, the main function is social, which is implemented by protecting the interests of ordinary citizens, by expressing social responsibility.

The construction of smart vessels demonstrates an open process of interaction between societies, where digital technologies play an important role in creating a favorable business environment.

23.4 Conclusion

The incorporation of artificial intelligence technology into the judicial process has allowed for substantial progress in the judicial system's information reform [8]:

1. Internet courts should be established to protect the rights and legitimate interests of a broad variety of people interested in electronic commerce, as well as to settle Internet-related disputes.
2. The emergence of "smart courts" has made it possible to process huge amounts of information and use modern digital technologies in the dispute resolution process.
3. The introduction of artificial intelligence technologies in legal proceedings takes place at all levels from filing a claim in court to making a decision.
4. The use of cloud technologies has allowed us to process a huge array of information that allows us to harmonize judicial practice and legal services, as well as other areas of knowledge. They are expected to be used not only in civil cases but also in criminal and disciplinary cases, reducing the amount of time the judge spends on the case since the software will suggest solutions for decisions based on the data provided.
5. The creation of intelligent legal proceedings provides various services for the search for information, its processing, and the provision of court services in real-time based on artificial intelligence technologies.
6. Creation of intelligent management of legal proceedings through the use of innovative technologies that allow processing court data and managing court cases [9]. In addition, the program offers the judge options for resolving the dispute, as well as prepares templates of court acts.

It is necessary to note the reverse side of this process, according to Judge Li Jianli, an intelligent trial management system that connects courts across the country, transmits information about all courts, all judges, and all cases, which is processed in real-time. According to this data, Chinese courts can analyze trials across the country in real-time, evaluate individual work of judges, as well as assess the socio-economic situation in a particular region [1].

In addition, such an active digital transformation of justice is aimed at establishing the rule of law on the Internet and ensuring cyber security. Electronic transactions also need a clear legal regulation by the state, since a huge number of people are involved in this area [10].

The process of automation moves by huge steps and makes the whole society think about how far it is ready to go for the sake of replacing a person with a robot, whose actions depend on the program installed in it, and most importantly about their consequences [11].

Acknowledgements This work was financially supported by the Grant of the President of the Russian Federation No.NSH-2668-2020.6 "National-Cultural and Digital Trends in the Socio-Economic, Political and Legal Development of the Russian Federation in the 21st Century".

References

1. Guodong, D., Meng, Y.: Big data, AI and China's Justice: Here's What's Happening. https://www.chinajusticeobserver.com/a/big-data-ai-and-chinas-justice-heres-whats-happening. Accessed 15 May 2021
2. Wen, X.: Robot Gives Guidance in Beijing Court (2017). https://www.chinadaily.com.cn/china/2017-10/13/content_33188642.htm. Accessed 15 May 2021
3. Dialani, P.: Artificial Intelligence Latest News (2021). https://www.analyticsinsight.net/ai-will-have-robot-judges-soon-what-about-human-judges. Accessed 13 May 2021
4. Rusakova, E.P., Frolova, E.E., Arzumanova, L.L.: Challenges of the Judicial Systems of the Russian Federation and People's Republic of China in the Era of the Pandemic. In: Popkova, E.G., Sergi, B.S. (Eds.) Modern Global Economic System: Evolutional Development vs. Revolutionary Leap. ISC 2019. Lecture Notes in Networks and Systems, vol. 198, pp. 1541–1549. Springer, Cham (2021). https://doi.org/10.1007/978-3-030-69415-9_170
5. Rusakova, E.P., Inshakova, A.O., Frolova, E.E.: Legal Regulation of Internet Courts in China. In: Popkova, E.G., Sergi, B.S. (Eds.) Modern Global Economic System: Evolutional Development vs. Revolutionary Leap. ISC 2019. Lecture Notes in Networks and Systems, vol. 198, pp. 1515–1521. Springer, Cham (2021). https://doi.org/10.1007/978-3-030-69415-9_167
6. Rusakova, E.P., Inshakova, A.O.: Industrial and manufacturing engineering in digital legal proceedings in the Asia-Pacific region: a new level of quality based on data, blockchain and AI. Int. J. Qual. Res. **15**(1), 273–289 (2021)
7. Dudin, M.N., Frolova, E.E., Protopopova, O.V., Mamedov, A.A., Odintsov, S.V.: Study of innovative technologies in the energy industry: nontraditional and renewable energy sources. Entrepren Sustain Iss **6**(4), 1704–1713 (2019)
8. Gaivoronskaya, Y.V., Miroshnichenko, O.I., Mamychev, A.Y.: The immodest charm of digitalization. Legal Concept **18**(4), 40–47 (2019)
9. Li, L., Li, Y.: Application of communication and information technologies in the "smart court". Commun. Inform. Technol. **3** (2019)
10. Tarakanov, V.V., Inshakova, A.O., Dolinskaya, V.V.: Information society, digital economy and law. In: Popkova, E.G. (Eds.) Ubiquitous Computing and the Internet of Things: Prerequisites for the Development of ICT. Studies in Computational Intelligence, vol. 826, pp. 3–15. Springer Science + Business Media, Cham (2019)

11. Rusakova, E.P., Frolova, E.E.: Procedural aspects of proof in China's Internet Courts: opportunities for receiving BRICS jurisdiction. In: Popkova, E.G., Sergi B.S. (Eds.) Modern Global Economic System: Evolutional Development vs. Revolutionary Leap. ISC 2019. Lecture Notes in Networks and Systems, vol. 198, pp. 1598–1605 (2021)

Chapter 24
New Technologies for Resolving International Energy Disputes

Elena P. Ermakova and Mahshid Yazdanimoghadam

Abstract To date, energy disputes between companies and states account for a fifth of international commercial disputes. In 2020, more than 50% of the cases accepted for consideration by the International Center for Settlement of Investment Disputes were disputes in the field of energy. In view of the objective doubts of investors in the ability of the host state to ensure independent judicial proceedings in recent years, the world scientific community has been actively exploring alternative mechanisms for resolving international energy disputes in an attempt to identify the most appropriate form that corresponds to modern realities and guarantees compliance with the interests of both parties. The purpose of this study is a comprehensive analysis of the existing forms and methods of resolving international energy disputes, identifying the optimal mechanism for their consideration, suitable for both the state and the investor. The authors formulated the definition of the concept of an international energy dispute, highlighted the advantages and disadvantages of the most common methods of resolving this type of dispute. The authors come to the conclusion that in the conditions of technology development, increasing global interest in renewable energy sources, increasing number of energy contracts, the leading role in the system of the most promising models for resolving international energy disputes belongs to international commercial arbitration. Arbitration is uniquely well-suited to deal with disputes relating to innovative technologies arising in the energy sector. As a process, arbitration is more readily able to adapt to the new status quo brought about by innovation than many national courts and can offer innovative dispute resolution solutions. Innovators in arbitration are already developing automated dispute resolution processes that operate in the same way that a smart contract does.

E. P. Ermakova (✉) · M. Yazdanimoghadam
Department of Civil Law and Procedure and International Private Law, Peoples' Friendship University of Russia (RUDN University), 6 Miklukho-Maklaya St., 117198 Moscow, Russia
e-mail: ermakova_ep@pfur.ru

A. O. Inshakova and E. I. Inshakova (eds.), *New Technology for Inclusive and Sustainable Growth*, Smart Innovation, Systems and Technologies 288,
https://doi.org/10.1007/978-981-16-9808-8_24

24.1 Introduction

Today, fuel and mineral raw materials account for almost 25% of the total volume of world trade in goods, while energy disputes between companies and states account for a fifth of international commercial disputes. Meanwhile, with the growth of international trade and the volume of foreign direct investment into the energy sector, the number and variety of disputes will also grow [14]. Thus, in 2020, 30% of the cases accepted for consideration by the International Center for Settlement of Investment Disputes (hereinafter, ICSID) relate to the oil and gas sector and the mining industry, 20%—to electricity and other types of energy. At the same time, the second—after bilateral investment treaties—the most applicable international act in establishing jurisdiction, according to ICSID, was the 1994 Energy Charter Treaty.

The supply of energy carriers still remains one of the main items of Russian exports [20]. With the development of public relations in the field of turnover of natural resources and electricity, energy disputes have turned into the largest and most expensive litigations, including for political reasons. Due to the dominance of state ownership of mineral resources and state management of energy pricing relations in most countries, the number of applications to non-state law enforcement bodies is growing, since the independence and competence of state courts remain in question [7].

Meanwhile, neither Russian legislation nor international law contains the very concept of an energy dispute or an international energy dispute, despite its almost universal use [13]. For the purposes of this article, we propose to define **an international energy dispute** as a dispute arising from relations related to the turnover of energy resources and complicated by a foreign element.

In the scientific literature, there are several types of disputes in the global energy sector divided by the composition of subjects:

- a company against a company;
- a state against a state;
- a company against a state;
- an individual or a group of persons against a company.

The most common type of disputes listed above is international commercial disputes between energy companies. Usually, such disputes affect relations between oil and gas project operators and service companies or arise from various bilateral agreements [2].

Before talking about the forms and methods of resolving energy disputes, it is necessary to determine how these concepts relate, what should be understood by form and method as legal categories. Since these concepts are not legal, i.e., the content of these concepts is not defined by law, different approaches to the correlation of these concepts have developed in Russian and foreign science [8].

24.2 Methodology

The study is based on the application of logical, dialectical techniques and methods of scientific knowledge, comparative-legal and legal-technical analysis of normative acts texts.

24.3 Results

The Civil Code of the Russian Federation defines ways to protect civil rights by listing them in Article 12. These include:

- recognition of the right;
- restoration of the situation that existed before the violation of the right, and suppression of actions that violate the right or create a threat of its violation;
- recognition of the disputed transaction as invalid and the application of the consequences of its invalidity, the application of the consequences of the invalidity of the void transaction;
- invalidation of the decision of the meeting;
- invalidation of an act of a state body or a local self-government body;
- self-defense of the right;
- award to the performance of duties in kind;
- compensation of losses;
- recovery of a penalty;
- compensation for moral harm;
- termination or change of legal relationship;
- non-application by the court of an act of a state body or a local self-government body that contradicts the law;
- other methods provided for by law.

It is worth noting that in the procedural sense, the concept of a method of protecting a right generally coincides with the concept of a form of its protection.

This idea is confirmed by the fact that the term "alternative dispute resolution methods" has developed and is widely used in the Russian legal doctrine. In this regard, the following practice has been formed: scientists refer to judicial and non-judicial forms of protection as forms of legal protection, and the term "method" is used in relation to alternative dispute resolution procedures [7].

In the law of foreign countries, a stable concept has developed—dispute resolution. Among the main forms of dispute resolution, foreign authors most often name:

(1) proceedings in state courts;
(2) arbitration;
(3) alternative dispute resolution methods;
(4) online dispute resolution.

Alternative Dispute Resolution (ADR) is a term denoting out-of-court procedures through which the parties have the right to settle a dispute that has arisen between them.

With all the variety of ADRs—negotiations, mediation, arbitration, mini-court, corporate ombudsmen, etc.—their goal is to resolve a disputable situation without applying to state courts [4]. Meanwhile, as Zdrok O.N. notes, today ADR refers to procedures that exist not only in parallel with the judicial system but also within it, as an alternative to a full-scale trial [21].

International energy disputes are commonly referred to as international investment disputes [12]. The procedure for their consideration and resolution is established by various bilateral and multilateral international treaties, while one of the main and system-forming documents regulating the legal relations in question is the Energy Charter Treaty (ECT), signed in Lisbon in 1994.

To date, the ECT is the only agreement related to intergovernmental cooperation in the energy sector that fully covers the entire energy supply chain, including a dispute resolution mechanism between states, as well as between states and investors, energy products, and energy-related equipment [1].

The issues of Russia's ratification and application of the Energy Charter Treaty still remain controversial and ambiguous.

In 1994, the ECT was signed by the Russian Federation, while Russia used the possibility of temporary application of the ECT "to the extent that such temporary application does not contradict its constitution, laws or regulations", from time to time returning to its consideration at the level of the State Duma [18].

In 2009, the Russian Federation withdrew from the list of states that used ECT on a temporary basis, and in 2018, it withdrew its signature under the treaty.

In accordance with Article 26 of the Energy Charter Treaty, disputes between a Contracting Party and an Investor of another Contracting Party may be sent for resolution:

(1) to the courts or administrative tribunals of the Contracting Party that is a party to the dispute;
(2) to an international arbitration or other conciliation body;
(3) to the International Center for Settlement of Investment Disputes (ICSID);
(4) to ICSID in accordance with the rules governing the Additional Procedure for the provision of proceedings by the secretariat of the Center;
(5) to the sole arbitrator or to the ad hoc arbitration court;
(6) to the Arbitration Institute of the International Chamber of Commerce in Stockholm.

Article 26 of the Energy Charter Treaty also specifically emphasizes the need to resolve disputes between a Contracting Party and an Investor of another Contracting Party concerning the latter's investments in the territory of the former in a "friendly manner".

It is interesting to note that in recent years the international community has been actively exploring mechanisms and ways of resolving international energy disputes, attempts are being made to identify the most effective models for their consideration that are suitable to modern realities.

One of the results of this process was the preparation of a Model Instrument on Management of Investment Disputes, approved following the Energy Charter Conference in Brussels in December 2018 (hereinafter referred to as the "ECT Model Instrument"). The document is of a recommendatory nature and does not imply mandatory implementation unchanged in the legal systems of the participating states. The prerequisites for its development were the analysis of the national legislation of the participating states conducted by the Energy Charter Secretariat, which revealed a number of problems and potential barriers in the settlement of disputes by non-judicial methods.

One of the main obstacles to the application of ADR by states was called the lack of a clear domestic legal framework, resulting in ambiguity in the distribution of powers to settle international investment disputes, concerns about possible accusations of corruption and abuse of power, as well as the lack of financial resources to conduct the process.

The "ECT Model Instrument" proceeds from the need to prevent international investment disputes and manage them before the need arises to use formal procedures for their resolution. Paragraph 2 of Article 3 of the "ECT Model Instrument" lists such procedures—arbitration, negotiations, mediation, conciliation, and others—with the proviso that the state or the state's bodies have clearly agreed to the use of these procedures in international investment agreements and investment contracts or if the parties to an international investment dispute have agreed on the use [6].

Taking into account the expediency of establishing a responsible body for resolving disputes arising from international investment agreements, the participants of the conference (which approved the "ECT Model Instrument" of December 23, 2018) identified as such a state body that negotiated or signed an agreement with a foreign investor on behalf of the state—a ministry, an interdepartmental commission, or another body providing the necessary coordination (Articles 6 and 9 of the "ECT Model Instrument").

The tasks of the responsible body include, in particular, coordination of work on the management of international investment disputes, including friendly dispute resolution, development of a strategy for dispute resolution, development and compilation of sets of documents for submission to international arbitration or international judicial institutions, communication with investors and response to the notice of arbitration, appointment of mediators, arbitrators or intermediaries in accordance with the chosen dispute settlement mechanism and other tasks, aimed at the consideration and resolution of the international investment dispute that has arisen [6].

A certain uniqueness of the "ECT Model Instrument" consists in the official recognition of the priority of alternative forms of dispute resolution—negotiations, conciliation regulation, and mediation—through which more effective and rapid conflict resolution is achieved. It is assumed that the adopted "ECT Model Instrument" will be useful both for states that do not yet have an internal legal framework for the consideration of investment disputes, and for those states that have such a framework, but would like to improve it.

Currently, almost any investment contract contains a clause on the transfer of disputes for resolution to international commercial arbitration [9]. The attractiveness of arbitration can be explained by a number of reasons. Thus, Gary Born, a well-known world expert in the field of arbitration, Chairman of the Arbitration Court at the Singapore International Arbitration Center (SIAC), reduced all the main reasons for preferring arbitration to state courts to the main 5 ones:

- arbitration is more impartial;
- arbitration has deeper expert knowledge;
- arbitration is time-efficient;
- arbitration is better executed;
- arbitration is more effective in general [3].

The practical convenience of arbitration is that the dispute resolution procedure can be built by the parties themselves, based on the specific needs of the parties to the dispute and its substance. The procedure is carefully tailored to each case, and therefore the consideration of the dispute becomes not only cheaper and more efficient for investors but also shorter in time.

Deeper expert knowledge of international arbitration is mostly determined by the fact that the parties themselves choose the decision-makers on a particular dispute, appointing a non-state judge for their case, based on his knowledge of the subject matter of the dispute. The advantage of arbitration also lies in the fact that the dispute can be considered not on the territory of its member states but on the territory of a neutral state—this shows the independence, impartiality, and objectivity of arbitration procedures.

The resolution of energy disputes—technically and organizationally complex—can take decades even in government agencies. It should be understood that the consideration of these categories of disputes a priori will not be either fast or cheap. In addition, the parties—if they have not agreed to resolve disputes within the framework of the international arbitration procedure—can participate in court proceedings not only in their own states but also in the courts of other countries.

Another attractive feature of all ADRs, including international arbitration, is their confidentiality. Unlike the judicial procedure for dispute resolution, the use of ADR is not accompanied by logging, and the hearing of the case itself is held in closed sessions [15]. This plays a significant role in the consideration of large and resonant energy disputes, which are sufficiently politicized. Thus, many energy companies, in an effort to ensure the protection of trade secrets, intellectual property, and high-tech information, apply for arbitration to resolve the dispute in order to minimize the illegal actions of their competitors [17].

Taking into account all these advantages, the popularity of international commercial arbitration as a form of resolving international energy disputes has grown significantly in recent years. Here are some examples [10].

One of the last cases considered by ICSID in 2020 was a dispute between the energy companies Petroceltic Holdings Limited, Petroceltic Resources Limited (Great Britain), and the Arab Republic of Egypt (9: ICSID Case No. ARB/19/7) about the alleged violation by Egypt of its obligations under several agreements on the sale of gas and the recovery of debt accumulated to investors. In its lawsuit, Petroceltic also pointed to the adoption by Egypt of unjustified and discriminatory measures against Petroceltic's investments in violation of the bilateral "Agreement on the Promotion and Protection of Investments" concluded between the Government of Great Britain and the Government of Egypt on February 24, 1976.

The dispute was registered by ICSID on April 4, 2019. In the process of forming the composition of the tribunal, the candidacy of arbitrator Bridget Stern on the part of the defendant was questioned by the plaintiff. As a justification for the disqualification proposal, Petroceltic representatives pointed to the "pro-state" reputation of the respondent's arbitrator, the lack of guarantees of independence, impartiality, and transparency of the case. The proposal to challenge the arbitrator was rejected by the co-arbitrators.

Nevertheless, presumably having reached an agreement on the dispute, on September 8, 2020, the parties filed a motion to terminate the proceedings. On September 15, 2020, the Tribunal issued an order to terminate the proceedings in accordance with Article 43 of the Arbitration Rules of the International Center for Settlement of Investment Disputes.

In 2014, a company UFG filed a lawsuit with ICSID against Egypt for damages due to a break in gas supplies under a sale and purchase agreement in the amount of $3.2 billion. In 2018, the ICSID tribunal found Egypt responsible for the failure to supply gas in accordance with the terms of the contract and ruled in favor of UFG, rejecting Egypt's arguments about corruption in the conclusion of the contract. The applicant was awarded $2 billion.

Later, UFG initiated another arbitration proceeding under the rules of The Cairo Regional Centre for International Commercial Arbitration (CRCICA), filing a lawsuit directly against Egyptian Natural Gas Holding Company (EGAS). The Arbitration Tribunal established in Madrid had to answer the question whether UFG had lost the right to file a claim against EGAS on the basis of the same factual circumstances, but against another person and guided by other legal grounds.

In the decision on the case, the Chairman of the Tribunal, Pierre Tercier, formed a majority with his colleague, the arbitrator Wolfgang Peter, for the first time applying Article 26 of the 1965 Washington Convention on the Procedure for Resolving Investment Disputes between States and Foreign Persons (hereinafter referred to as the Washington Convention) as a basis for prohibiting the filing of additional lawsuits arising from the merits of the previously considered case. At the same time, according to Article 26 of the Washington Convention, consent to transfer a dispute to ICSID means refusal to use other means of dispute resolution. Thus, the Tribunal concluded that UFG had lost the right to file a lawsuit against EGAS.

The German arbitrator Karl-Heinz Beckstiegel, who was a member of the Tribunal, disagreed with the majority of his colleagues and in his dissenting opinion pointed out that Article 26 of the Washington Convention excludes filing a lawsuit only if the legal grounds of the lawsuit and the parties to the arbitration proceedings in both disputes completely coincide. At the same time, the lawsuit filed by UFG with ICSID was against the State of Egypt and was based on a Bilateral Investment

Agreement on Mutual Attraction and Protection of Investments between Spain and Egypt, concluded on November 3, 1992.

Thus, this decision can become a precedent and can be used in the future to prevent the multiplicity of arbitration proceedings initiated by plaintiffs in order to increase the chances of winning.

24.4 Discussion

The current challenges faced by international commercial arbitration became the subject of discussion at the sixth annual conference held jointly by the Institute of Transnational Arbitration (ITA), the Institute of Energy Law (IEL), and the International Arbitration Court of the International Chamber of Commerce (ICC) in 2019 in Houston, Texas. The participants discussed, in particular, the issues of investing in the energy sectors of the Russian economy under sanctions, paying special attention to the problem of applying bilateral investment agreements of the Soviet period when applying to arbitration, issues of technology development, and cybersecurity in the legal field, changes in the settlement of disputes between investors and the state in the near future [5].

It was noted that the perception and use of technology in the legal space of arbitration may differ depending on the experience of lawyers and the involvement of the parties. For example, electronic data retrieval services e-discovery, used to detect information on electronic media that can act as evidence in a lawsuit, function differently in Europe and the United States, and the parties should have an idea of the content and format of the procedure at the earliest stages of the arbitration process.

In addition, the changes that have taken place in Europe in the legal regulation of the handling of personal data have indirectly affected the arbitration process in terms of the need for the arbitrators to provide guarantees of a "legitimate interest" in the collection of relevant private data.

The participants of the conference also raised questions about the expediency of the system of judicial settlement of disputes between investors and the state (ISDS), implying the investor's right of access to international courts in order to resolve investment disputes. According to Professor Marike Paulsson (Albright Stonebridge Group), the international investment court has always been a "pipe dream", and renewed interest in it is inappropriate. The speakers also noted the need to protect the energy sector from any attempts of nationalization but stressed that, despite the general reduction in the use of the ISDS system, some of its types will be preserved in the future. At the end of the meeting, the thesis was also voiced that in the nearest future we should expect an increase in the number of investment arbitrations based on contracts and, on the contrary, a reduction in arbitration proceedings based on bilateral investment agreements [5].

Another vulnerable point of arbitration is one of its strong sides—expertise. As noted above, the appeal to arbitrators seems to be preferable for investors because of, among other things, their experience and knowledge in the field of the subject of the proceedings, the ability of arbitrators to pay more attention to the dispute compared to overloaded judges of state courts. However, the problem lies in the fact that the specifics of energy law, as well as the high demand for competent arbitrators, creates an acute shortage of them, and therefore they have to devote time to developing an effective strategy for dealing with excessive workload. The limited number of available arbitrators also increases the risk of a conflict of interest, since it is highly likely that the arbitrators will face the same subject matter of the dispute, the parties, and the positions on which the claims are based [16].

The peculiarity of international investment law, as Rachkov notes, is that there are no uniform rules in it—neither substantive nor procedural. The result is unpredictability in the application and interpretation of the norms of international investment law [19]. At the same time, the development of the energy sector has caused a large number of disputes, which together have formed a case law in this area [20]. Thus, in particular, G. Born emphasized that one of the trends in the evolution of international arbitration is the tendency of national courts of different jurisdictions to focus on each other when interpreting conventions, to refer to international arbitral decisions, and also to use the reasoning that was used in other courts regardless of legal systems.

24.5 Conclusion

Despite some difficulties that international commercial arbitration is facing today, it still remains the most promising form of resolving international energy disputes, allowing to choose the procedure most suitable for a specific type of dispute. More and more previously conservative-minded states, in an effort to attract foreign direct investment in their energy sector, are reconsidering their attitude to arbitration: for example, Latin American countries in recent years have made changes to national and international policy programs to promote the development of international arbitration as an effective means of dispute resolution. China, the largest energy importer, which has so far restricted the use of arbitration procedure in bilateral investment agreements (BIT), has revised its position and allowed the use of international arbitration in investment agreements with Germany and the Netherlands on expropriation issues [17].

There is no doubt that with the development of technology, the rapid increase of interest in renewable energy sources, the increase in the number of contracts for the supply or transit of energy carriers in an effort to provide their country with energy potential, the growth in the number and diversity of energy disputes is only a matter of time [11]. It seems that being flexible and convenient, credible and adaptable to any circumstances, international arbitration is able to cope with any challenges of the future.

Acknowledgements This paper has been supported by the RUDN University Strategic Academic Leadership Program.

References

1. Anisimov, A.P., Matytsin, D.E.: The concept and types of environmental entrepreneurship. In: Inshakova A.O., Frolova E.E. (eds.) Smart Technologies for the Digitisation of Industry: Entrepreneurial Environment. Smart Innovation, Systems and Technologies, vol. 254. Springer, Singapore (2022). https://doi.org/10.1007/978-981-16-4621-8_25
2. Baturina, Y.: Pravovaya forma i pravovoe sredstvo v sisteme ponyatiy teorii prava (Legal form and legal means in the system of concepts of the theory of law). Dissertation. Academy of the Ministry of Internal Affairs of Russia (2001)
3. Born, G.: Kompanii otdayut predpochtenie mezhdunarodnomu arbitrazhu potomu, chto, khot' on i dorogostoyashch, al'ternativy eshche dorozhe (Companies prefer international arbitration because, although it is expensive, the alternatives are even more expensive). Zakon **5**, 201–207 (2019)
4. Chudinovskaya, N.: Uchastie advokata v tsivilisticheskom protsesse (The participation of a lawyer in the civil law process). Statute, Moscow (2020)
5. Cohen, A.: A Changing Landscape for International Energy Arbitration: A Report from the 6th Annual ITA-ICC-IEL Joint Conference. Kluwer Arbitration Blog (2019). http://arbitrationblog.kluwerarbitration.com/2019/03/31/a-changing-landscape-for-international-energy-arbitration-a-report-from-the-6th-annual-ita-icc-iel-joint-conference/. Accessed 26 Mar 2021
6. Decision of the Energy Charter Conference: Brussels (2018), 23 December 2018, p. 20. Energy Charter. https://www.energycharter.org/fileadmin/DocumentsMedia/CCDECS/2018/CCDEC201826_-_INV_Adoption_by_correspondence_-_Model_Instrument_on_Management_of_Investment_Disputes. Accessed 26 Mar 2021
7. Ermakova, E., Chumachenko, I.: Formy i sposoby razresheniya sporov v oblasti energetiki (Forms and methods of dispute resolution in the field of energy). Perot, Moscow (2017)
8. Feyzrakhmanova, D.: K voprosu o sootnoshenii ponyatiy "forma", "sposob" i "sredstvo" zashchity korporativnykh prav uchastnikov aktsionernykh obshchestv (On the relationship between the concepts of "form", "method" and "means" of protecting the corporate rights of participants in joint-stock companies). Lawyer **12**, 32–37 (2017)
9. Hausmann, R.: Investitsionnye dogovory s inostrannymi investorami (Investment agreements with foreign investors). In: Aliev, A., Krupko, S., Trunk, A. (eds.) Pravovye aspekty investitsionnykh dogovorov: sbornik statey (Legal aspects of investment contracts: a collection of articles), pp. 23–29. Norma, Moscow (2012)
10. ICSID Releases Fiscal Year 2020 Caseload Statistics. The World Bank Group. https://www.international-arbitration-attorney.com/icsid-caseload-statistics-for-fiscal-year-2020/. Accessed 1 Feb 2021
11. Inshakova, A.O., Frolova, E.E., Marchukov, I.P.: The general energy policy and ways of development of legal regulation of the foreign trade turnover of energy resources of the Russian Federation and the EU. In: Energy Sector: A Systemic Analysis of Economy, Foreign Trade and Legal Regulations. Lecture Notes in Networks and Systems, pp. 187–206. Cham, Switzerland (2019)
12. Inshakova, A.O., Anisimov, A.P.: Legal Regulation of the Development of Renewable Energy Sources in Russia, the BRICS, and EAEU Countries. In: Inshakova A.O., Frolova E.E. (eds.) Smart Technologies for the Digitisation of Industry: Entrepreneurial Environment. Smart Innovation, Systems and Technologies, vol. 254. Springer, Singapore (2022). https://doi.org/10.1007/978-981-16-4621-8_30
13. Inshakova A.O., Deryugina T.V., Tumakov A.V.: Legal incentives as a means of mediating the development of environmental entrepreneurship. In: Inshakova A.O., Frolova E.E. (eds.) Smart Technologies for the Digitisation of Industry: Entrepreneurial Environment. Smart Innovation, Systems and Technologies, vol 254. Springer, Singapore (2022). https://doi.org/10.1007/978-981-16-4621-8_26
14. Investitsii v TEK: (Investments in the fuel and energy complex). Energy Bull. **14**, 23–26 (2014)

15. Korovyakovsky, D.: O nekotorykh al'ternativnykh sposobakh razresheniya sporov (ADR gosudarstv, vkhodyashchikh v VTO (na primere SShA, Avstralii, Frantsii, Germanii, Velikobritanii, Rossiyskoy Federatsii) (On some alternative methods of dispute resolution (ADR of WTO member states (for example, the USA, Australia, France, Germany, Great Britain, Russian Federation)). Russian judge **11**, 16–17 (2014)
16. Le Bars, B.: Recent Developmentsin International Energy Dispute Arbitration. J. Int. Arbitr. **32**(5), 543–549 (2015)
17. Luki, B., Abubakar, N.: Dispute settlement in the oil and gas industry: why is international arbitration important? Int. J. Energy Technol **6**(4), 30 (2016)
18. Pominova, I.: Vsestoronnee issledovanie riskov i preimushchestv dlya RF ot uchastiya v Energeticheskoy Khartii. Sekretariat Energeticheskoy (Khartii Comprehensive study of the risks and benefits for the Russian Federation from participation in the Energy Charter. Energy Charter Secretariat). Knowledge Center, Moscow (2014)
19. Rachkov, I.: Tendentsii razvitiya mezhdunarodnogo investitsionnogo arbitrazha: voprosy po sushchestvu sporov (Trends in the development of international investment arbitration: issues on the merits of disputes). Int Just **1**(29), 93–117 (2019)
20. Rashchevsky, E.: Aktual'nye voprosy arbitrazhey v sfere postavok prirodnogo gaza (Topical issues of arbitration in the field of natural gas supplies). Arbitration Court **1**(2), 315–329 (2020)
21. Zdrok, O.: Primiritel'nye protsedury v grazhdanskom sudoproizvodstve: ponyatie, klassifikatsiya, tendentsii razvitiya (na primere Respubliki Belarus') (Conciliation procedures in civil proceedings: concept, classification, development trends (on the example of the Republic of Belarus)). Bull Civil Proced **1**, 180–223 (2012)

Chapter 25
Inclusion of Criminal Proceedings as a Factor of Sustainable Social and Economic Development

Valentina A. Lazareva, Natalya A. Solovyeva, and Aleksandr N. Pershin

Abstract The purpose of the study is to analyze the impact of inclusive criminal procedure activities on the stability of the development of society through the detection and suppression of crimes, compensation for damage caused by crime, protection of the rights and freedoms of individuals, and legal entities included in the economy. In this regard, the question of the goals and objectives of criminal procedure activity and its social purpose is raised again, which are not limited to the protection of the rights and freedoms of victims and accused persons, as it may seem with a cursory glance at article 6 of the Criminal Procedure Code of the Russian Federation. This article discusses the current problems of modern criminal procedure policy and its relationship with the purpose of criminal proceedings. The author substantiates the proposals aimed at strengthening the inclusion of the criminal procedure, including by improving its legal regulation. Using as the main system-historical method of scientific research, the authors trace the development of theoretical ideas about the role of the criminal procedure in the system of the mechanism of influence on public relations, analyze the controversial issues of modern criminal procedure science, and clarify and correct the previously formed position on the appointment of criminal proceedings. Result is based on the conducted research; the opinion is justified that there are no contradictions between Article 6 of the Criminal Procedure Code of the Russian Federation, taking into account the interpretation of the purpose of the criminal procedure proposed by the authors and the inclusion of the criminal procedure in the system of instruments of state crime control and national security protection.

V. A. Lazareva
Department of Criminal Procedure and Criminalistics, Samara National Research University Named After Academician S. P. Korolev, Akademika Pavlova Str., 1, 443011 Samara, Russia

N. A. Solovyeva (✉)
Department of Criminal Procedure and Criminalistics, Volgograd State University, Universitetsky Prospect, 100, 400062 Volgograd, Russia
e-mail: solovieva_na@volsu.ru

A. N. Pershin
Department of Criminalistics, Moscow State Law Academy Named After Kutafina O.E, Sadovaya-Kudrinskaya St, 9, 123242 Moscow, Russia

A. O. Inshakova and E. I. Inshakova (eds.), *New Technology for Inclusive and Sustainable Growth*, Smart Innovation, Systems and Technologies 288,
https://doi.org/10.1007/978-981-16-9808-8_25

25.1 Introduction

In the implementation of the state policy in the field of combating crime, which is formed by developing goals, setting objectives, as well as defining specific legal methods and means, an important role belongs to the criminal justice system and the criminal procedure activities carried out by it, the effectiveness of which depends on many interrelated factors. Among them, first of all, we should mention the purpose of the criminal procedure law, through which the state, realizing its will, directs the work of law enforcement agencies to achieve those goals that meet the needs of society [6]. The research subject is the question of what is the purpose of criminal proceedings because of their acute controversial nature. The answer to it allows us to understand whether it is legitimate to consider the modern criminal procedure as a state mechanism for crime prevention, if article 6 of the Criminal Procedure Code of the Russian Federation, which formulates the purpose of criminal proceedings, aims the criminal procedure at protecting the rights and freedoms of the individual.

25.2 Methodology

Using as the main system-historical method of scientific research, the authors trace the development of theoretical ideas about the role of the criminal procedure in the system of the mechanism of influence on public relations, analyze the controversial issues of modern criminal procedure science, clarify and correct the previously formed position. As a result of the study, the author's interpretation of the goals, objectives, and purpose of criminal proceedings is proposed, the opinion that there is no need for immediate amendments to Article 6 of the Criminal Procedure Code of the Russian Federation is justified, and the prospects for further scientific research are determined.

25.3 Results

The question of the purpose of the criminal procedure, which is relevant at all times, sounded in a new way in the post-perestroika period, when the previous ideas about the role of criminal procedure activity in the soviet (and socialist) society at that time seemed to contradict the idea of the supremacy of individual rights and freedoms guaranteed by the Constitution of the Russian Federation adopted in 1993. In theoretical studies of this period, the unsatisfactory state of the practice of defending the rights of the accused, the low quality and long terms of the preliminary investigation, the inequality of the parties, the accusatory orientation of the criminal procedure, red tape in the administration of justice, and, as a result, the dissatisfaction of society, disbelief in the possibility of achieving justice in court, etc. The reason for this was the idea of the criminal procedure as an instrument of public authority, a form of implementation of criminal law, and a means of combating crime [25]. Researchers have noted the danger of such a perception of the criminal procedure—the focus on fighting crime is fraught with disregard for human rights, which are sacrificed for a higher goal, and therefore does not guarantee against mistakes [14].

The new criminal procedure policy of the Russian Federation, which was embodied in the Concept of Judicial Reform approved on October 24, 1991, by the Supreme Soviet of the RSFSR, had a deep humanistic orientation, contributed to the widespread dissemination of ideas about the worth of the human person, the supremacy of his rights and freedoms, and gave hope for a fundamental change in the practice of functioning of the entire criminal justice system. In the course of numerous and lengthy discussions in the Russian Federation in 2001 was developed and adopted Criminal Procedure Code, based on different values and principles than before [10]. The unprecedented expansion of guarantees of the right of accused to defense, the presumption of innocence, adversary character of the judicial process and the equality of participants, the judicial independence, and other normative signs of changes in the criminal procedure policy of the Russian state were evaluated by contemporaries as a reflection of the revolutionary change not only in the relations of power and property but also of officially declared values [17]. It was received with approval and great enthusiasm due to the changes in the economic, political, and ideological spheres. It seemed that the priorities were set correctly that the Code of Criminal Procedure stood up for the main value—the human person, his life, rights, and freedoms in full compliance with the constitutional provisions that the whole meaning of the existence of the state, the bodies of all branches of government is to take care of the interests of the person. The criminal procedure has come to be seen as a way to protect human and civil rights and freedoms [15].

The question of the role and significance of the criminal procedure as an instrument of state power, as a way of its influence on public relations, continues to be debated [22]. The adoption of the value concept allowed us to consider criminal justice as an inclusive social system consisting of functionally and organizationally different bodies united by the subject and type of activity that is important for the successful functioning of other social systems, which does not allow us to reduce the role and importance of criminal proceedings solely to the protection of the rights and freedoms of its participants. Criminal proceedings are one of the forms of influence on reality as a type of state activity, the functioning of which is subordinated to the need to respond to the crimes committed [21]. The work of the investigator, the prosecutor, and the court ensures the implementation of the criminal law, the possibility of imposing criminal liability on the perpetrator of its violation, thereby fulfilling the role of a stabilizer of public relations and the function of general prevention.

A certain complexity in the solution of this issue is brought by the changed terminology of the criminal procedure law. When expressing their opinion, researchers often compare the purpose of criminal proceedings with its goals and objectives, while wrongly identifying criminal procedure with criminal proceedings, without taking into account the changed system of principles on which the legal regulation of criminal procedure activity is based, and the ambiguity of the initial concepts due to the richness of the Russian language [8].

There is no equal sign between the goals (objectives) and the purpose of the criminal procedure. The purpose of the criminal procedure (criminal proceedings) characterizes it as a structural element of the social system that affects its other elements. It is not by chance that is added to this term, in theory, the definition of "social" [18] or even "political" [16]. Article 6 of the Code of Criminal Procedure, indicating that the purpose of criminal proceedings is to protect the rights and freedoms of victims of crimes and persons accused of committing them, achieved as a result of the restoration of justice by the court (conviction and punishment of the guilty, acquittal of the innocent, compensation for the harm caused) as if it says that the court thereby contributes to the pacification of society, the restoration of the violated law and order, and the crime prevention. Such a result is ensured by competent legal regulation of the procedure for criminal proceedings. And is achieved by the efforts of all persons involved in the criminal procedure, including due to the difference in the goals of the participants in the criminal procedure and the tasks they solve.

Ignoring this circumstance became the main problem that had to be overcome in the course of judicial reform. And it was overcome by the consolidation in the Constitution of the Russian Federation, and then in the Criminal Procedure Code of the Russian Federation of the independence principles of the court, adversarial parties, the presumption of innocence, and ensuring the right to a defense. The unity of goals and objectives of the Soviet criminal procedure during the preparation and implementation of judicial reform rightly served as the main target for its critics. Moreover, all researchers are in Art. 2 of the Criminal Procedure Code of the RSFSR found grounds for asserting that the main purpose of the criminal procedure is to combat crime, prevent and eradicate it [23], later called by the authors of the Concept of Judicial Reform a "vulgar idea".

The origins of ideas about the criminal procedure as a tool for combating crime are in the natural connection of criminal and criminal procedure law, the application of which regulates the process of imposing criminal responsibility [11]. This means that the emergence of criminal procedure is not due to the fact of the existence of such a phenomenon as a crime but to the desire to regulate spontaneously developing social relations [5], so in the norms of criminal procedure law, in contrast to criminalistics, one should see not so much the rules of effective investigation of crimes to impose criminal liability on the guilty, but rather a system of restrictions established by law [19]. In other words, the State exercises its right to punishment through procedures that guarantee the avoidance of error, and the stronger the democratic and humanistic ideas in society, the more perfect the system of procedural guarantees it establishes. Criminal procedure law does not regulate the procedure for proceeding in a case of a crime, but establishes rules and restrictions that must be observed when performing procedural actions, indicates not the methods of searching and collecting evidence, but the limits of their admissibility, not the need for making procedural decisions, but the grounds for their adoption, etc. Therefore, we distinguish between the tasks of criminal procedure as a type of activity and the tasks of criminal procedure law, which directs this activity in a legal direction. In this sense, the Criminal Procedure Code of the Russian Federation, undoubtedly, is not only more perfect than its predecessor but continues to improve under the influence of the processes taking place in society, dictating new requirements and putting the criminal procedure in front of new challenges [12]. This is confirmed by the huge number of changes and additions made to the Code of Criminal Procedure of the Russian Federation during its validity. Although not always successful, these changes are directly or indirectly aimed at preventing unjustified restrictions on individual rights and freedoms.

The very dispute about the concepts of "goals" and "tasks" of the criminal procedure seems to be fruitless and meaningless, which was noticed by the authors who tried to seriously analyze and systematize the works of those who studied these concepts [2, 13]. Each researcher interprets in his own way not only the goals and objectives of the criminal procedure but also the concepts themselves. As a result, what some call goals, others are presented as tasks. It should also be noted that the difference in theoretical approaches to each of these concepts- task, purpose, including in application to criminal proceedings [4, 7], is largely predetermined, both by the lack of a clear distinction in theory between goals and objectives, and by the ambiguity of the concept of the criminal procedure itself, which denotes both criminal procedure law, and the proceedings regulated by it in a specific criminal case and the functioning of the entire criminal justice system. In this regard, bringing the perpetrators to criminal responsibility, assigning them a fair punishment, and compensating for damages are called by some authors the goals of the criminal procedure [20], while others are called the tasks [1, 3].

Understanding the goal as a mental conscious image of the expected result, which is aimed at achieving human activity, we see no reason to talk about the goal of a complex system of multidirectional actions of many people. The goals set by the investigator, the prosecutor, and the judge are dictated by the tasks that, by the function performed by the subject, are predetermined by law about each stage of the proceedings.

We do not consider it possible to consider the defense of the rights and freedoms of persons accused of crimes and persons who have suffered from a crime as the purpose, goal, or objectives of criminal procedure activities to initiate, investigate, and consider criminal cases. Given that the procedure activities carried out in violation of the rights of the accused and victims, the purpose of criminal proceedings, undoubtedly, does not correspond, it is legitimate to say that the protection of the rights and freedoms of all persons involved in criminal procedure activities, from the non-necessary restrictions, there is a sense of purpose of criminal procedure law, this activity is regulated [24]. However, the social and political significance of criminal procedure is not limited to the protection of the rights and freedoms of its participants; by providing guarantees to any potential participant in criminal procedure relations, criminal procedure law protects them from criminal infringement [9]. By ensuring fair punishment of those guilty of violating the Criminal Code, the criminal procedure protects the values protected by the Criminal Code of the Russian Federation. These are such values as property, public order and public safety, environment, the constitutional order of the Russian Federation, peace, and mankind security.

25.4 Conclusion

As social processes, crime and criminal procedure activities are interrelated and mutually dependent. The criminal procedure activity ensures the implementation by the state of the right to punish the person who committed the crime, contributes to the solution of the tasks of general and special prevention. Official activities aimed at solving crimes and exposing the perpetrators, ensuring fair punishment of the perpetrators, and protecting the interests of the victims, have a positive impact on the state and dynamics of crimes and harmonize public relations. Hence the conclusion: the purpose of criminal proceedings is not limited to the values mentioned in Article 6 of the Criminal Procedure Code of the Russian Federation, but putting them in the foreground, criminal proceedings performs its important role in regulating social relations that permeate all spheres of human society. Therefore, the search for legal means and ways to increase the effectiveness of criminal procedure activities inevitably involves strengthening the guarantees of individual rights.

References

1. Alekseev, A.I., Ovchinsky, V.S., Pobegailo, E.F.: Russian Criminal Policy: Overcoming the Crisis. M., RF, Norma (2006)
2. Alexandrov, A.S., Alexandrova, I.A., Kruglov, I.V.: Appointment of Criminal Proceedings and Punishment. N. Nizhny Novgorod Law Academy, Novgorod, RF (2006)
3. Azarov, V.A., Revenko, N.I., Kuzembaeva, M.M.: Functions of Preliminary Investigation in the History, Theory and Practice of the Criminal Procedure in Russia. RF: Publishing house of Omsk State University, Omsk (2006)
4. Barabash, A.S.: Public start of the Russian Criminal Procedure. SPb, RF: Legal Center-Press (2009)
5. Boikov, A.D.: The third power in Russia. Essays on Justice, Legality and Judicial Reform 1990–1996. RF: GUIPP "Kursk", Kursk (1999)
6. Cheltsov, M.A.: Criminal procedure. Textbook. M., RF: Jurid. Publish. USSR Ministry of Justice (1948)
7. Davletov, A.A., Azarenok, N.V.: Criminal Justice Program. M., RF: Yurlitinform (2009)
8. Fantrov, P.P., Bobovkin, M.V., Medentsov, S.A.: Improving the activities of the Prosecutor's Office for preventing crimes of extremism among minors. J. Legal Concept **20**(2), 74–81 (2021). https://doi.org/10.15688/lc.jvolsu.2021.2.10
9. Fantrov, P.P., Shinkaruk, V.M.: Smart technologies and prevention of extremism among young people. In: Lecture Notes in Networks and Systems, vol. 155, pp. 665–673. Springer (2021). https://doi.org/10.1007/978-3-030-59126-7_74
10. Gladysheva, O.V., Sementsov, V.A.: Modern criminal law policy to ensure economic security by criminal procedural means. Legal Concept **17**(2), 17–25 (2018)
11. Inshakova, A.O.: Extremism as a threat to national security: yesterday, today, tomorrow. Legal Concept **19**(2), 6–12 (2020). https://doi.org/10.15688/lc.jvolsu.2020.2.1
12. Inshakova, A.O.: The coordination of the law enforcement activities in the fight against crime safeguarding the national security of the Russian Federation. Legal Concept **20**(2), 6–11 (2021). https://doi.org/10.15688/lc.jvolsu.2021.2.1
13. Kasatkina, S.A.: Publicity and Dispositiveness in Russian Criminal Procedure. Stavropol (2006)

14. Kozyavin, A.A.: Social Purpose and Functions of Criminal Proceedings. M., RF: Yurlitinform (2010)
15. Lazareva, V.A.: Criminal procedure as a way to protect human and civil rights and freedoms (appointment of criminal proceedings). Lex Russica **3**, 540–550 (2010)
16. Lupinskaya, P.A.: High political significance of criminal justice. Lex Russica **2**, 277–298 (2008)
17. Mikhailovskaya, I.B.: Purposes, Functions and Principles of Criminal Proceedings (Criminal Procedure Form). M. RF: Prospect (2003)
18. Mikhailovskaya, I.B.: The social purpose of criminal justice and the purpose of the criminal procedure. State and Law **5**, 111–118 (2005)
19. Mizulina, E.B.: Criminal Procedure: The Concept of Self-restraint of the State. University of Tartu, Tartu, Estonia (1991)
20. Orlov, Y.K.: Problems of the Theory of Evidence in Criminal Proceedings. M., RF: Yurist (2009)
21. Perlov, I.D.: Judicial Investigation in the Soviet Criminal Process. RF: Gosyurizdat, Moscow (1955)
22. Shinkaruk, V.M., Fantrov, P.P., Solovyova, N.A.: The role of network technologies in preventing youth extremism in the context of integration of government and civil society's activities. In: Lecture Notes in Networks and Systems, vol. 155, pp. 507–513. Springer (2021). https://doi.org/10.1007/978-3-030-59126-7_57
23. Shpilev, V.N.: The Content and Form of the Criminal Procedure. BSU Publishing House, Minsk, USSR (1973)
24. Solovyova, N.A., Fantrov, P.P.: The integration of the state and civil society to prevent youth extremism in the context of legal socialization. Legal Concept **19**(2), 21–26 (2020). https://doi.org/10.15688/lc.jvolsu.2020.2.3
25. Yunoshev, S.V.: The history of the development and adoption of the Code of Criminal Procedure of the RSFSR 1960. Criminal Procedure Codes of the RSFSR. Educational and scientific aid. M., RF: Yurlitinform (2016)

Chapter 26
The International Value of Judicial Practice of International Court of Justice in the Context of Achieving Global Sustainable Development Goals (The Case of International Border Disputes on Delimination Disputes)

Aslan Abashidze and Marsel I. Garaev

Abstract The paper is devoted to the topical aspects of the implementation of the law enforcement of the ICJ—International Court of Justice of UN on International Border Disputes. The study reveals the practice of the International Court of Justice on the most significant cases, which has played defining importance in the termination of long interstate conflicts and stabilising international relations. Conclusions of research justify the significant importance of the role of ICJ in ensuring international peace and security. The given trends predetermine the dynamics of international legal regulation of relations in the field of international dispute resolutions. The factors causing the complication and complexity of legal means of international disputes by ICJ in the context of achieving sustainable development goals defined by the UN Charter are marked.

26.1 Introduction

The achievement of sustainable development goals approved by the UN [1] is of particular importance to maintaining peace and international security. Without the creation of a solid foundation for peaceful international relations, it is difficult to present the positive dynamics of all other areas of social life.

One of the most important guarantees of maintaining stability in international relations is the creation of legal instruments in order to prevent and suppress international disputes. As the practice of international relations shows, even local disputes may subsequently develop into large-scale military conflicts. In this view one of the key

A. Abashidze · M. I. Garaev (✉)
Peoples' Friendship University of Russia, Moscow, Russia

A. Abashidze
e-mail: abashidze-akh@rudn.ru

A. O. Inshakova and E. I. Inshakova (eds.), *New Technology for Inclusive and Sustainable Growth*, Smart Innovation, Systems and Technologies 288,
https://doi.org/10.1007/978-981-16-9808-8_26

tasks of International Law, especially in an unstable global political situation, is to lay down a system of differentiated legal means of settling disputes of different types and levels of complexity. In the context of achieving goals of sustainable development by the world community, the existence of an effective tool for dispute resolution is an indispensable condition for the functioning of the system of international relations and for ensuring the progressive development of society.

Claimed in subsection 3 of section 2 of the UN Charter, the principle of peaceful resolution of disputes [2] should be distributed absolutely to any kind of disputes, even if disagreements do not threaten international peace and security [3].

Among all legal means for the peaceful settlement of international disputes [3], the fundamental place occupies the International Court of Justice (hereinafter Court; Court of justice; The Court of Justice of the United Nations; UN Court). Its practice has gained significant importance in establishing peace in many long, most complex international conflicts.

The greatest number subject to the UN Court is related to territorial border disputes. The activities of the International Court in the field of the resolution of border disputes played a crucial role in the development of international law, the principles, methods, and special legal means of resolving territorial border disagreements.

The international border dispute is a multilevel, complex legal phenomenon, connected with a number of political, legal, social, economic aspects in international relations [4]. Non-solved border disputes significantly destabilise international cooperation in various spheres, impede integration processes, and undermine the foundations of sovereignty and national and international security.

The uncertainty of the legal status of territories and the regime of borders and the lack of legitimately established jurisdictions of states have a negative impact on the maintenance of global environmental security, the processes of nature management, the distribution of international rights and obligations in the sphere of the circulation of economic resources, which are undoubtedly the serious obstacles to sustainable global development.

In the context of the global problem caused by the spread of coronavirus infection (COVID-19), the "closure" of the borders between countries, and restrictions of movement of people, the issue of settlement of border disagreements acquires even greater social significance in terms of maintaining sanitary-epidemiological safety.

But the most negative consequences of cross-border territorial conflicts between states are growing social tensions, violations of human rights, lack of adequate legal protection for citizens and residents, low levels of well-being and livelihood of people in these countries, and a growing refugee problem.

To date, the International Court of Justice has had several non-solved boundary delimitation disputes before it, the existence of which creates an escalation in the territories concerned and destabilises international security. In particular, there are still non-solved border disputes between Nicaragua and Colombia over the delimitation of the continental shelf beyond 200 nautical miles [5] and between Bolivia and Chile over the status of the border territories in the Silala area [6].

On 18 March 2021, Court hearings were held in the complex boundary dispute between Somalia and Kenya over maritime delimitation in the Indian Ocean and the delimitation of part of the continental shelf beyond 200 miles [7]. Kenya did not accept the jurisdiction of the Court and did not participate in the hearing. Somalia reiterated its claims. Researchers speculate that a decision of the International Court of Justice on the dispute between Somalia and Kenya will only "formalise" some legal title to the territories, while not "concluding" the recurrent disputes between the countries. This is due to a multitude of factors that determine the nature of geopolitical relations between states [8]. The factors relate both to the historical aspects of the relations between the African states and to the problem of distribution of oil, water, agricultural, hydrocarbon resources in the disputed territory.

The dispute between Gabon and Equatorial Guinea over the implementation of maritime and land boundary delimitation is also currently before the Court. By order dated 7 April 2021, the Court fixed 5 October 2021 and 5 May 2022 as the respective time limits for the filing of Equatorial Guinea's memorial and Gabon's counter-memorial [9].

Today, the jurisprudence of the International Court of Justice in the field of international border disputes must be aligned with the conceptual priorities of sustainable development [10]. The Court's resolution of any, even a minor border dispute, should always be coupled with a focus on ensuring the basic human and civil rights, including the right to self-determination, preservation of one's historic place of residence, freedom of movement, as well as the maintenance of social and economic guarantees.

26.2 Methodology

The research carried out in this article is based on the application of a combination of general scientific and special scientific methods of knowledge. For example, the method of legal analysis has examined the most significant International Court of Justice cases for the development of international law in resolving certain categories of international border disputes, and has described the legal means of resolving them. The systematic method has allowed, first of all, to consider the international boundary dispute as the complex phenomenon, formation of various polysectoral elements; second, to define a place and value of law enforcement activity of UN Court in a system of legal means of the international disputes settlement; third, to reveal interrelation of practice of International Court of Justice and concept of international legal maintenance of steady development of a world community; fourth, to prove legal tools of increase of efficiency of procedures of disputes settlement. The formal-legal method was used to identify the essence and content of international legal norms and principles in the peaceful settlement of international border disputes.

26.3 Results

The variety of border disputes allows set up the differentiation into the following main types, depending on the subject matter and nature of the dispute: delimitation disputes (defining, changing, and clarifying the position of the border line); disputes related to the violation of the legal regime of borders. The different types of boundaries predetermine the following types of delimitation disputes: disputes over the position of land boundaries, maritime boundaries, and continental shelf boundaries.

The origins of the UN Court's enforcement activities in the field of border disputes and the conceptual principles of international law that have subsequently influenced the development of peaceful dispute settlement were laid down in the Burkina Faso/Mali border dispute [11]. The decision on the dispute became the determining factor for the establishment of the uti possidetis principle in judicial, arbitral, and international legal treaty practice of establishing the legal regime of borders and has generally had a positive impact on the resolution of many other territorial and border conflicts in Africa.

In the Burkina Faso/Mali case the role of the principle of equity in the interpretation of the law and the choice of methods of delimitation was also highlighted. The International Court of Justice took the regulation of border relations to a new level by prioritising the protection of property rights, territorial integrity, and the sustainability of borders. Moreover, it has established a presumption that the consideration of legal and national aspects in delimitation should be in accordance with the international legal concept of recta ratio.

In subsequent practice, the International Court of Justice's conclusions on the conditions of the uti possidetis principle have been supplemented by provisions on the subsidiary use of the principles of "effectivities" in the application of instruments confirming title to territory (Judgment in the Boundary Dispute concerning the delimitation of the Cameroon-Nigeria land and maritime boundary) [12]. Also, in that case, a significant contribution of the Court to the elimination of hostilities between the countries was its ruling that military and police forces and other administrative resources must be withdrawn from foreign territories, and the obligation of Cameroon to protect Nigerians living in Bakassi and the Lake Chad region.

One of the characteristics of the Burkina Faso-Niger boundary dispute (International Court of Justice judgment 16 April 2013) [13, 14] is the effective treaty regulation of the mutual rights and obligations of the parties to the dispute, as well as the obligations to the Court regarding the legal procedures for delimitation and demarcation (Special Agreement of 20 November 2009). Particular attention has been paid to the obligations to maintain peace and security in the disputed territories and to cooperate between the states.

It should be noted that on the African continent the problem of settling international territorial and border disputes has a long historical origin and is largely due to the nature of the political and legal system, the low level of economic prosperity of countries, and the lack of effective treaty-based legal tools to respond to inter-state disagreements. Many of the ongoing disputes within Africa relate to the processes of decolonisation and subsequent regional and local political struggles. Today, there are many non-solved territorial claims within Africa, all of which significantly destabilise social and economic relations and impede the sustainable development of the region [15].

Resolution of the border dispute between Somalia and Kenya by the UN Court is unlikely to ensure peace and agreement on all contentious issues, especially as Kenya has not recognised the jurisdiction of the Court. Here and in other similar relations, additional treaty-based international legal regulators are needed to specify the legal regime of the border territories, to establish the use of natural and other resources and to determine the legal status of citizens.

In the framework of research we also should note the border dispute between Honduras and El Salvador (International Court of Justice decision of 11 September 1992) [15]. The subject matter of the dispute was of a mixed character, combining the delimitation of six sectors of the land boundary and the determination of the legal status of the maritime space and the islands. The disagreement between the states lasted for about 90 years and often took the form of violent conflicts. As J. Dominguez notes, the territorial disputes between Honduras and El Salvador are among the most severe in the history of Latin American border disputes [16].

In this case, The International Court of Justice interpreted the content, purposes, objectives of implementation, the relationship of the principles of delimitation as they should be applied in the post-colonial Latin American continent. The complexity of the implementation of uti possidetis juris was due to the need to establish the credibility of the existence of rights originating and exercised several centuries ago. One of the regulators in this respect has been the principle of consent, embodied in particular in the confirmation of the interpretation of uti possidetis juris, the establishment of effectivities, the definition of situations of acquiescence or de facto recognition of governance.

The ruling in the Nicaragua-Costa Rica border dispute [17] was complicated by multiple legal relationships over delimitation, sovereignty determination, and environmental damage resulting from construction activities in the border area. By ruling in favour of Costa Rica to award appropriate compensation for environmental damage, the UN Court effectively broke new ground in the enforcement of border disputes by comprehensively addressing environmental concerns [4].

A separate place in the jurisprudence of the International Court of Justice is occupied by disputes on the delimitation of the continental shelf beyond 200 nautical miles. The Court's decisions related to the delimitation of the continental shelf, faced for the first time by the subjects of international law [18–20], contain provisions of great value for international law. In particular, questions on the division of the shelf beyond 200 nautical miles were systematised, the content and grounds for the states' rights to use the continental shelf were substantiated and the most important problems of sea pollution were touched upon. The Court identified the factors to be taken into account in reaching an equitable agreement on the delimitation of the continental shelf: the general configuration of the coast, the presence of special features, the geological structure and natural resources of the seabed, a certain proportion of reasonable proportionality, and the effect of the delimitation. The International Court of Justice applied the principle of equity and the method of taking into account special circumstances as determinative bases for the delimitation of the continental shelf and assessed the potential interests of third countries in relation to the disputed territories. In all this, the leading role of international agreements as the main legal means of delimiting disputed continental shelf territories was presumed.

An important contribution of the International Court of Justice has been the recognition of the concept of the continental shelf in the UN Convention on the Law of the Sea [21] as part of Customary Law. The Court has also brought legal certainty to the understanding of the right to an exclusive economic zone [22].

The boundaries of the continental shelf in many parts of the world remain uncertain. With the increasing depletion of resources and the growing interest of the leading world powers in the richest reserves of shelf zones, there is a risk of new hotbeds of boundary disputes related to the division of the continental shelf. The ruling of the ICJ in this matter constitutes the major legal and instrumental importance for International Law. In addition, the role of the UN Court of Justice in maintaining the environmental security of the seabed in resolving international delimitation disputes and addressing issues of allocation of responsibilities for the use of the shelf should now come to the fore.

The effectiveness of the International Court of Justice's legal procedures for dispute settlement is greatly enhanced when the relationship between it and the parties is reinforced by contractual regulation. Bilateral regulators outline the subject matter of the proceedings, define the system of mutual rights and obligations of the parties to the dispute, the interaction with the Court, establish special procedures, and set out the consequences of the execution of the judgment.

It is interesting to note that in referring the territorial boundary dispute between Equatorial Guinea and Gabon to the ICJ, the States defined in detail the nature of the maritime and land delimitation required, raising before the Court a number of questions concerning the application of international law [23]. It is also noteworthy that the parties have established detailed legal procedures for the provision of evidence. Such a contractual model of enshrining the mutual rights and obligations of the parties in a judicial proceeding would seem to be necessary for the resolution of any international disputes submitted to the UN Court, as it increases the level of legal security for the timely delivery and enforceability of procedural acts.

The enforcement of the Court's case law in the field of boundary delimitation disputes has developed a conceptual framework that defines the dynamics of the system of all means of peaceful settlement of such disputes. In particular, the Court's work has grounded the legal concept of the interrelationship between territory and population. A key feature of international border disputes is the impact on citizens' interests and the realisation of their rights in various spheres of life. As Sumner B.T. pointed out the symbiosis of population and territory is one of the main factors to be taken into account in border disputes [24]. In view of this, the most important task of the International Court of Justice in today's sustainable global development goals is to find a balance between the interests of citizens and states based on law and justice. No enforcement decision in a border dispute would be sustainable and enforceable without comprehensively safeguarding the interests of citizens and a resolution of a series of issues to build cooperative inter-state relations in cross-border activities.

26.4 Conclusion

The International Court of Justice has played a leading role in resolving international border disputes. The practice of the Court is a point of reference for various international and national judicial bodies. The institutional specificity of the ICJ manifests itself in the fact that it is a universal international means of dispute settlement, acting exclusively within its competence and through special legal procedures. It is also significant that the Court is not influenced by any regional integration associations. One of the guarantees of ensuring peaceful settlement of international disputes is a clear procedural sequence of actions of participants of disputes, reflecting the order of implementation of their procedural rights and obligations, enshrined in the UN Charter, the Statute of the International Court of Justice [25], the Rules of the International Court of Justice [26].

The Court of Justice of the United Nations today implements the idea of applying the legal means of settling international disputes, maintaining the principles of consistency and stability in the interpretation of the law, providing legal certainty and predictability. At the same time, the individualisation of international disputes creates a unique set of legal instruments to be applied in the resolution of a particular dispute and the particularities of the use of settlement procedures. The multistructure of international border disputes consists of not only legal regulators but also a whole set of historical, ethnographic, geographical, natural, and other circumstances. All this predetermines the unity and differentiation of the Court's enforcement activity.

The dynamics of the judicial mechanism in the area of delimitation disputes follows the diversification of international legal regulation of the relevant relations and is associated with the development of new legal regulators and tools to prevent disputes at an early stage. According to A.Kh. Abashidze, a new trend in international law is the development of norms aimed at preventing international disputes [3]. In this regard, it is worth noting that in the absence of a system of special legal regulators reflecting current problems of international relations arising over territorial-border issues, strengthening the preventive function of the International Court of Justice, in particular, should be carried out by creating a control mechanism at the stage of preliminary agreement of parties to a dispute to prevent the escalation of conflicts and violation of rights and interests of citizens and society.

However, it must be understood that the decisions of the International Court of Justice are not always the final point in the resolution of conflicts between states. Among the Court's diverse jurisprudence, there are also decisions with a "positive" outcome, when the conflict has indeed been settled, and there are also "problematic" disputes which, despite the judicial decision, have not ceased to exist. For this reason, it is useful to perceive models of "post-conflict" contractual interaction between the parties regarding mutual rights and obligations during the execution of the judicial act.

As researchers point out, The International Court of Justice today is considerably limited in the exercise of its judicial powers in cases between states, unless they are subject to the Court's jurisdiction. There is also the problem of the "politicization" of the Court. In particular, political confrontations between permanent members of the Security Council have in some cases undermined the efforts of the International Court of Justice to secure the peaceful settlement of conflicts [27]. As long as these kinds of political issues are not effectively resolved between international legal institutions, the role of the International Court of Justice in securing global peace may be weakened.

Also noteworthy is the problem of the very conservative approach of the International Court of Justice in reflecting environmental issues in the resolution of international disputes. Although the jurisprudence in recent decades has been characterised by a certain increase in the proportion of litigation on environmental legal relations, including in the aforementioned Costa Rica-Nicaragua border dispute, the Court's "environmental" decisions are still in single digits and, for the most part, based on well-established principles [28]. Meanwhile, today the Court should "extend" the system of legal instruments to secure the most significant environmental legal relations for the world community.

The improvement of the international legal mechanism of judicial settlement of international border disputes seems to be built systematically, taking into account the latest social, economic, and political trends in international relations. Factors contributing to the effective resolution of international disputes in the framework of law enforcement activities of the International Court of Justice to date are:

(1) The widespread introduction of treaty mechanisms setting out the prerequisites and conditions for the referral of international disputes to The International Court of Justice
(2) An increase in the number of states accepting the jurisdiction of the UN Court
(3) Strengthening the UN's role in facilitating the enforcement of judgments in international border and other disputes
(4) Increasing the visibility of the International Court of Justice as an instrument for the settlement of international disputes, including through the widespread use of digital technologies to ensure transparency in judicial proceedings
(5) Improving legal procedures for international border disputes through institutional and functional redesign of the International Court of Justice, as well as reducing the time for dispute resolution
(6) Reforming the legal framework for the interaction of the International Court of Justice with various international legal institutions and bodies, reducing the risks of "politicization" of the Court's activities
(7) Establishment of specific international legal treaty mechanisms to enforce the Court's judgments.

References

1. Transforming our World: The 2030 Agenda for Sustainable Development. UN, New York (September 2015)
2. The Charter of the United Nations of June 26, 1945 (adopted in San Francisco on 26.06.1945). Collection of existing treaties, agreements and conventions concluded by the USSR with foreign states, Issue XII, M., pp. 14–47 (1956)
3. International law: Peaceful settlement of disputes: studies. In: Abashidze, A.Kh., Solntsev, A.M. (eds.) Manual for Undergraduate and Graduate Studies, 3rd ed., corrected and supplemented, p. 221. Yurayt Publishing House, Moscow (2018)
4. Shaikhutdinova, G.S., Garaev, M.I.: The usage of international court of justice in the field of border delimitation disputes. Int. J. Criminol. Sociol. **9**, 979–983 (2020). (November 2020)
5. International Court of Justice. Reports of judgments. Advisory opinions and orders. Judgment of 17 March 2016. Question of the delimitation of the continental shelf between Nicaragua and Colombia beyond 200 nautical miles from the Nicaraguan coast (Nicaragua v. Colombia) Preliminary objections. https://www.icj-cij.org/public/files/case-related/154/154-20160317-JUD-01-00-EN.pdf. Accessed 30 Mar 2021
6. Application instituting proceedings filed in the registry of the Court on 6 June 2016 dispute over the status and use of the waters of the Silala (Chile v. Bolivia). https://www.icj-cij.org/public/files/case-related/162/162-20160606-APP-01-00-EN.pdf. Accessed 15 Apr 2021
7. Maritime Delimitation in the Indian Ocean (Somalia v. Kenya). Conclusion of the public hearings held from Monday 15 to Thursday 18 March 2021. https://www.icj-cij.org/public/files/case-related/161/161-20210318-ORA-01-00-BI.pdf. Accessed 03 Apr 2021
8. Onguny, P.: Why the Kenya-Somalia Maritime Dispute is far from over, regardless of the international court of justice ruling. Africa Policy J., 41–51 (2020). (Cambridge, Spring 2020)
9. Land and maritime delimitation and sovereignty over Islands (Gabon/Equatorial Guinea). https://icj-cij.org/public/files/case-related/179/179-20210407-ORD-01-00-EN.pdf. Accessed 05 Apr 2021
10. Inshakova, A.O.: Determining factors of the development of jurisdictions: the evolution of legal policy and legal culture. Legal Concept **19**(4), 6–16 (2020). https://doi.org/10.15688/lc.jvolsu.2020.4.1
11. Case concerning the Frontier Dispute (Burkina Faso/Republic of Mali) Judgment of 22 December 1986. https://www.icj-cij.org/files/case-related/69/069-19861222-JUD-01-00-EN.pdf. Accessed 05 Apr 2021
12. Case concerning the Land and Maritime Boundary between Cameroon and Nigeria (Cameroon v. Nigeria: Equatorial Guinea intervening) Judgment of 10 October 2002 [Electronic resource]. https://www.icj-cij.org/files/case-related/94/094-20021010-JUD-01-00-EN.pdf. Accessed 05 Apr 2021
13. Frontier Dispute (Burkina Faso/Niger). Judgment of 16 April 2013 [Electronic resource]. The International Court of Justice official website. https://www.icj-cij.org/files/case-related/149/149-20130416-JUD-01-00-EN.pdf. Accessed 03 Apr 2021
14. Abashidze, A.Kh., Amirova, A.A. (2020). Africa. Sustainable Development Goals (SDGs) and Africa. In the book: Africa and International Law. Ser. Library of the Department of International Law, pp. 29–42. Moscow
15. Land, Island and Maritime Frontier Dispute (El Salvador/Honduras: Nicaragua intervening). Judgment of 11 September 1992. https://www.icj-cij.org/files/case-related/75/075-19920911-JUD-01-01-EN.pdf. Accessed 01 Apr 2021
16. Jorge, I.D.: Boundary Disputes in Latin America. United States Institute of Peace, p. 30 (2003)
17. Certain Activities Carried Out by Nicaragua in the Border Area (Costa Rica v. Nicaragua). Question of compensation Summary of the Judgment of 2 February 2018. http://www.icj-cij.org/files/case-related/150/150-20180202-SUM-01-00-EN.pdf. Accessed 15 Apr 2021
18. North Sea Continental Shelf cases (Federal Republic of Germany/Denmark; Federal Republic of Germany / Netherlands / Judgement of 20 February 1969. http://www.icj-cij.org/files/case-related/52/052-19690220-JUD-01-00-EN.pdf. Accessed 12 Apr 2021

19. Case Concerning the Continental Shelf (Tunisia/Libyan Arab Jamahiriya), Judgment of 24 February 1982. http://www.icj-cij.org/files/case-related/63/063-19820224-JUD-01-00-EN.pdf. Accessed 12 Apr 2021
20. Case concerning the Continental shelf (Libyan Arab Jamahiriya/Malta) Judgment of 3 June 1985. https://www.icj-cij.org/files/case-related/68/068-19850603-JUD-01-00-EN.pdf. Accessed 15 Apr 2021
21. United Nations Convention on the Law of the Sea (1982). https://www.un.org/depts/los/convention_agreements/texts/unclos/unclos_e.pdf. Accessed 15 Apr 2021
22. Birni, P.: Trends in the activities of the International Court of Justice in the consideration of cases related to the delimitation of the continental shelf. Soviet Yearbook of Maritime Law, pp. 94–95 (1991)
23. Special agreement between the Gabonese Republic and the Republic of Equatorial Guinea of 15 November 2016. https://www.icj-cij.org/public/files/case-related/179/179-20210305-SPE-01-00-EN.pdf. Accessed 15 Apr 2021
24. Brian Taylor Sumner: Territorial disputes at the International Court of Justice. Duke Law J. **53**, 1783–1790 (2004)
25. Statute of the International Court of Justice. https://www.icj-cij.org/en/statute. Accessed 15 Apr 2021
26. Rules of court. Adopted on 14 April 1978 and entered into force on 1 July 1978. http://www.un.org/ru/icj/rules.shtml. Accessed 15 Apr 2021
27. Herbert, E.B.: The role of the International Court of Justice in actualising global peace. Indian J. Int. Law **59**, 323–354 (2021)
28. Bodansky, D.: The role and limits of the international court of justice in international environmental law (May 20, 2020). The Cambridge Companion to the International Court of Justice, Esposito, C., Partlett, K. (eds.) (Cambridge University Press, Forthcoming) (2020). https://ssrn.com/abstract=3606268

Chapter 27
Impact of UP-TO-DATE Technologies on Efficiency of Corporate Disputes Examination in the Russian Federation

Louise N. Voznesenskaya

Abstract Research and scientific purpose of the article drawn up is to reveal and analyze impact of up-to-date technologies on efficiency of corporate disputes examination in states courts of arbitration of the Russian Federation. Generating of new digital means of the state arbitration courts functioning, digital systems and databases (including the state ones) has allowed to improve efficiency of corporate disputes examination, that, undoubtedly, increases attractiveness of doing business in the form of "joint enterprise" or, in other words, a "corporation", and, respectively, levels up the volume of sales in goods and rendering services. Despite the procedural legislation has been reformed in the segment of corporate disputes solution, some procedural peculiarities of corporate disputes examination require further improvement.

27.1 Introduction

The procedural legislation of the Russian Federation takes into account and stipulates a wide variety of peculiarities of corporate disputes examination, for instance, as follows: exclusive jurisdiction, arbitrability of corporate disputes, multilateral structure of parties to such disputes, numerous specific features related to proof provision procedures, absence of mandatory pre-trial examination of corporate disputes, peculiarities of interim measures to secure corporate disputes, reduced procedural terms, impossibility to examine corporate disputes through summary judgement, closed-session concept of the disputes ensued from corporate agreements/shareholders' agreements examination and many others [4].

L. N. Voznesenskaya (✉)
Peoples' Friendship University of Russia (RUDN University), Moscow, Russia
e-mail: louise@zakon.ru

Postgraduate of the Department of Civil Law and Procedure and International Private Law, 6 Miklukho-Maklaya St, 117198 Moscow, Russia

A. O. Inshakova and E. I. Inshakova (eds.), *New Technology for Inclusive and Sustainable Growth*, Smart Innovation, Systems and Technologies 288,
https://doi.org/10.1007/978-981-16-9808-8_27

All the aforesaid procedural specific features have arisen as a result of corporate interrelations specifics and the necessity to keep equilibrium between the examination operativeness of a corporate dispute and the necessity to secure corporate interest with a view to prevent actual freezing of its operations.

Any bona fide party to a corporate dispute is concerned with its prompt settlement and normalizing of a corporate's operations mostly depends upon operativeness of such dispute's examination.

It is not enough for prompt and efficient examination of the aforesaid disputes to simply adopt reduced terms for examination of corporate disputes in the procedural legislation. Up-to-date technologies are paramount in actual acceleration of corporate disputes examination process and, as a consequence, in increase of their efficiency.

27.2 Methodology

There have been applied the following general scientific methods of cognition: (analysis, synthesis, inducing, and deducing), special methods (dialectical, system, etc.), and specific scientific methods (legal comparative, statistical, legalistic, etc.). Applying of such methods allowed to reveal impact of up-to-date technologies on efficiency of corporate disputes examination in state courts of arbitration of the Russian Federation.

27.3 Results

The fourth industrial revolution in the world, or the digital revolution, has a significant impact on the legal systems of all countries of the world, including Russia. The settlement of legal issues concerning the digital transformation of society has become a vital factor in the progressive development of national and cross-border trade, the services sector and the movement of capital [6, 10].

27.3.1 Claim Drafting

A corporate dispute, as any other dispute to be examined in trial, is to be drafted for court proceedings.

Due to complicated nature of such type of cases, the legislator has stipulated a requirement to corporate disputes proxies as follows: they must have higher legal education degree. A proxy is to be allocated with authorities to represent interests of the parties. There is supposed a multilateral structure of parties to such disputes in the types of disputed being regarded herein, in relation to which a notarized Power of Attorney might be required (for instance, the one legalized by a natural person acting as a corporate shareholder). Meanwhile, corporate disputes mostly imply extremal urgency that might be expressed in necessity to file a claim or an application for interim measures urgently (the last may be submitted either along with the claim or prior to the claim). A claimant (e.g., a shareholder) can be located in other region than a defendant corporation. Along with it, it is necessary to consider exclusive territorial jurisdiction: corporate disputes are to be examined at the region of the defendant corporation location (as per Art. 38, P. 41 of the Arbitration Procedure Code of the Russian Federation (hereinafter referred to as APC RF)) [14].

Thus, a Power of Attorney is to be legalized very promptly, but the Principal can be located at a significant distance from the proxy (a representative), therefore, an issue concerning legalization of a digital Power of Attorney throw into sharp relief. This procedure is relatively new for Russia but has already proved itself as a convenient and efficient way of legalization of a Power of Attorney in case a Principal and a Proxy are separated by a significant distance. Legal significance of such procedure is the following: notaries are eligible to certify equal legal force of a digital and a hard-copy instrument and vice versa. Firstly, a notary converts a hard-copy Power of Attorney into a digital file, then, another notary converts the file into a hard-copy Power of Attorney. As a result, such instrument has equal legal force as the original document and is not a copy of the latter (Art. 77, 103.8, 103.9 of the Fundamental Principle of Legislation Concerning Notarial Services) [5].

Thus, availability of such opportunity as a "digital Power of Attorney" might significantly reduce terms of drafting corporate dispute for court of arbitration examination.

Comprehension of actual court practice and evidence gathering is a significant issue for any lawyer drafting a case for court examination at the initial stage.

Up-to-date technologies are very helpful for a lawyer and there are exist various informational legal systems (e.g., Consultant+, Caselook [13], Garant) and various reference informational systems (e.g., Spark [2], Kontur.Focus).

Informational legal systems help to learn legislation actual for issues under examination and study court practice thereon as well as to choose similar court practice to affirm a legal position.

Reference informational systems allow to obtain high-quality and complete data on corporations and carry on comprehensive assessment of counterparties as well as search for interrelations between them. There is data on payment behavior of corporations and on foreign legal entities and publications in mass-media among the exclusive SPARK information. SPARK provides access to credit bureaus, notaries, and the Russian State Register databases in a single-window mode.

Corporate disputes imply substantial expenses (including counter injunction while applying for interim measures which, in fact, very rarely can be avoided while corporate disputes are examined). A potential claimant can by no means always afford such unexpectable expenses. Due to this fact, various court investment services have become of special importance and actuality. Claimants and investors are congregated at such sites. Claimants publish the information concerning their potential claim against a defendant, and investors are capable to choose a claim that appeals to them and pay court fees for such claimant, herewith, in case a win in action is achieved, the claimant shall pay the investor 30–50% of funds obtained in compliance with a court order. The digital services depicted above (refer, for instance, to the "Platform" service) [1] are also a derivative of up-to-date technologies, a type of investment digital platforms that directly affect impact on efficiency of corporate disputes examination in arbitrary proceedings.

A lawyer today cannot even fancy his or her professional activity without the aforesaid up-to-date technologies. They help not only to save elaboration time dedicated to the issue, but also significantly increase the quality of drafting a claim to court proceedings.

27.3.2 Evidence

A corporation itself must be attracted to court proceedings while examining a corporate dispute. In relation to this, it is necessary to attach an instrument confirming the corporate's location. Such instrument is for a while an extract from the Unified State Register of Legal Entities (hereinafter referred to as USRLE). Superior courts have asserted the statement that it is the corporate indicated in the USRLE to which all notifications addressed to the corporation and concerning judicial facts (including commencement of a court examination) are to be sent. Earlier, it took quite a long time to order such extract from USRLE and it was issued by an authorized body (Federal Tax Service) in a hard copy. Since 2018 the aforesaid extract is provided for free in real-time mode in digital format signed by a digital electronic signature of an authorized authority. There has been generated the public state digital source—"PROVISION OF INFORMATION FROM THE USRLE/USRIE IN DIGITAL FORMAT" [9]. Now, it takes less than a minute to obtain the above-mentioned digital document that spurs the procedure of filing a claim.

Besides that, the aforesaid extract from the USRLE on a corporation registered as limited liability company is at the same time a proof for filing such claim for the corporation itself as well as for corporate participants or the corporate's director.

The document confirming the right to file a claim for joint-stock companies' shareholders is an extract from the shareholder register which also can be submitted in digital format [8].

The documents upon which a claimant substantiates its position can be regarded as evidence. In relation to development of digital documents circulation software, many documents are signed by an electronic digital signature and are transferred via digital telecom channels with a fact of sending and receiving a document registration by software means, without personal handwritten signature on a hard copy. The aforesaid item (the relevant software) has speeded up the process of document circulation and simplified substantiation of the relevant facts.

Even in case the digital document circulation is unavailable, the opportunity to sign a document with a digital electronic signature has significantly simplified and skyrocketed many procedures.

Legislation in relation to this is keeping up with up-to-date technologies and regulates legal relations associated with digital signatures trying to suppress possibilities to commit malfeasances and fraud.

It is necessary to highlight, while depicting the procedure of providing evidence within the framework of corporate disputes examination, the convenient procedure of submitting new procedural instruments and evidence via "My arbitr", a special state system.

Sending of documents is to be effected as per "The Procedure of Submitting Documents to the Courts of Arbitration of the Russian Federation in digital format." The aforesaid procedure is posted for public access or is located in the following reference section https://my.arbitr.ru/#help/2/0 [12].

We will analyze the system below.

The recent trend to be outlined separately is acceptance of correspondence held in messengers as case evidence. Much earlier, in order to deliver evidence that, for instance, letters were sent to e-mail addresses, lawyers had to carry out notarized check of a mailbox that made the procedure of proving substantially more difficult and more expensive. The recent court practice demonstrates that it is sufficient for a court to be provided with a screenshot containing one or another correspondence or a page in the Internet network; the court requests the relevant aggregators or messengers to confirm or refute the documents submitted only in case the other party to a process states unreliability of these evidences.

27.3.3 Filing a Claim, Court Sessions, Familiarization with Materials of a Case

In connection with the specifics of these relations, and for dispute resolution, standard arbitration regulations are not always applicable, so now arbitration institutions tend to follow the way of including separate regulations with regard of the specifics of these disputes [3].

It became possible since 2016 to submit documents to court of arbitration in digital format via "My arbitr" state system. Judicial recourse and the documents attached thereto might be submitted in digital format signed by enhanced qualified digital signature of an applicant (or the applicant's proxy) or in digital images of the documents format authorized by simple digital signature or by enhanced qualified digital signature of an applicant.

Initially, it was possible to submit documents to a court of arbitration via "My arbitr" system through registration on the relevant portal. Relatively recently, it has become impossible to submit documents to a court of arbitration via "My arbitr" system without authorization on "The Public Services Portal". "The Public Services Portal of the Russian Federation" is a unified portal for provision of state services and is a state federal informational system. It is necessary to input reliable personal data to confirm an account in the Service center to register on the aforesaid portal.

In other words, in case one has a verified account on "The Public Services Portal", it is possible for him or her to submit any procedural instruments to a court of arbitration in digital format. Some procedural appeals may be submitted with enhanced qualified digital signature only (we will provide details of such procedural appeals below).

Thus, it is allowable to file a corporate claim to court of arbitration in digital format, meanwhile, as we have highlighted above, obligatory documents to be attached may also be collected in digital format. Evidently, a state fee for a corporate claim examination can also be paid digitally.

The impact of up-to-date technologies is not limited by these achievements. Under the restriction caused by COVID-19 pandemic spread, the Supreme Court and the Council of Judges presidia have recommended courts to carry on sessions using video-conference communication systems since April, 2020 [12].

Hearings via video-conference systems are to be initiated by the parties independently through submitting a digital application via "My arbitr" system.

Video-conference communication system was in operation until 2020 only if a party (or the party's proxy) mandatory attends the court of arbitration session at the party's (or the party's proxy) region of location. Thus, for instance, in case a claimant is located in St.Petersburg, but the case is being examined in the Court of Arbitration of Moscow, the claimant might have submitted the relevant appeal and, in case the appeal is satisfied, might attended the Court of Arbitration of St.-Petersburg and Leningrad region and to be connected via video-conference system with the Court of Arbitration of Moscow.

At present, in order to participate in court session via video conference, it is unnecessary to attend the court of arbitration that is a significant time-saving factor for court proxies and, as a consequence, cut down on costs born by the hearings participants. Meanwhile, it is necessary to keep in mind technical factors and possible malpractices in this sphere, in case participants to a dispute are inclined to protract a dispute examination. So, we repeatedly faced references provided by our opponents to "technical issues" in relation to which a court session intended for examining a corporate dispute was postponed.

It is necessary to separately highlight the procedure of digital notification of the parties to examination about date and time of the court session effected by courts.

So, as per the explanations provided in p. 16 of the Supreme Court Plenum Decree dated 12/26/2017 No 57 "On some issues concerning application of the legislation regulating usage of digital documents in general jurisdiction courts' and courts of arbitration activities", in case a court has information confirming that the persons identified in P. 1 of Art. 121 of the APC RF, are aware about the initiated examination, the above-mentioned persons might be notified by a court about date and place of the court session or for carrying out some procedural activities, including appeal, cassation, and supervisory courts via posting the relevant information on the court's official website in the Internet network.

Thus, in case a court of arbitration possesses evidence that the participants to the examination or other arbitration proceeding parties have received the court decision about acceptance of claim application or filing and commencement of proceedings application, the information on date and time of the first court session, judicial acts that appoint time and date of subsequent court session or carrying out separate procedural activities, are to be sent to the parties to the proceedings and other participants to the arbitration examination via posting such judicial acts at the official website.

The possibility to familiarize with case materials in online mode has become a significant facilitation of a court lawyer's work. Until 2020, a lawyer, in order to familiarize with case materials had to attend the relevant courts and, applying photofixation means, got familiarized with the documents attached to the case. Currently, courts of arbitration (including the ones examining corporate disputes) make scan copies of all case materials on their own and send a link for downloading them to an applicant.

27.3.4 Interim Provisional Measures

Interim measures in an arbitration proceeding are urgent interim activities aimed at securing a claim or property rights of parties to the case (provisional measures), including in case the court proceedings are postponed with a view to find a settlement of the dispute (p. 1, Art. 90 of the APC RF). It is necessary to outline, that the aforesaid interim measures institution is the one of the ultimate significance while examining corporate disputes because it mostly predetermines the practical sense of holding court proceedings aimed at examining a dispute. In most cases, if the interim measures are not adopted promptly, restitution of claimant's rights will be impossible within the frameworks of a single court hearings due to the impossibility to execute the court order.

It should be noted that submitting of applications on interim provisional measures in digital format via "My arbitr" system is regulated in special way, as follows: the documents enumerated below are to be signed with enhanced qualified digital signature:

- application on provision of evidence (P. 3, Art. 72 of the APC RF),
- application on provision of claim (Art. 92 of the APC RF),
- application on provision of property interests (Art. 99 of the APC RF),
- application on provision of a judicial act execution (Art. 100 of the APC RF),
- appeal on suspension of a court order execution (P. 1, Art. 265.1, 283 APC RF),
- statement of claim, application, petition for appeal, cassational appeal containing a petition to undertake interim measures (P.1, Art. 125, 265.1, 283 of the APC RF).

The aforesaid special regulation poses some difficulties while submitting interim measures application in digital format because the enhanced qualified digital signature is issued to an applicant by the authorized body upon condition of the applicant's corporal appearance. However, in comparison with material submitting of the relevant application, its submitting in digital format signed by the enhanced qualified signature speeds up the procedure significantly.

It is necessary to point out that a court of arbitration, allowing provision of a claim upon defendant's appeal, is entitled to demand from an applicant seeking interim measures to ensure provision of the defendant's possible losses reimbursement or propose the applicant to do it on his or her own initiative (counter injunction), through enrollment to the court's deposit account funds in the amount proposed by the court, or provision of a bank guarantee, surety or other financial provision which covers the same amount of funds (p. 1, Art. 94 of the APC RF).

Due to up-to-date digital technologies, legalization and provision of counter injunction upon the interim measures application is also possible to be effected digitally.

Practically all credit bodies (banks) allow to submit a request for provision of a bank guarantee distantly, in online mode, through fulfilling special form on their Internet official websites. In case the request has been approved, the bank guarantee might be provided as a digital document signed by the bank's digital electronic signature.

While choosing the counter injunction in the form of enrolling funds to a court's deposit account, an applicant is entitled to effect this transaction via Client-Bank digital system or through online payments in the relevant personal accounts registered in banks.

Due to the fact that the promptness while drafting, submitting and examining the interim measures is the factor of ultimate significance, the digital systems and software directly affects the procedure of corporate disputes examination.

It is especially important that court orders concerning interim measures are to be promptly executed. As per Art. 96 of the APC RF, a court of arbitration order concerning securing a claim is to be executed immediately and according to the procedure stipulated for execution of arbitration courts orders. As per Art. 93 of the APC RF, copies of orders concerning securing a claim are to be sent to the parties to the case and to other parties held liable by the arbitration court to execute the interim measures not later than the next day after such order has been adopted, as well as, depending upon the type of the adopted measures, to the state authorities, or other bodies effecting state registration of property or the relevant property rights. On our point of view, such provision is ought to be comprehended in a wider manner: those might be different bodies—registrars, credit bodies, depositories, and other entities that hold funds of a borrower, etc. Meanwhile, it is necessary to highlight that it is currently impossible to be sent by a court of arbitration its own order via digital telecom channels directly to the above-mentioned bodies, and in relation to this fact, orders are sent in hard copies that protracts notification of these state authorities and bodies about the interim measures adopted by a court, including for corporate disputes orders.

27.3.5 Court Order

By virtue of P. 9.5 of the Rules for Proceedings in Courts of Arbitration of the Russian Federation (the 1st, appeal and cassational instances) approved by the Higher Court of Arbitration of the Russian Federation Plenum dated at 12/25/2013 No 100, texts of all judicial acts excluding the ones containing information classified as state secret or other legally protected data are to be placed at the "The Arbitration Court Case Files" informational source, "Arbitration Courts Awards" automated system in the Internet network within 24 h since the moment of their signing in the judicial proceedings automated system.

Judicial acts of courts of arbitration are to be signed by digital electronic signature of a judge and are to be published in the system with a specified publication date.

In case a corporate dispute has been examined in the frameworks of closed session, as per P. 1, Art. 177 of the APC RF, the court order made in digital format is to be sent to the parties to the case through its placing if the court of arbitration official website in informational and telecommunication system "Internet".

The aforesaid procedures significantly speed up a procedure of obtaining judicial acts related to a case by the persons concerned and exclude risk of non-receipt of paper postal correspondence due to technical failures or a human factor.

27.3.6 Appeal from Judgements

APC RF requires that the persons appealing from judicial acts in the form of appeal, cassation, or supervising complaints attach the relevant judicial acts to the relevant complaints. This requirement has long been an inconvenient and complicated one because the applicant had to receive original version of the appealed judicial act in hard copy. The possibility to attach the selected digital copy of the judicial acts being contested to appeal, cassation, and supervising complaints, selected from "The Arbitration Court Case Files" informational system has significantly facilitated the procedural "red-tape" for the parties to the examination. The contested judicial acts can be attached to appeal, cassation and supervising complaint provided in digital format and signed by enhanced qualified digital signature(s) of judge(s) adopted the judicial act, or in digital image format of the contested judicial acts signed by enhanced qualified digital signature of the presiding judge for the case (or the deputy presiding judge) or an authorized court officer.

Considering that corporate disputes are distinguished by procedural regulations as the ones with reduced terms, the above-mentioned peculiarity is rather necessity than just convenience.

27.4 Conclusion

Up-to-date technologies have significantly changed procedures of corporate disputes examinations. From the first stage (a dispute drafting for court proceedings) to contesting and executing a court order, the procedure of a corporate dispute examination is tightly associated with digital technologies. Efficiency of, particularly, corporate disputes examination directly depends upon the aforesaid technologies because, namely, corporate disputes so tightly depend upon promptness of evidence collecting and their examination by courts of arbitration.

The statistical information points out that the newly generated mechanisms of corporate disputes settlement increase the interest of businessmen in the protection of their legal rights and interests in corporate disputes in states courts [11]. As a consequence, business partners will be more interested in building up corporations that stipulates pooling of their assets. The aforesaid fact increases the attractiveness of Russian jurisdiction for doing business.

References

1. About the Platform (2021). https://platforma-online.ru/. Accessed 01 Oct 2021
2. About the SPARK system (2021). https://spark-interfax.ru/about. Accessed 01 Oct 2021
3. Ermakova, E.P., Frolova, E.E., Sitkareva, E.V.: New trends in developing alternative ways to resolve financial disputes. J. Polit. Law **13**, 280–286 (2020)
4. Frolova, E.E., Inshakova, A.O., Dolinskaya, V.V.: The concept and legal framework for judging corporate conflicts on the US market. In: "Conflict-Free" Socio-Economic Systems: Perspectives and Contradictions, pp. 101–108. Bingley, West Yorkshire (2019)
5. Fundamentals of the legislation of the Russian Federation on notaries dated February 11, 1993, No. 4462-1. Bulletin of the Congress of People's Deputies of the Russian Federation and the Supreme Soviet of the Russian Federation-March 11, 1993, No. 10, Art. 357
6. Inshakova, A.O., Frolova, E.E.: Preface. In: Inshakova A.O., Frolova E.E. (eds.) Smart Technologies for the Digitisation of Industry: Entrepreneurial Environment. Smart Innovation, Systems and Technologies, vol. 254 (2022). Springer, Singapore
7. JUDICIAL DEPARTMENT at the Supreme Court of the Russian Federation: official site (2021). http://www.cdep.ru/index.php?id=79&item=5257. Accessed 01 Oct 2021
8. Matytsin, D.E.: Digital Technologies of the Bank of Russia for regulating investment relations. In: Inshakova A.O., Frolova E.E. (eds) Smart Technologies for the Digitisation of Industry: Entrepreneurial Environment. Smart Innovation, Systems and Technologies, vol. 254. Springer, Singapore. https://doi.org/10.1007/978-981-16-4621-8_19
9. Providing information from the Unified State Register of Legal Entities. EGRIP in electronic form. (2021). https://egrul.nalog.ru/index.html. Accessed 01 Oct 2021
10. Rusakova, E.P., Frolova, E.E., Gorbacheva, A., Kupchina, E.V.: Implementation of the smart-contract construction in the legal system of Russia. In: Materials of the 6th international conference on education, social sciences and humanities, pp. 748–753 (2019)
11. Supreme Court of the Russian Federation: official site (2021). http://www.supcourt.ru/press_center/news/28875/. Accessed 01 Oct 2021

12. System "My Arbitr" (2021). https://my.arbitr.ru/#help/4/0. Accessed 01 Oct 2021
13. System for searching and analyzing judicial practice Caselook (2021). https://about.caselook.ru/#search. Accessed 01 Oct 2021
14. The Arbitration Procedure Code of the Russian Federation of July 24, 2002 No. 95-FZ. Collected Legislation of the Russian Federation of July 29, 2002, No. 30, Art. 3012

Chapter 28
Procedural Documents in the Era of New Technologies: How Legal Design Has Changed the Legal World

Boris I. Bashilov and Alisa M. Berman

Abstract This article reviews the impact of new technologies through Legal Design tools on the established approach to the preparation of procedural documents and the legal community as a whole. What is Legal Design and what are the reasons for its emergence and development? What is the Concept of perfect procedural document? What are the main challenges that can be solved with Legal Design? How to work with the tools of Legal Design and when it is appropriate? These and many other questions are answered by the authors in this article. As a result, the article is not only of general theoretical nature, but also has an applied value, and therefore will be useful and interesting to practicing lawyers.

28.1 Introduction

Litigation has two sides. The flip side of the coin is that we are faced with a large volume of procedural documents, the content of which is difficult to comprehend. The reason for this can be caused by many factors: the documents are poorly structured; written in complex language; contain a lot of redundant information; contain many facts, events, and circumstances that are difficult to perceive in textual format, etc. All this leads to the fact that the document becomes useless and the judge has to independently search for information from the evidence in the case.

The other side of that coin is that judges have a very heavy workload and an absolute shortage of time to consider the case. If we turn to statistics, only in 2020, which was a year of multiple restrictions on the judicial system functioning due to the spread of coronavirus infection [10], the Arbitration Court of Moscow has heard

B. I. Bashilov (✉) · A. M. Berman
Department of Civil Law and Procedure and International Private Law, Peoples' Friendship University of Russia (RUDN University), 6 Miklukho-Maklaya St., 117198 Moscow, Russia
e-mail: bashilov_bi@pfur.ru

A. M. Berman
e-mail: berman_am@pfur.ru

A. O. Inshakova and E. I. Inshakova (eds.), *New Technology for Inclusive and Sustainable Growth*, Smart Innovation, Systems and Technologies 288,
https://doi.org/10.1007/978-981-16-9808-8_28

234,635 cases, which corresponds to the workload per judge to hear 266 cases per month [4].

For comparison, in 2018, when the courts were still hearing cases in the regular mode, the Arbitration Court of Moscow heard 415,967 cases [8] (almost twice as many as in 2020), which corresponds to the workload of one judge to hear about 40–60 cases per day. At the same time, the average volume of case files is 450 sheets, and one court hearing usually lasts about 10 min, of which only half, due to the need to comply with the mandatory procedural aspects, is left for the judge to hear the parties and study the case.

As a result, we have the judge has little time to study the case and lawyers have little time to present arguments during the court session, but the quality of the average procedural document does not allow to facilitate the work of either the court or the lawyer. At the same time, given the described critical lack of time of the judge, procedural documents should be the tool, which will have a decisive influence on the perception of the information in the case and decision-making by the judge [5]. And Legal Design will help to achieve this goal.

What is Legal Design and what are the reasons for its emergence and development? What is the Concept of perfect procedural document? What are the main challenges for which Legal Design is a solution? How to work with the tools of Legal Design and when it is appropriate? In the course of this study through the answers to these and other questions, the authors will consistently disclose the concept of Legal Design and its impact on procedural documents and the legal community as a whole.

28.2 Methodology

Due to the fact that Legal Design is an interdisciplinary phenomenon that exists at the junction of such areas as design, technology and law, in the course of this study the whole range of general, general scientific and particular scientific methods and knowledge techniques were used, including systemic, functional, logical, comparative and other scientific research methods. To solve private research problems such modern general scientific methods as explanation, analysis, synthesis, analogy, abstraction, deduction were widely used.

Due to the applied, practice-oriented specifics of the present study, only comprehensive use of these methods allowed to identify existing problems and formulate practical proposals for the application of Legal Design in civil and arbitration proceedings.

28.3 Results

28.3.1 Legal Design

In today's legal education, the key focus is on developing the skills necessary to form the content of a procedural document. At the same time, in order to make the content understandable to the reader (be it the judge, the trustee, the counterparty, etc.) and to have the desired effect, the form of communication of this content is important.

And Legal Design is responsible for enhancing the visual component of the document. Legal Design is a modern approach to document preparation, in which information is presented not only in text, but also by layout and data visualization.

This definition was first introduced by Legal Design Lab Director and Stanford Design Institute faculty member Margaret Hagan.

Margaret is also the author of "Law by Design" [6]. The book tells about how to form a creative mindset, what are the steps for creating a product in legal design sphere, and also the techniques needed in the work. The book describes many ways to apply legal design to your work, which can be roughly divided into two groups:

1. How to improve the lawyer's work, which Margaret, by analogy with programming, calls the "backend"?
2. How to improve the trustees' impression of the legal services provided to them?

If we look at Legal Design from a practical point of view, it helps to solve the following tasks, which can be clearly seen in Fig. 28.1:

- task: to describe the facts' sequence, reconstruct the chronology of events, etc.; solution: to use a timeline, which will allow you to get rid of a large amount of

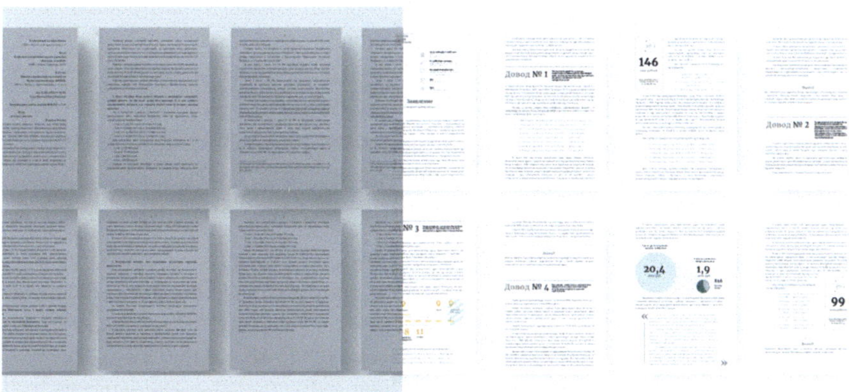

A procedural document without Legal Design A procedural document with Legal Design

Fig. 28.1 Comparison of procedural documents with and without Legal Design [2]

text, the need to spend a lot of time reading it, and unnecessary inconvenience, setting out all the information in a convenient picture, understandable in a minute;
- task: show ratios, differences in numbers, etc.; solution: to draw on a diagram that emphasizes the significant differences;
- task: to understand the structure, show the subordination and links between staff; solution: to use a block-scheme, which will allow to look at the substance of the dispute through the interrelationship of its elements;
- task: to hold a coherent thought, to prove a thesis, to convince; solution: text structuring and editing, accents, inserts with large figures;
- task: to attract attention, to provide a competitive advantage; solution: layout and icons, which will get rid of monolithic text without structure and allow you to create a clear structure with intuitive "anchors".

Despite the essential importance of the form of communication, this category of skills is underdeveloped in lawyers and is not given sufficient attention in legal education. This leads to the fact that many chartered lawyers are unable to communicate information, albeit with quality content, in a comprehensible form [7]. The lack of this skill becomes a barrier to achieving their objectives. So why does it become a barrier?

28.3.2 Channels of Perception

This is due to the fact that there are several channels of information perception.

The first of these is auditory. It is influenced by the lawyer's ability to speak (oratory). This skill is actively used by lawyers in practice and is universally studied in various educational programs.

The second channel of information perception is visual. Visual perception is activated through Legal Writing and Legal Design.

Legal Writing, like public speaking, is common in use and study among lawyers.

But Legal Design, which is no less effective tool for influencing the visual channel, is undeservedly ignored.

In most cases, people make the mistake of thinking that the information they originally intended to convey to the interlocutor, when talking it out in their head, and the information eventually perceived by the interlocutor, are not the same at all. This is due to the fact that during the process of verbal transmission of information, when putting a thought into a verbal form, a part of the transmitted information is lost or modified.

The amount of information conveyed is reduced even further at the stage of perception by the other person, taking into account their individual characteristics as well as their personal views of the situation. This becomes especially significant due to the fact that court proceedings often extend over several months or even years. So how does the human brain perceive information transmitted in different forms?

It is scientifically proven [11] that when information is received through the auditory channels of perception, a person remembers only 36% of this information the next day after receiving it. Three days later, this volume is reduced to 10%. What to say about months and years.

However, when information is received through visual channels of perception, a person remembers 85% of this information the day after receiving it, and 65% of it three days later. Thus, when we need to convey information in the long term, it is more effective to convey information through visual channels of perception.

Considering the information received about the channels of perception and the key problems of court proceedings, what then consists of a perfect procedural document?

28.3.3 *The Concept of a Perfect Procedural Document*

The concept of a perfect procedural document is that a procedural document should combine three elements (Fig. 28.2):

1. Legal Expertise—knowledge of the law and the ability to apply it in practice;

Fig. 28.2 The concept of a perfect procedural document [9]

2. Legal Writing—the ability to present an argument in clear language;
3. Legal Design—skills of visual presentation of information in a document.

The combination of all three elements is the ideal to strive for.

In fact, Legal Design in this triad is responsible for taking care of the legal information reader. It is a kind of empathy to him: when you compose a procedural document imagine how a judge will study it. A well-designed text always works more visually than solid paragraphs without any visible structure or emphasis.

As a result, it turns out that Legal Design is an interdisciplinary phenomenon at the junction of law, design, and technology.

Legal Design requires legal skills, design thinking, and technological capabilities.

In this regard, work with Legal Design can be structured through two models:

1. When a lawyer simultaneously combines the qualities of a lawyer and a designer and becomes a Legal Designer;
2. When a lawyer collaborates with a designer.

28.3.4 Working as a Designer or Working in Conjunction with a Designer

Let's first take a closer look at the lawyer-designer situation.

As the term "Legal Designer" makes clear, it is a combination of the qualities and skills both of a lawyer and a designer. The Legal Designer must combine the following skills: legal assistance skills; visual communication skills; research skills.

Research skills are necessary because design is not a core area for a lawyer, and with the constant evolution of tools and technology in Legal Design, a lawyer must have the research skills to master them.

Thus, a Legal Designer must: think like a lawyer; think like a designer; be able to work with his hands like a lawyer; be able to work with his hands like a designer.

Now let's take a closer look at the lawyer-designer cooperation situation.

In such a model, the role of the lawyer is reduced to the following: drafting the procedural document; identifying the aspect in the procedural document that requires visualization; assigning the task to the designer; checking the final visualization for compliance with the goals pursued.

The designer should also not take a passive role, his responsibilities include selecting the visualization method that most effectively meets the goals pursued; knowledge of major trends and new technologies in design.

What should you pay attention to when working with a designer? The following are some of the reference points to look out for:

- put the task to the designer as early as possible, so that there would be time to coordinate and finalize it;
- tell the designer not just the passage requiring visualization, but the entire case in context;

- provide an outline to the designer, because visualization is sometimes easier to understand through drawing rather than explaining in words;
- make sure that you and the designer understand each other and find common ground;
- be open to the designer's suggestions, even if they don't match your initial ideas about visualization;
- establish the deadline clearly;
- quickly respond to the designer's clarifying questions;
- give the designer feedback after receiving the rendering project: this will help with the current project as well as future projects.

It should not be forgotten that the skill of seeing aspects in a procedural document that require visualization is developed like any other skill and comes solely with practice.

28.3.5 Double Diamond: Design Process Model

In order to increase the efficiency of interaction with the designer, it is necessary to understand the essence of the design process model, which is also applicable to Legal Design. The easiest way to do this is with the Double Diamond model (Fig. 28.3).

"Double Diamond" is the name of the design process model popularized by the British Design Council in 2005 and adapted from the divergence-convergence model proposed in 1996 by Hungarian-American linguist Béla H. Banati [1].

The two rhombuses are a manifestation of divergent thinking and represent a process of more detailed and complex study of a problem. On the basis of which, due to convergent thinking, purposeful actions are taken [13].

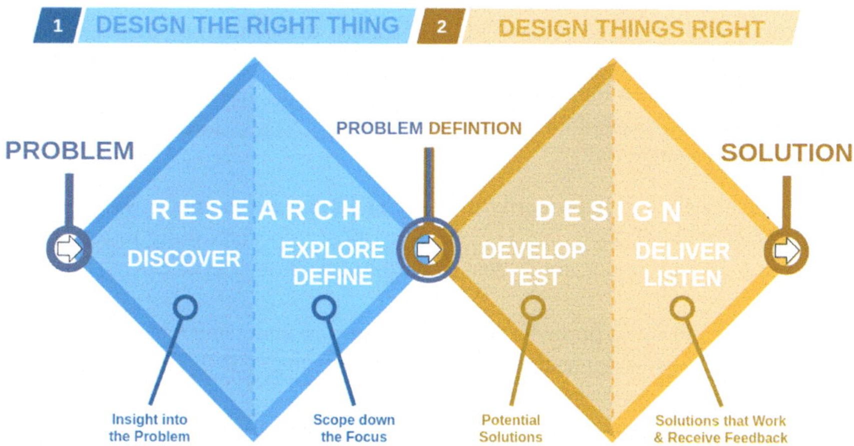

Fig. 28.3 Double diamond (design process model) [3]

The design process is supposed to be shaped by four steps:

1. Research: find out exactly what the problem is—assumptions alone are not enough;
2. Synthesis: try to look at the problem differently—the understanding gained during the research phase will help to assess the problem from a new perspective and synthesize a new vision of it;
3. Idealization: formulate different answers to a clearly defined problem, being inspired by different ideas and interacting with people with different points of view;
4. Implementation: Try different solutions to the problem—reject those solutions that do not solve the problem and improve those that are useful.

28.3.6 Assessing the Appropriateness of the Visualization

To summarize, the answer to the question, "How to understand that a procedural document requires visualization?"—is as follows, as often the same elements in a procedural document require visualization:

- factual circumstances of the case (chronology of events, corporate structure, etc.);
- information that requires focus of the judge's attention (evidence, figures, etc., supporting the main thesis);
- a large volume of information requiring structuring (contracts' register, list of transactions, etc.);
- complex constructions, if it is presumed that the judge, due to the specifics of the case, is not familiar with them (mechanism of operation of oil production equipment, mechanism of mining cryptocurrency, etc.);
- weaknesses of the procedural opponent's position (unproven totality of circumstances, use of an improper legal provision, etc.);
- comparative information (quantitative indicators, distribution of shares, volume ratio, etc.);
- the violation of court logic (the court's decision should have been based on one set of circumstances, and the court established only a part of them or established other circumstances, etc.).

Thus, the answer varies primarily from the tasks you are facing, as well as from the type of procedural document [12]. But the most important thing to remember is that Legal Design is not about beauty or simplification. Legal Design is about usefulness.

Legal Design makes a procedural document easy to understand and comprehensible. If Legal Design is not useful in a procedural document, it should not be placed in it.

As mentioned earlier, Legal Design is all about empathy for the reader. You must first understand the reader, get on his or her side, and identify his or her needs.

28.4 Conclusion

From the above it can be concluded that the correct, thought-out form of a procedural document, which is achieved through the tools of Legal Design, helps to overcome the following problems: the complexity of the legal text and the case as a whole; the large volume of case materials; lack of the judge‘s interest to study all the details of the case; lack of the judge‘s time to consider the case.

Due to the fact that Legal Design solves the indicated problems, the benefit of Legal Design for all participants of a legal process becomes obvious: a lawyer solves the problems of the principal in court, where he faces the fact that the judge has many cases and little time to study documents, but Legal Design helps a lawyer to make an understandable document and save time of the judge, and the more understandable to the judge are the documents, the more chances the lawyer have to win litigation of the principal. By using Legal Design's tools, everyone wins.

The very first question lawyers ask when encountering Legal Design is the following: "What is the reaction of judges to Legal Design?".

Judges rarely express their impression of a procedural document openly, although there are cases where judges can be seen to be pleasantly surprised by a procedural document. However, let us build a logical chain. If the judge has studied the procedural document and Legal Design has helped him or her to understand it, there can be no negative reaction. If the judge has looked at the procedural document and the Legal Design happened to be just a meaningless picture that has not helped in any way to understand it, then the judge may well give a negative impression. Therefore, as has been said before, Legal Design should be helpful in the first place.

Also one of the most common questions about Legal Design is, "Statistically, how many cases does Legal Design help to win?".

Such statistics does not and cannot be formed at all, because the use of only one means cannot lead to a victory. It is always a combination of means, among which, for example: a bright and structured speech; presentable appearance of a lawyer; neat, priorly submitted to the court procedural documents; use of Legal Design tools in a procedural document; etc.

It should not be forgotten that a judge is also affected by many other factors that have nothing to do with a particular trial and are beyond the lawyer‘s control, such as the percentage of reversed decisions of a particular judge that affect his approach to decision-making in the case; family and professional circumstances of the judge, the conflict he encountered at the beginning of the day, and among other things—Legal Design in a procedural document.

So if one asks the question: "Can the effectiveness of Legal Design be measured in numbers?"—the answer is unequivocally: "No".

The outcome of litigation can be influenced by many of the factors and tools outlined above. The effectiveness of each of them cannot be measured individually. However, the more tools are involved, the higher the chance of winning a dispute. Therefore, Legal Design tools should not be neglected.

Acknowledgements The article was prepared within the framework of the project implementation by the winner of the Master's program faculty grant competition 2020/2021 of the Vladimir Potanin fellowship program.

References

1. Banathy, B.H.: Designing Social Systems in a Changing World. Plenum Press, New York (1996)
2. "Bashilov, Noskov & Partners" Law Offices in Moscow (2021). https://bnplaw.ru. Accessed 13 Oct 2021
3. Double Diamond (design process model) (2021). https://en.wikipedia.org/wiki/Double_Diamond_(design_process_model)#/media/File:Double_diamond.png. Accessed 13 Oct 2021
4. Extended meeting of the Presidium of the Arbitration Court of Moscow, dedicated to summarizing the results of the court in 2020 (2020). https://msk.arbitr.ru/node/16082. Accessed 13 Oct 2021
5. Frolova, E.E., Ermakova, E.P.: Utilizing artificial intelligence in legal practice. In: Inshakova A.O., Frolova E.E. (eds.). Smart Technologies for the Digitisation of Industry: Entrepreneurial Environment, pp. 17–27 (2022). https://doi.org/10.1007/978-981-16-4621-8.
6. Hagan, M.: Law by Design (2021). https://www.lawbydesign.co. Accessed 13 Oct 2021
7. Inshakova, A.O., Frolova, E.E.: Preface. In: Inshakova, A.O., Frolova, E.E. (eds.) Smart Technologies for the Digitisation of Industry: Entrepreneurial Environment. Smart Innovation, Systems and Technologies, vol. 254 (2022). Springer, Singapore
8. Key performance indicators of the Arbitration Court of Moscow for 2018 (2018). https://msk.arbitr.ru/node/15676. Accessed 13 Oct 2021
9. Noskov, I.Yu.: Legal Design in judicial disputes. Materials of the scientific-practical seminar "Technique, design and aesthetics in legal education" VIII Moscow Legal Forum "Socio-economic development and quality of the legal environment" (2021). https://msal.ru/news/myuf-tekhnika-dizayn-i-estetika-v-yuridicheskom-obrazovanii/. Accessed 13 Oct 2021
10. Rusakova, E.P., Frolova, E.E., Arzumanova, L.L.: Challenges of the judicial systems of the Russian Federation and People's Republic of China in the Era of the Pandemic. In: Popkova E.G., Sergi B.S. (eds.) Modern Global Economic System: Evolutional Development vs. Revolutionary Leap. ISC 2019. Lecture Notes in Networks and Systems, vol. 198, pp. 1541–1549 (2021). https://doi.org/10.1007/978-3-030-69415-9_170
11. Ruttenberg, K.L.: Images with Impact: Design and Use of Winning Trial Visuals, pp. 10–11. American Bar Association, New York (2017)
12. Tarakanov, V.V., Inshakova, A.O., Dolinskaya, V.V.: Information society, digital economy and law. Stud. Comput. Intell. **826**, 3–15 (2019)
13. What is the framework for innovation?: Design Council's evolved Double Diamond (2021). https://www.designcouncil.org.uk/news-opinion/what-framework-innovation-design-councils-evolved-double-diamond. Accessed 13 Oct 2021

Part IV
Stream of Commerce in the Conditions of the Fourth Industrial Revolution: Financial and Information Support

Chapter 29
Private-Legal Control as a Factor of Trade Turnover Stability

Oleg V. Grabko

Abstract The paper describes private legal control as a determining factor affecting the stability of trade turnover. The concept of the definition of private legal control, its features are defined and a specific characteristic is given. The author determined the methodological basis of the study, justified the relevance of the study and made the appropriate conclusions.

29.1 Introduction

The need for the study of private legal control in modern conditions of economic development is explained by a number of reasons.

Private law is the main legal form of the modern market economy, which is characterized by free economic relations, decentralization, the minimum necessary influence of the state on participants in civil turnover, economic entities, broad autonomy, independence, initiative of the latter.

At the same time, the civilized market imposes certain requirements on the behavior of subjects of civil law: conscientiousness, reasonableness, discipline, ability to self-regulation and self-organization. A large role in compliance with these requirements is assigned to the implementation of private legal control, based not only on equality and free expression of will, but also on the mutual interest of the controlling and controlled person, their interaction and mutual support. The main purpose of such control is to correct the behavior of the controlled person, the preventive desire of the participants in the control legal relationship to avoid the use of state-coercive measures.

The very concept of control in civil law is used in two different senses (as a check, and as an influence on decision-making) without much fixation on the fundamental differences between these legal phenomena.

O. V. Grabko (✉)
Judge of the Twelfth Arbitration Court of Appeal, Saratov, Russia
e-mail: federal64@yandex.ru

A. O. Inshakova and E. I. Inshakova (eds.), *New Technology for Inclusive and Sustainable Growth*, Smart Innovation, Systems and Technologies 288,
https://doi.org/10.1007/978-981-16-9808-8_29

The study of the legal nature and specific features of civil law control is of great practical importance, primarily for the stability of trade turnover, as well as the further development and improvement of all civil legislation and the practice of its application.

29.2 Methodology

To achieve the objectives of the study and determine the effect of private legal control on the stability of trade turnover believe define the private legal control as part of a decentralized civil law system of self-regulation based on the law or the contract validation of the actual behavior of the controlled entity mandatory requirements and (or) interest, controlling persons, carried out the last:

(a) at its sole discretion,
(b) in the field of powerless, coordinating relations between him and the controlled entity,
(c) in the private interests of the controlling and (or) controlled person,
(d) in a non-standardized manner,
(e) mainly to correct the behavior of the controlled person, and not to punish him,

Failure to exercise control constitutes the risk of the controlling person.

29.3 Content

We emphasize that from the etymological and general theoretical positions, control is a test. However, in recent years, the word "control" has been used in a different sense in law, as the ability of one person to influence the decision-making of another person. For example, the category "corporate control" has firmly entered scientific circulation after the publication of a number of judicial acts adopted by the Supreme Arbitration Court of the Russian Federation, in which for the first time the highest court used the term "restoration of corporate control" [1–3, 6]. It is not surprising that these judicial acts were followed by a number of publications devoted specifically to the "restoration of corporate control" [2–4, 6–8, 10–12].

Obviously control-verification and control-influence are different legal institutions that have independent implementation mechanisms, as well as goals, objectives, principles, functions of implementation, and also have different effects on civil turnover.

It is worth noting that civil science has not developed a classification of types of civil control, which actualizes the need for its implementation. Despite the absence of a deep doctrinal study of the concept of "civil control", as well as its individual types, the study of this category seems to us expedient from both practical and theoretical positions, since through civil control failures in the mechanism of industry regulation are identified, adjustments and changes in the legal norms regulating trade turnover, are made. The main purpose of civil law control is to facilitate the fulfillment of civil law obligations within the prescribed period, to ensure the effective and unhindered operation of the entire mechanism of civil law regulation and the high-quality functioning of civil turnover as a whole. The most important task of control, in this regard, is the prevention of undesirable consequences.

Taking into account the above, the classification of types of civil law control makes it possible to systematize knowledge about this legal phenomenon, facilitate the clarification and understanding of its essence, improve the theoretical and legal apparatus in this area, as well as increase the effectiveness of law enforcement activities in the subject of the study.

The classification of types of civil law control can be based on various criteria that make it possible to understand the essence of this type of control, its features and scope of application in civil law.

Summarizing the above, we consider it possible to propose the following basis for classification and types of civil control: contractual and non-contractual civil control. It is worth emphasizing that this classification, which is based on such a feature (or basis) as the basis for the emergence of civil law control in the sense of the theory of legal facts, is very conditional and has no direct connection with the well-known division of civil law obligations into contractual and non-contractual (or protective).

The allocation of such types of civil law control is primarily due to the fact that of all the many manifestations of control in civil law, its most striking and frequently occurring case should be recognized as contractual control, which is inherent in absolutely any contractual legal relationship in the field of civil law.

At the same time, there are other manifestations of civil law control in civil law, which are not based on a contract, but are based on other legal facts [5].

Taking into account the fact that civil law control is an auxiliary civil legal relationship existing within the framework of the main legal relationship, which determines the actual content, nature, and form of the implementation of control functions, it is necessary to distinguish internal control existing "inside" the legal relationship arising between its direct participants (the controlling entity and the controlled person) and manifested during the implementation of their rights and obligations, and external control implemented outside the framework of the main public legal relationship, however, it has an impact on his further legal fate. Internal control is carried out by the parties to the contract, since it is implemented within their legal relationship and does not affect third parties.

Civil law control is expressed, in particular, in giving consent (approval) for transactions or other legally significant actions. Civil control, expressed in the form of approval or prohibition of a transaction, allows the controlling person to "predict" the onset of certain legal consequences and fully manage them. Such "control" functions create additional guarantees for the authorized person for the timely protection and protection of subjective civil rights from possible encroachments, and also ensure a balance of interests of subjects of civil legal relations [9].

Taking into account the signs of civil law control and the variety of forms of its implementation, control in private law can be classified according to such criteria as the content of control, information, and activity. The essence of information control is manifested in the right of the controlling person to receive the information necessary for control from the controlled entity in a timely manner. At the same time, activity control is expressed in the form of specific actions (or inactions) aimed at realizing the legitimate rights and interests of subjects of civil legal relations. For example, in accordance with paragraph 2 of Article 1037 of the Civil Code of the Russian Federation, the copyright holder is obliged to control the quality of goods (works, services) produced (performed, rendered) by the user on the basis of a commercial concession agreement. It seems that in this case, information control by the copyright holder alone is clearly not enough, since the complexity of legal relations arising from a commercial concession agreement requires specific control measures aimed at establishing the quality of goods, works, or services produced by the user [10]. The need for activity-based civil control in the implementation of a commercial concession agreement is due, as noted in the scientific literature, "the desire of the parties to the commercial concession agreement to carry out the business activities provided for by the agreement according to uniform standards, i.e. to level the differences between them in the implementation of the relevant business activities" [8].

29.4 Conclusion

In order to clearly determine the effectiveness of influence on civil and commercial turnover, it is necessary to build a theory of civil law control, systematize knowledge about this legal phenomenon, clarify and understand its legal essence, form a theoretical and legal categorical apparatus in this area, as well as increase the effectiveness of law enforcement activities in the field under study.

References

1. Aleshchev, I.O.: On the concept of "restoring corporate control". EG-Yurist, **36** (541) (2008).
2. Frolova, E.E., Belikova, K.M., Badaeva, N.V., Akhmadova, M.A., Dudin, M.N.: General approaches to the market structure control in BRICS countries. J. Adv. Res. Law Econ. **9**(1), 96–105 (2018)
3. Frolova, E.E., Zankovsky, S.S., Dudin, M.N., Zinkovsky, S.B., Kirsanov, A.N.: Studying concepts of the breakthrough economic reforms in selected developed and developing countries and regions of the world: Economic and legal aspects. J. Adv. Res. Law Econ. **9**(4), 1236–1242 (2018)
4. Guseva, A.A.: An object of vindication: problems of law enforcement. Actual Probl. Russ. Law **16**(4), 76–93 (2021)
5. Inshakova, A., Dolinskaya, V., Frolova, E.: Classification criteria: defining the specific features of corporate conflicts. In: Popkova, E.G. (ed). "Conflict-Free" Socio-Economic Systems. Emerald Publishing Limited, pp. 89–99 (2019)
6. Inshakova, A.O., Frolova, E.E., Marchukov, I.P.: TNCs as subjects of economic activity and lawmaking in the sphere of Foreign trade in energy resources. Lecture Notes in Networks and Systems, vol. 44, pp. 151–171 (2019)

7. Khokhlov, V.A.: Restoration of corporate control. J. Entrep. Corp. Law **4**, 9–14 (2016)
8. Lomakin, D.V.: General characteristics of rights and obligations of limited liability company members. Civil Law **2**, 3–12 (2021)
9. Matytsin, D.E.: Retail investment financing by means of a remote digital computer algorithm. Legal Concept = Legal paradigm, vol. 20, pp. 150–158 (2021)
10. Nazimov, I.A.: Restoration of corporate control. Russ. Law J. **2**, 111–119 (2014)
11. Rainikov, A.S. (2009). Commercial concession agreement, p.14. Moscow, Russia: Statut.
12. Sarbash, S.V.: Restoration of corporate control. Bull. Civil Law **4**, 70–79 (2008)

Chapter 30
Concept and Types of Distance Digital Investment Transactions in the Information Space

Denis E. Matytsin

Abstract In this chapter of the monograph, the author develops the concept and types of investment transactions in the information and telecommunications network Internet. Transactions are made and executed, using computer algorithms based on the autonomy of the will of subjects, who are located at a significant mutual distance. In this regard, investment transactions in the information space can be called remote. It is proved that for these transactions, both at the preparatory stage and at the stage of video negotiations, and at the stage of subsequent execution by the parties of the agreements reached, the use of computer programs by the participating entities is mandatory. The programs provide the transmission of data and commands by working out algorithms, the movement inside of which is carried out at the digit 1, not carried out at the digit 0. In this regard, investment transactions in the information space can be called digital. According to the author, a remote digital investment transaction in the information space is a remote interaction on the Internet between an investor and (or) an investment intermediary and (or) an investment recipient regarding the investment of funds in an object of civil rights (property) that has a monetary valuation. This interaction is carried out in order for the investor to acquire the right of temporary ownership of the object and terminate it after a certain period to make a profit. The author proposes a classification of investment transactions in the information space on several grounds.

30.1 Introduction

In the legal literature, sufficient attention has always been paid and is currently being paid to the topic of transactions, including the level of doctoral monographic research. At the same time, colleagues in civil studies of the dissertation format have

D. E. Matytsin (✉)
Reasearcher of Insitute of Law, Volgograd State University, Volgograd, Russia
e-mail: Dmatytsin@ya.ru; gimchp@volsu.ru

Department of Civil Law and Procedure, Volga Branch of the International Law Institute, Volzhsky, Russia

A. O. Inshakova and E. I. Inshakova (eds.), *New Technology for Inclusive and Sustainable Growth*, Smart Innovation, Systems and Technologies 288,
https://doi.org/10.1007/978-981-16-9808-8_30

not considered remote digital investment transactions in the information space until recently. The remote sign of the studied transactions is a positive achievement of computerization and internetization of investment relations. First of all, the circle of potential private investors is expanding indefinitely, who do not need to come to the Capital from distant places of the vast Russian Federation to meet with a professional investment intermediary and discuss the terms of an agreement on services that will be provided to him.

Participation in investment relations of a professional business entity-intermediary of special legal capacity does not at all reduce the risks and responsibility for the decisions made for the investor. A private investor is not legally an entrepreneur and does not become a de facto entrepreneur even in cooperation with a licensed broker and/or asset manager. The key substantive feature that reveals the definition of investments is the property that has a monetary valuation. Moreover, its dynamics of involvement in the transaction is mandatory in order to increase the cost (profit) in the process of implementing a certain entrepreneurial idea and executing the transaction itself. In addition, the immanent characteristic of investing is risk.

30.2 Methodology

The development of the content of this chapter is based on the materialistic worldview and the universal scientific method of historical materialism. General scientific methods of cognition are applied: dialectical, hypothetical-deductive method, generalization, induction and deduction, analysis and synthesis, and empirical description. The research also uses private scientific methods: dogmatic, comparative-legal, hermeneutic, structural–functional, etc.

30.3 Results

Justifying the scientific interpretation of remote digital investment transactions is necessary first to identify the distinctive features characteristic of them.

Firstly, in order to carry out the transaction and the occurrence of this legal fact in the legal field, special devices are needed to ensure the contact of interacting entities: computers or smartphones that must necessarily have a cable or wireless connection to the Internet, as well as a connection to an energy source (stationary AC network or battery). In addition, special programs must be installed on some devices that allow you to communicate with a certain website on which the transaction will be carried out and recorded. Such a special program is the browser. It allows you to find, view, read, listen to text, audio, video information on websites on the Internet, and download files (texts, images, audio, and video) from websites to your computer (smartphone). The browser is the main link between the user subject and the global information space.

The most widely used browsers are Google Chrome, Opera Mini, Firefox, Safari, Yandex, and Internet Explorer. The mechanism of working on the Internet is the same for all browsers. 1. The subject-the user opens the browser on his computer (smartphone) and enters the unique name of the website. 2. The browser searches for the server on the Internet by IP address and domain name, which are individual for each website, referring to the Domain Name System (DNS). 3. When the browser has found the IP address, it establishes a connection with it via a special TCP/IP protocol that provides data transmission on the Internet. 4. The browser sends an http://-request to the server. 5. The server processes this request and sends a response to the browser. 6. The browser, in turn, processes the response received from the server, as a result, the subject-user sees on the monitor of his computer a loaded page of a certain website with text and (or) a video image of the counterparty of the investment relationship.

Here, it is necessary to clarify that computer programs function on the basis of processing units of information (called bits) and binary code, in which there are only two digits 1 (one) and 0 (zero). Moreover, 0 means that there is no command, 1 means that there is a command (signal). The computer perceives any number in a binary system (for example, the number 2,000 = 0,000,011,111,010,000). The computer also translates letters into numbers, then turns the number into signals and writes them down, as well as numbers (for example, the letter B-193-11,000,001). Similarly, an image is formed on a computer monitor, on a smartphone. Each dot, depending on the color, has its own code (for example, black dot 0, 0, 0; brown: 153, 102, 51), transmitting colors in three bytes (1 byte is equal to 8 bits); you can encrypt more than 16 million colors. Sound and video information are also translated into bits and bytes (that is, into signals consisting of two digits, 1 and 0). It is in this technically correct sense that the correct term digitization exists. However, the derivative expressions "digitalization", "digital rights", and "digital economy" are nonsense. Since digits (zero and one) in computers, algorithms are only code tools of the programming language currently used.

Secondly, a private investor (an individual) how does an initiative entity look for opportunities to increase the available capital. He must have physical and intellectual abilities reflecting his satisfactory health (capacity) and a high level of training in order to conduct an analysis on the Internet investment segment and identify an acceptable investment option for himself in a certain investment object. This training includes, first of all, sufficient vision and hearing, knowledge, skills, and skills of an individual in handling a modern computer (smartphone). As well as his intellectual abilities in terms of understanding the order and conditions of functioning of the Internet investment segment to choose from a mass of investment proposals a promising investment object for himself.

The Indication of the Bank of Russia dated April 29, 2015, No. 3629-U (hereinafter referred to as the Indication) establishes the procedure for recognizing persons as qualified investors and maintaining a register of persons recognized as qualified investors [1]. If an individual meets one of the five requirements of the Indication, he can be recognized as a qualified investor. For example, requirement 2.1.2: a citizen has worked for 3 years or more for an organization that made transactions with securities and (or) concluded contracts that are derivative financial instruments. Such a requirement looks very strange if a citizen worked, for example, as a driver or electrician in this organization. For example, requirement 2.1.4: a citizen owns property worth 6 million rubles or more (money in bank accounts, securities, etc.). This is also a very strange requirement. The question immediately arises, why exactly 6 million rubles, and not 8 and not 10 million rubles? Further, suppose a citizen who has been working as an installer on a construction site all his life receives a 10 million inheritance in the form of a bank deposit—this random event automatically makes him a qualified investor.

A natural person receives recognition as a qualified investor upon his application from a broker, asset manager, and a management company of a mutual investment fund, who disclose the Rules of such recognition on their websites on the Internet. It should be clarified that according to paragraph 3.4 of the Indication, the Decision to recognize a person as a qualified investor records in respect of which types of securities, and (or) derivative financial instruments, and (or) types of services the applicant is recognized as a qualified investor. Such a very narrow qualification, in our opinion, rather represents the admission of a private investor to a closed list of investment objects. In 2020, amendments concerning qualified investors were made to Federal Law No. 39-FZ of April 22, 1996, "On the Securities Market" [2]. Currently, we can say that modern Russian society contains a special social stratum of people—qualified investors with a special legal position in investment relations. At the same time, we believe that the qualification of investors is based not on the amount of money in the bank, not on the length of work experience in a certain organization, but on professional education, which is not easy and cannot be quickly obtained in higher educational institutions of the Russian Federation. In addition, in our opinion, investment transactions in the information space are a type of complex and risky entrepreneurial activity, and high qualifications (specialized higher education) are mandatory for this.

Thirdly, it is quite possible for an individual to invest sums of money without intermediaries, for example, when purchasing a package of issuer's bonds under an individual transaction. Similarly, it is also possible, without intermediaries, to acquire a block of shares by one of the shareholders of his business company. However, it should be borne in mind that such cases are isolated; there is no systematic turnover of investment objects here. In accordance with the requirements of the legislation, individuals with the participation of professional intermediaries mainly in electronic form, using electronic signatures, carry out the absolute majority of investment transactions. Intermediaries—legal entities of special legal capacity (having licenses and permits of the Bank of Russia) have been conducting business activities in the field of investments for many years, on organized platforms provided for this purpose—exchanges, financial platforms, and investment platforms. A private investor will not be able to independently enter the auction, for example, Moscow exchanges in order to make a deal with any object of investment. On the exchange's website, there is information about the very professional intermediaries admitted by this exchange to participate in the auction, where each individual can implement his plans for investing in cooperation with such an investment intermediary. Interaction is carried out based on an agreement between the parties, as a rule, with a broker, with an asset manager. Modern information technologies have saved the mentioned parties from the need to organize and hold meetings and negotiations in the office premises at the location of the legal entity. A video call has become completely ordinary, which is used easily and gratuitously even in the simplest e-mail services: @gmail.com; @mail.ru; @yandex.ru. During the pandemic of the new coronavirus infection, during 2020–2021, video services Zoom; Skype; Teams, Discord, etc. have become familiar, the use of which by the widest segments of the population has been brought almost to automatism.

In our opinion, the remote feature of the studied transactions is a positive achievement of computerization and internetization of investment relations. First of all, the circle of potential private investors is expanding indefinitely, who do not need to come to the Capital from distant places of the vast Russian Federation to meet with a professional investment intermediary and discuss the terms of an agreement on services that will be provided to him. On the contrary, a video call on the same occasion can be made even 10 times, while the costs of the parties in both time and material resources will be minimal. Let us clarify here that the participation in investment relations of a professional business entity—an intermediary of special legal capacity does not at all reduce the risks and responsibility for the decisions made for the investor. A private investor is not legally an entrepreneur and does not become a de facto entrepreneur even in cooperation with a licensed broker and/or asset manager.

Fourth, we believe that the remote version of the video communication of a private investor with the "sharks" of the investment sphere has a positive effect on the process of forming his will, which should be completely free and conscious. We agree with Yu. V. Kholodenko, according to whom the expression of will is an external expression of will in the form prescribed or permitted by law, making the will of the subject accessible to the perception of other persons [3, p. 37]. A private investor, being at a distance and negotiating with an investment intermediary in a video way, undoubtedly has a more balanced mental state, because "houses and walls help". He (she) can, without hesitation and false modesty, ask many times more questions of interest to him (her) and put forward favorable conditions for himself than being in the premises of the negotiation office of the organization. Modern Internet technologies provide for recording the process of video communication of the parties. This allows you to create an individual video in which to record that the visible expression of will accurately and completely reflects the inner will of each of the subjects, that the subjects saw it and took note of it. By agreement between them, the specified video can be recognized as proof of the free and conscious expression of the will of the subjects at the conclusion of the transaction.

Fifth, the legal purpose of the transaction in the information space is to ensure that in the process of translating an entrepreneurial idea into reality and bringing the investor's funds into a dynamic state, a certain investment object comes into a dynamic state and, following the results of the transaction, becomes the property of the investor. The recipient of the investment becomes an obligated person to him, the investor who owns the investment object during the planned period becomes a creditor, a business participant, and a founder of a trust management. Thus, the planned legal result of the transaction is achieved: establishment, possible agreed modification, termination within the stipulated period, or termination of civil rights and obligations agreed before the deadline.

We believe that investing in the information space is the implementation of practical actions in the information and telecommunications network "Internet" to invest capital, namely on the use of objects of civil rights (property) that have a monetary value for profit and (or) increment of value in the process of implementing a certain entrepreneurial idea that has an external expression of the transaction.

The term investment transaction does not exist in Russian legislation. In Federal Law No. 211-FZ of July 20, 2020, "On Financial Transactions using a Financial Platform", as part of the basic concepts of the law, the legislator discloses the definition of "financial transaction". These are transactions specified in the law for the provision of financial services provided for by the rules of the financial platform (banking, insurance, securities market services, transactions with financial instruments) made between financial organizations or issuers and consumers of financial services. The consumer here is an individual who has joined the service agreement, which is provided by the operator of the financial platform for making financial transactions with financial organizations and issuers using the financial platform [4].Unfortunately, in the specified normative act, the legislator does not fully disclose the key definition. Since the explanation "financial transactions are transactions for the provision of certain services" reflects only the nature of the process during which services are provided to the consumer. Nevertheless, what exactly these actions of the parties are, what their legal purpose and basic conditions are for implementation are not disclosed.

Differentiation by types of remote digital investment transactions in the information space is expedient for several reasons.

Firstly, transactions differ depending on the level of preparation of investors; it is necessary to distinguish transactions of qualified investors and those of unqualified investors. In December 2020, the Bank of Russia recommended professional investment intermediaries to not make transactions with complex and risky financial instruments with retail investors before the introduction of mandatory testing of unqualified investors. According to the alarming monitoring of the Bank of Russia, investment intermediaries offer citizens to invest in various bonds, the amount of payments which depends on specific circumstances (for example, on changes in the price of several underlying assets). They also offer to invest in other complex or structural instruments, as well as combined products, where an investment instrument is purchased simultaneously with the conclusion of a bank deposit agreement. Some intermediaries deliberately mislead individuals by presenting distorted characteristics of these investment objects, hiding the consequences of their acquisition or early sale. Thus, citizens replace savings products with investment objects contrary to their risk profile and the purposes of placing funds.

It is true that the Bank of Russia considers it unacceptable to offer complex investment instruments with high investment risk or difficult conditions for determining the amount of income on them, the presence of advantage, or containing a foreign element to people who do not have special knowledge and have no investment experience. In the list not recommended for unqualified investors are derivatives and complex bonds, trust management strategies with investments in such objects, trading with advantage, and combined deposits.

Implementing a preventive approach to possible problems in the investment sphere, the Mega-Regulator supplemented Federal Law No. 39-FZ of April 22, 1996, "On the Securities Market" with Article 51.2–1. From April 1, 2022, mandatory testing of an individual—an unqualified investor will be introduced in accordance with the procedure established by the basic standard for protecting the rights and interests of investors. We regard the decision on testing as an event "for show", which in the process of a formal and ostentatious educational program does not at all ensure the appearance of knowledge about the investment sphere and the qualifications of an investor in an individual. In paragraph 4 of Article 51.2 of Federal Law No. 39-FZ of April 22, 1996, "On the Securities Market", the legislator discloses the characteristics of a qualified investor—this is an individual who meets one of the five requirements specified in the law. It should be noted with satisfaction that in 2019, in particular, requirement five was changed; now a citizen is recognized as a qualified investor by the level of his education: namely has an education or qualification in the field of financial markets established by the regulations of the Bank of Russia, confirmed by a certificate of qualification.

Secondly, transactions differ depending on the counterparty; it is necessary to distinguish transactions with commercial banks, transactions with investment intermediaries—non-financial organizations, and transactions with recipients of investments. At the beginning of 2021, there were 464 professional participants in the financial market of the Russian Federation. The number of investors in brokerage services in 2020 increased 2.3 times compared to 2019, to 9.9 million people (8.4 million people—commercial banks; 1.5 million people—non-financial organizations). In this increase, credit institutions added 85% of new investors (in 2019—73%). It is significant that the majority (about 70%) of clients in brokerage services are active users of mobile applications younger than 40 years [5]. At the same time, the leading operators, for example, on the Moscow Exchange in terms of the volume of client transactions are investment intermediaries—non-financial organizations: LLC "Brokerage company REGION"—more than 4,144 trillion rubles; Financial group "BrokerCreditService"—more than 2.174 trillion rubles; Renaissance Broker LLC—more than 1.125 trillion rubles, etc. VTB Bank is in the 4th place, SBER is in the 8th place, and VEB.RF is in the 10th place [6].

In the conditions of total immersion of a modern person in the information space, both through a computer monitor and through a mobile smartphone, it is natural to appeal to him as a private investor, a variety of persons—initiators of development projects. For example, on the website of Maxim Arkadyevich Seryakov's sole proprietor, we see a sentence: "Get 22–30% per annum of fixed passive income with monthly payments. Invest from 500,000 rubles in profitable businesses operating for 3 years with an annual revenue of 15–500 million rubles, verified by Seryakov/Investments. It should be noted that without a license (an individual entrepreneur and cannot have one), the said individual entrepreneur acts de facto as an investment intermediary. Consider his message on the website: "investors invest money directly in businesses, and companies pay us for completed transactions—either a share in the business or a commission from completed transactions. Today we have 7 companies in our portfolio, including a network of 47 women's clothing stores with an annual turnover of more than 500 million rubles and a 9-year history; clothing and beverage companies; dental network; commercial real estate with its own assets of 200 million rubles; vegetable supplier in "Vkusville" and "Magnit". Every month we receive more than 200 applications for financing from businesses and in total there are more than 6000 projects in our "funnel" looking for investments" [7]. In our opinion, such proposals may well be relevant, reasonable, and profitable, but investing, as noted above, is a type of difficult and risky business activity. Moreover, suggestions like "make a money transfer to us in debt, we will return everything to you on time and with interest" without deep professional study to make a positive decision are unacceptable.

Thirdly, transactions differ depending on the object of investment. We distinguish two groups of objects in this regard. Group I delimits the real objects of the material world. Investment recipient presenting one or another legally justified instrument for the investor put up received money in producing material goods, performing works, providing services, granting property rights. We include 13 such investment objects in the composition of the most common and most attracting the attention of private investors. Government securities of the Russian Federation (subjects of the Russian Federation); municipal securities; shares and bonds of Russian issuers; government securities of foreign states; shares and bonds of foreign issuers; Russian depositary receipts, as well as foreign depositary receipts for securities; investment units of mutual investment funds, as well as units (shares) of foreign investment funds; mortgage certificates of participation. For example, in March 2021, MTS PJSC fully placed bonds of the 001P-18 series with a 3-year maturity of 4.5 billion rubles. As part of the state contract, MTS must connect 4995 socially significant objects to the Internet—the cost budget is about 4.9 billion rubles. MTS will equip communication centers in settlements and provide the opportunity to connect everyone to them. According to MTS PJSC, the volume of the bond issue will not exceed the project cost budget. All funds from the bond issue will be used to finance the costs of connecting schools, institutions of secondary vocational education, medical and obstetric centers, state and local government bodies, military enlistment offices, election commissions, police stations, and fire stations in the Russian Federation to the Internet [8].

Group II distinguishes between electronic-virtual and crypto-objects [9]. The recipient of the investment, providing one or another legally sound instrument to the investor, invests the money received in certain ideas (contracts concluded at organized auctions that are derivative financial instruments), in cryptographic records—monetary surrogates (computer fiduciary "money") [10]. For example, the Moscow Exchange has permanent services for investing in contracts that provide for profit in the event of a change in the value of the so-called basic asset. The investor can choose to invest money in several ideas, for example: (1) the Moscow Exchange Index will increase/decrease by some amount; (2) the Blue-Chip Index will increase/decrease;

(3) the degree of volatility of the Russian market will increase/decrease; (4) the value of ordinary shares of PJSC Aeroflot will increase/decrease; (5) the value of ordinary shares of PJSC MTS will increase/decrease; (6) the value of "fifteen-year" federal loan bonds will increase/decrease; (7) the Euro/US dollar exchange rate will increase/decrease; (8) the credit MosPrime interest rate; (9) the price of Brent crude oil will increase/decrease; (10) the price of natural gas will increase/decrease; (11) the price of grade 4 wheat will increase/decrease, etc. [8].

30.4 Conclusion

A distance digital investment transaction in the information space is an intellectual and volitional behavioral act of capable, legally capable subjects (investor, investment intermediary, recipient of investments), forming a legal fact in their mutual contact and expression of will in the information and telecommunication network "Internet" by means of special software and hardware, expressed in an electronic form regulated by law. The legal purpose of their contact and expression of will—the receipt of the object of investment into the investor's property, funds to the recipient of the investment—is achieved and forms legal consequences: establishment, possible agreed change, within the stipulated time or termination of civil rights and obligations agreed before the deadline The essence, the internal content of a remote digital investment transaction, is a certain entrepreneurial idea, without which investing in the information space is impossible.

The types of remote digital investment transactions in the information space should be differentiated into three types: First, depending on the level of professional training of investors (qualified and unskilled investors); second, depending on the counterparty (commercial banks, investment intermediaries—non-financial organizations, direct recipients of investments); third, depending on the object of investment (real objects of the material world, electronic-virtual and crypto-objects). This typology should be taken as a basis for the development of similar subject blocks of an educational program for the training of private investors. The final mandatory certification of a citizen with a positive result should be completed by assigning the OKVED-2 code 64.99.1 (investments in stocks, bonds, promissory notes, securities of joint-stock funds and mutual investment funds, etc.), as well as granting the right of state registration in the status of an individual entrepreneur admitted to investing (with the issuance of an electronic document stored in his personal account on the website of GOSUSLUGI). It is necessary to make additions to the relevant legislation.

Acknowledgements The work was supported by Russian Science Foundation (project No. 20-18-00314).

References

1. "On the recognition of persons as qualified investors and the Procedure for maintaining the Register of persons recognized as qualified investors": Indication of the Bank of Russia. 29.04.2015 No. 3629-U. Bulletin of the Bank of Russia 15.06.2015. No. 51 (2015)
2. "On the Securities Market": Federal Law No.39-FZ of 22.04.1996. Rossiyskaya Gazeta. 25.04.1996. No. 79 (1996)
3. Kholodenko, Yu.V.: Invalidity of transactions with defects of will: problems of legal regulation Lambert Academic Publishing Saarbrücken 277 (2011)
4. "On Making Financial Transactions using a Financial Platform": Federal Law No.211-FZ of 20.07.2020. Rossiyskaya Gazeta. 24.07.2020. No. 162 (2020)
5. Overview of key indicators of professional participants of the securities market. Information and analytical material, 4 (2020). http://www.cbr.ru/collection/collection/file/32068/review_secur_20.pdf
6. The official website of Moscow Stock Exchange: Customer statistics for 2021 (2021). https://www.moex.com/s719
7. Individual entrepreneur Maxim Arkadyevich Seryakov (2021). https://www.seryakovinvest.ru/
8. The official website of Moscow Stock Exchange. Information on futures contracts and options (2021). https://www.moex.com/ru/derivatives/select.aspx
9. Inshakova, A.O., Goncharov, A.I., Ershova, I.V.: Digital blockchain registration of legally significant stages of complex good's export-import supplies by business entities of the EAEU and BRICS jurisdictions. In: Inshakova A., Inshakova E. (eds.) Competitive Russia: Foresight Model of Economic and Legal Development in the Digital Age. CRFMELD 2019. Lecture Notes in Networks and Systems, vol. 110, pp. 328–336. Springer, Cham (2020). https://doi.org/10.1007/978-3-030-45913-0_37
10. Inshakova, A.O., Goncharov, A.I., Salikov, D.A.: Electronic-digital smart contracts: modernization of legal tools for foreign economic activity. In: Lecture Notes in Networks and Systems, vol. 91, pp. 3–13 (2020)

Chapter 31
Remote Registration of International Mortgages in the Context of Digitalization of Banking Services

Olesya P. Kazachenok

Abstract The dynamics of scientific and technological development of modern society is characterized by high rates of development, accompanied by digital algorithmization of the economy and acts as a catalyst for the emergence of new social relations and the transformation of the existing ones. The paper examines the digitalization of banking services and the possibilities of concluding international mortgage transactions in a remote format in the context of the transition to Industry 4.0, as well as the gaps in the current legislation in the field of identification that do not allow the implementation of remote banking services in full. It is concluded that in order to ensure the possibility of concluding international mortgage transactions in a remote format, it is necessary to form new approaches to identification, which, on the one hand, must comply with the international principles of countering money laundering and terrorist financing, and on the other hand, increase their accessibility and simplicity for users, including individuals. In this area, the use of biometric data and the formation of a bank of identifiers in "digital profiles" is seen as promising.

31.1 Materials

The scientific development of the material is carried out on the basis of a set of normative and doctrinal sources. The paper uses the federal laws and other normative acts of the Russian Federation: Federal Law No. 115-FZ "On Countering the Legalization (Laundering) of Criminally Obtained Incomes and the Financing of Terrorism" of August 7, 2001, Federal Law No. 476-FZ "On Amendments to Federal Law "On Electronic Signature" of December 12, 2019, Decree of the Government of the Russian Federation No. 566 "On Approval of the Rules for reimbursement to credit and other organizations of lost income on housing (mortgage) credits (loans) issued

O. P. Kazachenok (✉)
Department of Civil and Private International Law, Volgograd State University, 100 Prosp Universitetsky, Volgograd 400062, Russia
e-mail: o.kazachenok@yandex.ru

A. O. Inshakova and E. I. Inshakova (eds.), *New Technology for Inclusive and Sustainable Growth*, Smart Innovation, Systems and Technologies 288,
https://doi.org/10.1007/978-981-16-9808-8_31

to citizens of the Russian Federation in 2020" of April 4, 2020, and the information letters of the Central Bank of the Russian Federation.

As the policy documents of a strategic and advisory nature, dedicated to civil law in the digital economy, we studied Doctrine of Information Security of the Russian Federation (appr. by Decree of the President of the Russian Federation No. 646 of December 5, 2016); the Strategy for the Development of the Information Society in the Russian Federation for 2017–2030 (appr. by Decree of the President of the Russian Federation No. 203 of May 9, 2017); the State program of the Russian Federation "Information society (2011–2020)" (appr. by Resolution of the Government of the Russian Federation No. 313 of April 15, 2014); the Program "Digital Economy of the Russian Federation" (appr. by Decree of the Government of the Russian Federation No. 1632-r of July 28, 2017), etc.

The doctrinal sources are represented by the scientific publications of the domestic lawyers, including Goncharov A. I., Inshakova A. O., Khomenko E. G., Ruzakova O. A., and others.

31.2 Methods

The content of the paper is developed on the basis of the materialistic worldview and the general scientific method of historical materialism. The general scientific methods of cognition are applied: dialectical, hypothetical-deductive method, generalization, induction and deduction, analysis and synthesis, and empirical description. The study also uses specific scientific methods: normative-dogmatic, comparative-legal, structural–functional, etc.

31.3 Introduction

A powerful impetus to the transition to Industry 4.0 economy was the so-called "coronavirus crisis" caused by the spread of the COVID-19 coronavirus infection in early 2020. The resulting isolation and significant restrictions on the movement led to a sharp increase in demand for remote satisfaction of public and individual needs, which prompted the operators of a wide range of services that were previously not provided without a full-time presence, to urgently introduce digital technologies and reorient business processes.

Banking services were no exception and in the context of the pandemic, banks began to actively force a significant expansion of the range of services that can be obtained in a remote format.

Within the territory of the Russian Federation, the information letters from the Central Bank of the Russian Federation provided an opportunity to issue a remote mortgage by simplifying the procedure for identifying the borrower, which contradicted the current legislation regulating the fight against money laundering and terrorist financing.

Despite the fact that the possibility of remote mortgage registration appeared within the territory of individual states, the conclusion of international mortgage transactions in the context of the coronavirus crisis was actually stopped.

In the transition to Industry 4.0., the need for a remote banking service that allows you to conclude the transaction, including across jurisdictions, becomes obvious, but for the development of the current mechanism one has to resolve the question of online identification of a person entering into a transaction which, on the one hand, would allow obtaining international mortgages remotely and, on the other hand, ensure complying with the international standards of legalization (laundering) of incomes obtained in a criminal way and financing of terrorism (AML/CFT).

31.4 Results

In the conditions of a market economy, the rapid development of civil turnover, increasing the movement of financial flows both within the country and abroad, there is a need to find the optimal tool that contributes to increasing the pace of socio-economic development of this country. One of these tools is the mortgage. In the context of the pandemic, the demand for mortgage products did not decrease, but there were significant problems of legal registration associated with the need for the repeated personal presence of the borrower in the bank, which in some cases became an insurmountable obstacle to the registration of mortgages, especially the international ones.

31.4.1 The Legal Basis of International Mortgage Lending

The credit resources guaranteed by real estate collateral provide an inflow of investments into the production sector, form a multi-level fictitious capital in the form of mortgages and mortgage-backed securities, increase payment demand in the real estate market, expand the opportunities for business development, and contribute to solving the problem of providing citizens with housing. This is the economic function of a mortgage.

A significant role in providing the population with affordable housing is played by mortgage housing lending, which is an alternative to the previously existing possibility of obtaining public housing.

Since the 2000s, mortgages have been rapidly developing and gaining popularity among the population.

The single development institution in the housing sector is the Agency for Housing Mortgage Lending (AHML), whose tasks include supporting the development of mortgage lending at the state level, as well as the secondary mortgage lending market.

In the ranking of the leading mortgage banks in Russia, three banks hold leading positions: Sberbank, VTB 24, and Rosselkhozbank.

Mortgage lending is one of the most important forms of bank credit. It should be noted that in this context we proceed from a broad interpretation of the term mortgage lending.

Despite some positive trends in the development of the institution under consideration, the need for its reform does not fade into the background. This is especially true in the sphere of legislative regulation. Thus, the practice of foreign countries indicates an increased interest in this issue [1].

In this part of the study, we will turn to various legislative approaches implemented in the regulatory framework of mortgage lending in foreign legal systems for comparison with the methodology of the national legislator.

If the relationship under the mortgage agreement is complicated by a foreign element, we should already talk about international mortgage, which is a promising area of banking practice.

The essence of international mortgage lending is as follows: the borrower, receiving funds under the loan agreement, acquires a real estate object outside of his country, while he is burdened with a mortgage, and another real estate object that is already owned by the borrower can also be provided as collateral.

International mortgage lending is a rather controversial topic. Even in the statistics of foreign countries, international mortgages do not stand out as an independent segment.

The attractiveness of buying real estate abroad is also due to the high interest rates on mortgage loans in Russia. Despite the fact that some banks, and in particular, the leading bank of mortgage lending Sberbank, make decisions to reduce interest rates, we believe that this measure will not completely reduce the attractiveness of buying foreign real estate with a mortgage.

In addition to the above, this fact is also explained by the fact that real estate and infrastructure abroad and, in particular, in Europe are of higher quality than the similar ones in the Russian Federation.

Currently, depending on various factors that characterize both the borrower and the subject of the mortgage, the initial payment can reach up to 60%, and credit funds are provided on average for 10–15 years, no more, provided that the borrower confirms his solvency and provides information about his regular expenses (rent, alimony, payments on other loans, etc.).

The European countries can be divided into several groups depending on the possibility of providing real estate to non-resident Russian citizens and the complexity of the procedure for obtaining a mortgage loan:

(1) countries where obtaining a mortgage is not possible for Russian citizens (Bulgaria, Romania, Montenegro, Switzerland, etc.);

(2) countries where a mortgage is available, but the procedure for obtaining it for Russian citizens is much more complicated (England, Italy, and Austria);
(3) countries where Russian citizens are granted a mortgage if certain requirements are met (Germany, France, Turkey, and Latvia);
(4) countries where it is difficult for Russian citizens to get a mortgage due to the unstable economic situation (Italy, Greece, and Croatia);
(5) countries where it is possible for Russian citizens to obtain a mortgage loan on terms similar to those established for citizens of the country (Spain, Cyprus, Portugal, Finland, Israel, the United States, and Belgium).

In our opinion, international mortgage lending is a type of mortgage lending that assumes that the main participants in the transaction and the real estate object fall within different jurisdictions.

31.4.2 The Classification of International Mortgage Lending Transactions

We will classify international mortgage lending transactions vertically, using such a criterion as the number of jurisdictions:

- a one-tier transaction in which the borrower and the bank are located in one country, and the subject of the mortgage in another one [2];
- a two-tier transaction (the borrower has a permanent residence (or citizenship) in one country, and the bank and the subject of the mortgage are located in another country);
- a three-level transaction (the borrower, bank, and mortgage object are located in different countries).

Of course, each of the countries in which a potential borrower wants to buy real estate has special requirements for them, which in turn complicates the conclusion of transactions and leads to an increase in costs.

In general, the mortgage market is a segment of the financial market, in particular, the loan capital market; it is obvious that the international mortgage lending market is also part of the global loan capital market.

At the moment, international mortgage lending is developing dynamically. Unlike Russia, the G7 countries have considerable experience in supporting transactions with foreign real estate.

In our opinion, international mortgage lending is characterized by a number of advantages.

The first of them is a long-term loan. The second advantage is due to the fact that mortgage lending is not subject to the market conditions, since credit resources are formed by attracting the savings of future borrowers on the principle of mutual assistance cash, regardless of the stock markets.

It seems that the latter should be applied in the Russian banking practice.

The main direction of using the funds received by the borrower under an international mortgage will be the purchase of real estate for personal residence. However, this is not the only direction of using the borrowed funds.

The second direction is investment, when the borrower purchases real estate abroad in order to invest money for further income generation, for example, by renting it out or using it for tourist purposes. According to research, 63% of the surveyed realtors believe that among Russian-speaking investors, apartments and houses for rent are in the greatest demand.

31.4.3 Remote Mortgage as a New Banking Product

Industry 4.0 involves the remote control of processes, making transactions, receiving not only the state and municipal, but also banking services.

One of the significant innovations that reflects the general trends of the digitalization of banking services, but did not actually exist before the coronavirus crisis, was remote or online mortgage.

The rapid pace of digitalization of this sphere in the Russian Federation is confirmed by the fact that just three years ago, in 2018, Elena Chaikovskaya, the adviser to the First Deputy Chairman of the Bank of Russia, speaking about remote mortgages, pointed out that "at the moment it looks close to fiction" [3].

In October 2019, the Ministry of Communications of Russia prepared a Draft Decree of the Government of the Russian Federation on the experiment on the remote use of a reinforced qualified electronic signature (QES), which allows, in particular, issuing the electronic agreements of purchase and sale of real estate with a mortgage, while using the funds of the credit institution using the platform "Masterchain", based on the technology of distributed registries.

It should be said that earlier Russian banks announced the provision of services for issuing online mortgages, however, in its essence, this service could not be called "remote" in the full sense of the word.

The process of issuing the so-called "remote mortgage" took place in two stages:

- a preparatory one, which takes place in remote format and is carried out in the client's personal account on the bank's website or in a mobile application, including filing an application, calculating payments, and scoring the borrower.
- a legal one, consisting, in fact, of the conclusion of the transaction, carried out in the traditional legal form.

Thus, previously, it was the organizational, not legal, actions that were transferred to a remote format. The number of visits to the bank by the borrower was reduced to one, but it was necessary. That, in the conditions of border closures and mass lockdowns, made it impossible to issue an international mortgage.

With the onset of the pandemic and sharp currency fluctuations, Russian banks recorded an increased demand for housing loans in the conditions of self-isolation and, in this regard, began to maximize the possibilities of the remote mortgage registration service.

The demand of citizens for mortgage lending also increased in connection with Decree of the Government of the Russian Federation No. 566 of April 2020, which established an extremely low interest rate on housing loans in the Russian Federation—6.5% per annum.

31.4.4 The Identification of Persons Entering into Mortgage Transactions

It should be noted that granting remote mortgages is still limited by the current legislation.

The Federal law "On counteraction to legalization (laundering) of incomes obtained in a criminal way, and terrorism financing" does not allow opening accounts (deposits) of clients without the personal attendance of the individual opening the account (Deposit) or the client's representative.

However, the Central Bank of the Russian Federation has taken significant steps to mitigate this requirement with a number of information letters.

By Information Letter of April 10, 2020, it is established that for the period up to July 1, 2020, to apply penalties against the credit organizations for violations by the credit organizations of the requirements of sub-paragraph 3 of paragraph 5 of Article 7 of the Federal Law "On Countering the Legalization (Laundering) of Criminally Obtained Incomes and the Financing of Terrorism", regarding the prohibition to open accounts (deposits) to customers without the personal attendance of the individual opening an account (deposit), or a representative of the client, only if the Bank of Russia reveals the facts of opening such accounts by the credit institutions on the basis of false identification information.

On April 17, 2020, by the Information Letter, the Central Bank of the Russian Federation introduced similar eases for small and medium-sized businesses, but with a targeted restriction: they receive a loan for urgent needs to support and maintain employment [4].

The banks reacted quickly to the message of the regulator, in particular, the bank DOM. The Russian Federation has announced the beginning of electronic transactions with mortgages, when all documents are signed by customers with an enhanced qualified electronic signature using a smartphone. According to the bank, at the moment such transactions are carried out in Moscow, St. Petersburg, Samara, Nizhny Novgorod, and Chelyabinsk. Sberbank has also announced the granting of the first remote mortgages [5].

The Information letters of the Central Bank of the Russian Federation do not have the force of regulations, since by virtue of Article 7 of the Federal Law "On the Central Bank of the Russian Federation", the Bank of Russia issues regulations only in three forms: instructions, regulations, and directions [6].

The coronavirus crisis motivated the Central Bank of the Russian Federation to go to extreme measures. No changes were made to the Law on Legalization, that is, the legislative ban on opening accounts without personal attendance still remains. But its effect was actually canceled by a strong-willed decision of the Central Bank of the Russian Federation, which explained that the measures of responsibility for these violations will not be applied and de facto legalized the remote opening of accounts and the establishment of the will of the recipients of banking services [7].

The recognition of remote will will take place in the event that the technical means used for expressing will allow you to reproduce the content of the expression in a tangible medium, and determine the person who expresses the will, including using biometric identification, electronic signature, and so on [8].

Thus, in making any remote banking transactions, the most important is the issue of proper identification. The priority is to establish a balance between the desire for development in the information society and the need to ensure proper identification to eliminate the risks of conducting banking operations for illegal purposes and committing fraudulent actions [9].

Thus, the cases of fraud in the real estate sector using a qualified electronic signature are already being recorded. The fraudsters re-registered the ownership of the apartment through the portal of Public Services, using a fake electronic signature [10].

At the present stage, the legislator in the Russian Federation has implemented the possibility of electronic interaction in the digital environment between all participants in legal relations by introducing such a concept as a single enhanced qualified electronic signature (CEP) into the regulatory framework [11].

An enhanced qualified electronic signature is confirmed by a certificate from an accredited certification center and in all cases is equated to a paper document with a "live" signature, except in cases where the law explicitly provides for the obligation to comply with the written form of the document.

But in practice, not only individuals and legal entities, but even public authorities have faced the problem of using an electronic digital signature in the information systems of various departments in connection with the requirement for a single certificate of the key for verifying the enhanced qualified electronic signature of the object identifiers (OID) of certain powers assigned to the user within a specific information system.

Due to the lack of the required identifiers in individual information systems, certificates issued by accredited certification centers (CC) may not always be used to verify an electronic signature. This restriction has led to the fact that certification authorities are forced to offer users qualified certificates designed to work with specific organizations [12].

On April 10, 2020, the Central Bank issued recommendations designed to overcome this limitation. Starting from April 2020, if the client has a valid qualified certificate that was issued by an accredited CC in force at the time of issue, then the credit institutions should not issue a qualified certificate from the servicing bank for their client [13].

Since today the statutory regulation of the use of electronic signature does not correspond to the current stage of development of this technology, all this systematically led the legislator to make changes to the federal legislation on the electronic digital signature, which are necessary for civil turnover.

On July 1, 2020, Federal Law No. 476-FZ of December 27, 2019 "On Amendments to the Federal Law "On Electronic Signature" came into force in the Russian Federation.

The amendments are designed to create a single space of trust for electronic signatures [14].

The Federal Law contains a number of provisions designed to overcome the problem faced by individuals, legal entities, and authorities by regulating the procedures for establishing and verifying the authority of these entities using electronic signatures in various information systems and ensuring the reliability of the identification of these entities using an electronic service that would provide electronic interaction [15].

The legislator introduced the concept and mechanism of using the so-called trusted time stamp, that is, reliable information in electronic form about the date and time of signing an electronic document with an electronic signature.

It also provides for the introduction and use of machine-readable powers of attorney for the purpose of confirming the authority in the digital environment [16].

A proposal to introduce a new legal institution—the institution of trusted third parties—is also under discussion. Such an institution will be the organizations authorized by the state authorities to verify electronic signatures in electronic documentation in the digital environment, as well as to implement confirmations in an electronic format of the results of such verification [17].

The introduction of biometric authentication technologies, which is a global trend, is actively discussed. The biometric platform uses two parameters for identification—a voice profile and a photo image. According to the expert estimates, it is the banking biometrics market that will show the highest growth rates for the period 2018–2022 [18]. Currently, there is an active development of the regulatory and technical base of biometric technologies [19].

It is quite possible that biometric identification technologies will take a dominant position, replacing the digital ones, with the accumulation of data and the improvement of technical systems for recognizing biometric parameters; biometric identifiers will become increasingly important, capturing new niches in banking services [20].

Since 2019 under consideration there has been the Draft Decree of the RF Government "On introducing amendments to the Statute on the Federal state information system "Single portal of public and municipal services (functions)", providing for the formation of the bank identifiers in the "digital profiles" of the citizens (and later probably and legal persons) that will give them the opportunity to trade by means of the infrastructure of the Single portal of public and municipal services.

31.5 Conclusion

Summing up, it should be noted that the coronavirus crisis has become a catalyst for the earlier structural changes in the banking system caused by the transition to Industry 4.0, significantly accelerating the process of digitalization of banking services, which included even the services that were previously impossible to obtain remotely, in particular, obtaining a mortgage loan.

However, it is currently impossible to obtain an international mortgage in a remote format, which, given the limited international communication in the context of the coronavirus crisis, has actually stopped the system of international mortgage lending.

International mortgage lending is a rather promising banking product. For a person who wants to purchase a property abroad with a mortgage, it is necessary to meet a number of certain requirements imposed on both the borrower and the purchased property, depending on the specific country.

International mortgage lending transactions can be structured as one-, two-, and three-level. In general, international mortgage lending has significant advantages, acts as an effective mechanism for solving housing issues of citizens, is and a way to obtain additional income in the case of purchasing foreign real estate for investment purposes.

The realities that the banking sector of the Russian Federation faced in the first half of 2020 pointed to the urgent need to introduce a broader form of information technology application to meet the needs of citizens of the Russian Federation and support the country's economy.

The Central Bank of the Russian Federation in the crisis conditions took a number of urgent measures, including temporary blocking of the legislative restrictions on opening accounts without the personal attendance of citizens. The legal requirement that, where there is a legal prohibition and punishment of the person authorized for monitoring its compliance, their willful decision to announce that the liability in a particular period will not apply, can be justified by the need for urgent action, however, creates a precedent, unacceptable in a legal state.

In the Russian Federation, it is necessary to amend the current legislation, namely Federal Law No. 115-FZ "On Countering the Legalization (Laundering) of Criminally Obtained Incomes and the Financing of Terrorism" (AML/CFT) of August 7, 2001, regarding the possibility of opening accounts (deposits) to customers without the personal attendance of an individual.

It is also necessary to review the legislation in the field of AML/CFT with the establishment of a certain level of powers granted to the authorized bodies for rapid changes in the regulation in this area, including in order to form and provide a legal basis for the inevitable and rapid process of digitalization of banking services. It is possible to establish the boundaries of the legal field within which the regulation will be carried out by the regulatory acts of the Central Bank of the Russian Federation, which will allow for a more flexible and timely response to changing circumstances, which does not contradict the current legislation.

The formation of the legal regulation of digitalization of banking services is impossible without fixing new approaches to identification, which, on the one hand, should protect as much as possible from fraudulent actions, and, on the other hand, increase its accessibility and simplicity for users, including individuals. In this area, the use of biometric data and the formation of a bank of identifiers in "digital profiles" are seen as promising.

Acknowledgements The study was carried out with the financial support of the Russian Science Foundation (project No. 20-18-00314).

References

1. Grishina, Y.S.: Public guarantee as a legal means of property support and ensuring the obligation to bank credit of the subject of social entrepreneurship. Bank Law **5**, 16–22 (2015)
2. Kazachenok, O.P.: Legal concept of international mortgage lending in the Russian Federation and European countries. Vestnik Volgu. Series 5. Jurisprudence **3**(32), 38–44 (2016)
3. Mekhtiev, E.: New regulatory norms for the development of e-finance. Bank Rev. **9**, 46–49 (2018)
4. Financial Pulse. Weekly Analytical Review of April 10. http://www.cbr.ru (2020). Accessed 01 Sept 2021
5. Sberbank Began to Issue Mortgages at Home. https://realty.rbc.ru/news/5e959e969a7947348a7e384c (2020). Accessed 03 Sept 2021
6. Sandalova, V.A., Kochubey, T.A. Mescheryakova, Y.V.: Civil turnover of objects by the intellectual property in conditions of digital intensification. In: Competitive Russia: Foresight Model of Economic and Legal Development in the Digital Age. Proceedings of the International Scientific Conference in Memory of Oleg Inshakov (1952–2018), pp. 531–536, Springer Nature, Switzerland (2020)
7. Tarakanov, V.V., Inshakova, A.O., Dolinskaya, V.V.: Information society, digital economy and law. In: Ubiquitous Computing and the Internet of Things: Prerequisites for the Development of ICT, pp. 3–15. Springer Science + Business Media, Luxemburg, Cham (2019)
8. Ruzakova, O.A.: The issues of digitalization of rights in the banking sector in accordance with the federal law "on amendments to parts one, two and article 1124 of part three of the civil code of the Russian Federation. Banking Law **4**, 29–35 (2019)
9. Kazachenok, O.P.: Non-compliance of the current legislation with the needs of the information society on the example of a loan of electronic money. Laws Russia Exp. Anal. Practice **10**, 47–50 (2017)
10. Electronic Theft: In Russia, for the First Time, an Apartment was stolen via the Internet. https://news.mail.ru/economics/37375662/?frommail=1 (2020). Accessed 01 Sept 2021
11. Kazachenok, O.P., Lavrentyeva, A.V.: New digital payment facilities: an analysis of institutional and legal regulation. In: Ubiquitous Computing and the Internet of Things: Prerequisites for the Development of ICT, pp. 333–340. Springer Science + Business Media, Luxemburg (2019)
12. Inshakova, E.I. Ryzhenkov, A.Y., Inshakova, A.O.: Neo-industrialization of the russian economy: technological and digital development. In: Ubiquitous Computing and the Internet of Things: Prerequisites for the Development of ICT. Studies in Computational Intelligence, vol. 826, pp. 239–250. Springer Science + Business Media, Luxemburg (2019)
13. Chedzhemov, S.R., Zoloeva, Z.T.: Biometric personal data of a person—a means of protecting his rights or total control of the State over the individual. Const. Munic. Law (11), 39–43 (2019)
14. Khomenko, E.G.: From traditional money transfers in the national payment system to the digital ones: features of the legal regulation. Business Law **3**, 17–19 (2019)

15. Inshakova, A.O., Goncharov, A.I., Ershova, I.V.: Digital blockchain registration of legally significant stages of complex good's export-import supplies by business entities of the EAEU and BRICS jurisdictions. In: Competitive Russia: Foresight Model of Economic and Legal Development in the Digital Age: Proceedings of the International Scientific Conference in Memory of Oleg Inshakov (1952–2018), pp. 328–336. Springer Nature, Switzerland (2020)
16. Matytsin, D.E., Rusakova, E.P.: Strategy of quality management in industry 4.0 and formation of cognitive economy based on industrial and manufacturing engineering in the Russian Federation and countries of the EU. Int. J. Qual. Res. (2020). https://doi.org/10.24874/IJQR15.04-03
17. Matytsin, D.E.: Digital technologies for the implementation of civil law transactions: a repo agreement for investment assets of the securities market. Pravo i praktika **2**, 136–140 (2020)
18. Tarasenko, O.A.: digital transformation of banking and payment systems of Russia. Business Law **3**, 3–10 (2019)
19. Lisachenko, A.V.: Identification of subjects of civil law relations: new problems and some solutions. Russ. Law J. **5**, 91–99 (2019)
20. Kazachenok, O.P.: Legal regulation of the legal aspects of computing circulation in the list of measures to increase the competitiveness of the Russian economy in the information society. In: Competitive Russia: Foresight Model of Economic and Legal Development in the Digital Age: Proceedings of the International Scientific Conference in Memory of Oleg Inshakov (1952–2018), pp. 407–413, Springer Nature, Switzerland (2020)

Chapter 32
Investing in Utilitarian Digital Rights as a Legal Mechanism for Improving the Welfare of the Population and a Factor of Inclusive Growth

Andrey N. Sadkov, Marina V. Goncharova, and Elvira O. Osadchenko

Abstract The authors of this chapter of the monograph explore the problems of the turnover of utilitarian digital rights. These rights are considered as an investment mechanism that allows for the inflow of free funds of the population into the real sector of the economy. This inflow of money contributes to achieving the target of protecting the savings of the population from inflation, and also contributes to improving the welfare of the population. According to the authors, utilitarian digital rights are an important element of improving the efficiency of small business development, allowing individuals and legal entities to gain access to markets and resources without discrimination. Such opportunities, in turn, are one of the factors of inclusive growth. The authors formulate proposals for improving Russian legislation aimed at increasing the attractiveness of using the investment structure through investment platforms.

32.1 Introduction

The progressive development of civilization is impossible without the use of new technologies that reduce organizational, time, and material costs to achieve a high standard of living in society. The search for optimal ways to determine the comfort of people's lives has necessitated the rejection of assessing the level of development of countries based on the Gross Domestic Product index for each individual person. In 2017, at the World Economic Forum in Davos, as the most correct indicator

A. N. Sadkov (✉) · E. O. Osadchenko
Volgograd State University, Volgograd, Russia
e-mail: sadkov@volsu.ru

E. O. Osadchenko
e-mail: osadchenko@volsu.ru

M. V. Goncharova
Plekhanov Russian University of Economics, Volgograd branch, Volgograd, Russia
e-mail: reu-kafedra-ekonomiki@mail.ru

A. O. Inshakova and E. I. Inshakova (eds.), *New Technology for Inclusive and Sustainable Growth*, Smart Innovation, Systems and Technologies 288,
https://doi.org/10.1007/978-981-16-9808-8_32

determining the real level of development of a country, it was proposed to use a new criterion—the Inclusive Development Index (IDI). This index takes into account not only the volume of Gross Domestic Product for each individual person, but also a number of indicators. In particular, the level of poverty, median income, the coefficient of stratification of society by income, employment of the population, the coefficient of demographic load, etc. are taken into account [1].

In the modern sense, the Inclusive Development Index (IDI) is a complex indicator calculated based on the analysis of 12 criteria combined into three groups: The first group: growth and development; the second group: continuity of generations and sustainability of development: The third group: inclusivity [2]. ID allows you to determine the degree of availability of material goods and services for various categories of the population. It is proposed to consider it as a tool used to form an economy with high employment, ensuring social and territorial cohesion [3].

Improving the level of comfort of people's lives in a particular country is impossible without the modernization of the management and production sector, the use of the latest technologies and innovations. We are in solidarity with Dmitry Medvedev's position that the more intelligent and efficient the economy, the higher the level of well-being of citizens, the freer, fairer, and more humane not only the political system but also society as a whole [4].

There is no doubt that the modernization of the economic and production sphere requires the involvement of certain resources, and above all, money, which is achieved in the process of investment. In our opinion, the use of the construction of "utilitarian digital rights" in order to invest the free funds of the population in the real sector of the economy can be considered as a legal mechanism for improving the welfare of citizens and one of the driving forces of inclusive growth.

32.2 Methodology

The conducted research is based on the application of the universal scientific method of materialistic dialectics. The authors also used general logical, general scientific, and special methods of scientific cognition. Among them are analysis, synthesis, deduction, induction, method of system research, method of structuring material, formal legal method, comparative legal method, etc.

32.3 Results

As a legal basis for the turnover of utilitarian digital rights and investment activities in Russia, first of all, the Civil Code of the Russian Federation should be noted. We will also highlight the Federal Law "On Attracting Investments Using Investment Platforms and on Amendments to Certain Legislative Acts of the Russian Federation" dated August 2, 2019, No.259-FZ (Law No.259-FZ, 2019). Of course, the regulation of investment issues using the construction of utilitarian digital rights is not limited only to the above-mentioned regulatory legal acts but should also be highlighted within the framework of the stated topic.

In accordance with the norms of Articles 8 and 9 of Law No.259-FZ, 2019, the legislator, revealing the essence of utilitarian digital law, highlights the right of an authorized person to a claim fixed in a digital certificate. This is the right to demand the transfer of a thing(s), the transfer of the exclusive right to the result of intellectual activity and (or) the right to use the results of intellectual activity, the right to demand the performance of works, and (or) the provision of services. In accordance with the norms of the current legislation, these rights are recognized as utilitarian digital rights. A condition is established here—these rights originally arose as a digital right based on an agreement on the acquisition of a utilitarian digital right concluded using an investment platform [5].

Unfortunately, the legal definition of the term "digital law", which is contained in Article 141.1 of the Civil Code of the Russian Federation [6], does not allow to clearly distinguish this right from other objects of civil rights. The analysis of public relations arising in connection with "digital rights" allows us to conclude that this is more about the form of fixing certain property rights. Therefore, we agree with the authors who consider the term "digital rights" to be incorrect and propose to introduce the phrase "digitized rights" into the legal lexicon as its replacement [7].

Inshakova and Goncharov [8], describing the specifics of utilitarian digital rights, argue that practically a crypto record is carried out in the database of the investment platform, which certifies the requirements reflected in the digital certificate. It should be noted that the legislator has imposed restrictions on the certification of certain requirements that exist as utilitarian digital rights. These are requirements concerning the transfer of property, the rights to which are subject to imperative accounting in state registers. These are the requirements concerning the transfer of property in transactions for which a mandatory notarial form is established. Alternatively, in accordance with the provisions of the law, state registration is provided for such deals.

As a person can be attracting investments using the construction of utilitarian digital rights only a legal entity whose national law is Russian law. Such a person can also be an individual entrepreneur. These persons are obliged to conclude an agreement on the provision of investment attraction services with the operator of the investment platform. On the basis of this agreement, the person attracting investments is granted access to the information platform for concluding an investment agreement. In practice, access is carried out with the help of information technologies and technical means of the corresponding information platform.

Any individual or legal person who has the appropriate amount of legal personality can act as an investor. The investor enters into an agreement with the operator of the investment platform on the provision of investment assistance services. In accordance with this agreement, the investor is granted access to the investment platform for signing the investment agreement. This agreement is signed with the help of information technologies and technical means of the investment platform.

In our opinion, the investment construction of utilitarian digital rights is an attempt by the legislator to streamline investment relations common in economically developed countries, which are called "Internet investing" [9], or "crowd-funding" [10]. The Russian legislator refused to introduce the term "crowd-funding" into the legal national legal lexicon. The Russian legislator refused to introduce the term "crowd-funding" into the legal national legal lexicon. At the same time, crowd-funding is a technology of collective financing that allows attracting a wide range of investors for the implementation of a certain project [11]. Kick-starter is considered one of the most popular international crowd-funding platforms, which has been successfully operating since 2009 [12]. Since 2012, such collective financing platforms as Boomstarter [13] and Planeta [14] have been successfully operating in Russia.

The use of the investment structure by acquiring property rights, referred to in paragraph 1 of Article 8 of the Law No.259-FZ, 2019, as "utilitarian digital rights" [5], is a variant of the legal regulation of attracting money from a wide range of retail investors to the real sector of the economy. The undoubted advantage of such an investment for an investor is its accessibility. The investor only needs to have a device that allows him to work on the Internet and identify the user. Such an investment provides a wide range of objects for investment. There is an obligation of the person attracting investments to transfer things, the results of work, to provide services, to grant exclusive rights to the results of intellectual activity, or the rights to use the results of intellectual activity. As advantages of this type of investment for the person attracting investments, it is necessary to highlight first of all the possibility of attracting a wide audience of investors on terms more comfortable than bank lending. This is especially important for economic entities classified in accordance with the norms of the Federal Law "On the Development of Small and Medium-sized Enterprises in the Russian Federation" dated July 24, 2007, No.209-FZ [15] as "small and medium-sized enterprises".

The Strategy for the Development of Small and Medium-Sized Enterprises in the Russian Federation for the period up to 2030 notes that small and medium-sized enterprises provide for the creation of about 1/5 of the Gross Domestic Product of the Russian Federation. In many subjects of the Russian Federation, the share of such subjects in the creation of the gross regional product reaches 1/3 or more. The Strategy approved by the Decree of the Government of the Russian Federation No. 1083-r dated June 2, 2016 [16]. In 2021, more than 5.5 million small and medium-sized businesses operated in the field of small and medium-sized business economic entities providing employment for more than 18 million people.

The authors [17], characterizing the problems of the development of small and medium-sized entrepreneurship at the present stage, reasonably note its features. These subjects of civil turnover react most quickly to market changes, adapt to new conditions in a mobile way, and are dynamic in choosing areas of activity and ensuring employment of the population. However, along with the positive aspects characterizing the participation of small and medium-sized businesses in economic activity, it should be noted that the risks for these entities are also great. E. Petrova, relying on the opinion of the expert community, notes that in the conditions of a pandemic COVID-19 in Russia, the number of small and medium-sized businesses is rapidly declining. This is due to a number of factors, including the complexity of using the bank financing mechanism for the implementation of new projects [18].

In the context of a pandemic COVID-19, when small and medium-sized businesses face not just optimization issues, but survival, the popularity of platform solutions is growing. Using the potential of P2P platforms allows not only to reduce transaction costs but also to attract funds for development. According to our estimates, unfortunately, when it comes to a startup, bank lending "for an idea" looks completely unrealistic. Then "people's financing" often becomes the only opportunity to enter the market of companies with high growth potential in the absence of a credit history and collateral property. At the same time, investors have the opportunity to place free cash savings in current investment projects and save them from inflation. There is no doubt that the development of promising projects and mutual satisfaction of the interests of businesses and consumers contribute to the growth of the welfare of the population, inspire people with confidence in the future, and help overcome the negative consequences of the economic downturn in the conditions of the COVID-19 pandemic.

However, with all the advantages of using investment mechanisms through the acquisition of utilitarian digital rights, it should be noted that crowd investing and crowd lending could not be considered as an alternative to a bank deposit. Although Internet investing assumes profitability at the level of 12–15% per annum, which is significantly higher than the income offered by banks on ruble deposits at the level of 5–7% per annum, the risk of unsuccessful investment through investment platforms is quite high.

This is due to a number of factors. Firstly, the operator of the investment platform is not liable for the obligations of the person attracting investments. Secondly, the legislator does not provide for mandatory insurance of the risk of liability of the person attracting investments. Thirdly, the objectivity of the information contained in the investment offer of the person attracting investments often remains on the conscience of such a person himself and is not confirmed by an independent audit.

Of course, the Russian legislator has provided measures aimed at reducing investment risks when investing using investment platforms. First, the responsibility of the operator of the investment platform for violating the rules of the investment platform and for providing inappropriate information about the investment platform and its operator has been established. Secondly, generally for an individual investor, the maximum investment volumes using investment platforms within one calendar year should not exceed 600,000 rubles. Third, during one calendar year, one business entity using investment platforms does not have the right to attract investments worth more than 1 billion rubles. Fourth, there are quite serious requirements for the operators of the investment platform in terms of personnel potential and the size of their own funds (at least 5 million rubles), etc. In our opinion, such "protective" measures should be strengthened with the help of business risk insurance structures of the person attracting investments. We also believe that it is necessary to publish on the investment platform's website information about the expected effectiveness of the investment project identified by an independent audit.

32.4 Conclusion

The legal system of the Russian Federation is actively searching for solutions the effectively regulate public relations arising within the framework of a new industry—digital financial technologies. One of such measures is the adoption of Law No. 259-FZ, 2019. One of the directions of legal regulation in this law is the regulation of the turnover of utilitarian digital rights in the process of investment by individuals of their monetary savings. There is no doubt that the new investment opportunities established by this law will contribute to the development, first of all, of small and medium-sized businesses. They will also contribute to the implementation of new achievements, primarily in the field of intellectual activity, and thereby improve the welfare of the population.

However, it is necessary to protect citizens as unqualified investors from the adverse consequences of a failed investment. In order to increase the financial security of new digital investment objects—utilitarian digital rights—we propose to supplement the norms of Law No. 259-FZ, 2019. The law needs to be supplemented with provisions concerning compulsory insurance of the entrepreneurial risk of the person attracting investments. In addition, a preliminary independent audit is required when placing information about an investment project on an investment platform. We believe that the legislator will be able to achieve an optimal combination of private and public interests, which will contribute to a noticeable improvement in the quality of life of the broad masses of the population and civilization as a whole.

References

1. Davos came up with an alternative to the GDP indicator. And who is the world leader now? http://news.mail.ru/card/117/ (2021)
2. Medvedev, D.A.: Russia, go forward! http://www.kremlin.ru/events/president/news/5413 (2021)
3. European strategy for smart sustainable and inclusive growth. http://ec.europa.eu/research/era/docs/en/ec-understanding-era-13.pdf (2021)
4. Official website of the project Boomstarter. https://boomstarter.ru/ (2021)
5. On Digital Financial Assets, Digital Currency and Amendments to Certain Legislative Acts of the Russian Federation. Federal Law No. 259-FZ of 31.07.2020. Rossiyskaya Gazeta. 06.08.2020. No.173
6. Part One of the Civil Code of the Russian Federation. Federal Law No. 51-FZ of 30.11.1994. Rossiyskaya Gazeta. 08.12.1994. No.238–239
7. Strokova, A.A.: Crowd-funding in Russia: the essence and prospects of development. In: Problems and Prospects of Economics and Management: Materials of the V International Scientific Conference, pp. 58–60. St. Petersburg, Publishing House "Svoe izdatelstvo" (2016)
8. Kotenko, D.A.: Crowd-funding is an innovative investment tool. Law **5**, 140–145 (2014)
9. Galets, L.A.: Internet investing as a remote algorithm for retail investment financing. Int. J. Human. Nat. Sci. **5**(2), 86–88 (2021). https://doi.org/10.24412/2500-1000-2021-5-2-86-88
10. Krasnokutsky, P.A., Zmiyak, S.S., Yarovaya, N.S.: The concept of inclusive growth: new opportunities or a conscious need? Theor. Econ. **6**(48), 77–86 (2018)
11. Tropskaya S.S.: Financial market in the conditions of digital economy development (financial and legal aspect). Financial law. 8. pp 28–33 (2018)

12. Tropskaya, S.S.: Financial market in the conditions of digital economy development (financial and legal aspect). Finan. Law **8**, 28–33 (2018)
13. Official website of the project Planeta. https://planeta.ru/ (2021)
14. Petrova, E.: Experts predict a decrease in the number of small and medium-sized enterprises in the Russian Federation. Rossiyskaya Gazeta. 07.05.2020. https://rg.ru/2020/05/07/eksperty-prognoziruiut-snizhenie-chisla-malyh-i-srednih-predpriiatij-v-rf.html (2020)
15. On the Development of Small and Medium-sized enterprises in the Russian Federation. Federal Law No. 209-FZ of 24.07.2007. Rossiyskaya Gazeta. 31.07.2007. No.164
16. On approval of the Strategy for the Development of Small and medium-sized Enterprises in the Russian Federation for the period up to 2030. Decree of the Government of the Russian Federation No.1083-R of 02.06.2016. http://www.consultant.ru/document/cons_doc_LAW_199462/ (2021)
17. Bykovskaya, Yu.V., Ivanova, L.N., Samokhina, E.A.: Small and medium-sized entrepreneurship in modern Russia: status, problems and directions of development. Bull. Eurasian Sci. **5** https://cyberleninka.ru/article/n/maloe-i-srednee-predprinimatelstvo-v-sovremennoy-rossii-sostoyanie-problemy-i-napravleniya-razvitiya (2018)
18. Sadkov, V.A.: Digitized rights as an electronic-virtual fiction for the legal turnover of subjective claims. Legal Concept **2**(20), 159–163 (2021)
19. Inshakova, A.O., Goncharov, A.I.: Legal regulation of investment platforms in the conditions of intensification of digital technologies in the securities market. Law Practice **1**, 107–115 (2021)

Chapter 33
Improving Legislation in the Field of Application of Key Technologies of the Fourth Industrial Revolution: Distributed Registries in Online Commerce

Alexander I. Goncharov, Agnessa O. Inshakova, and Elvira O. Osadchenko

Abstract The modern development of the Internet space is characterized by the fact that in most spheres of public relations, this space becomes indispensable. Currently, for the sustainable development of business, ensuring its competitiveness, it is necessary to use digital technologies in the activities of organizations. The technology of data processing and storage based on decentralized distributed registries (blockchain) is highly effective in commercial applications. Unfortunately, in the Russian Federation, the technology of decentralized distributed registries is not given sufficient attention. There is no legal framework regulating this technology, as well as relations that arise within the framework of remote purchase and sale transactions carried out by legal entities and individuals on the Internet.

33.1 Introduction

When referring to official documents, it can be noted that the technology of distributed registries is defined as a register of data blocks, each of which is an addition to each other, as a chain of blocks containing information and cryptographic links [1, 2].

Colleagues Professor Karelina S.A. and Professor Frolov I.V. give a similar definition, pointing out that the blockchain is a registry that consists of a database and

A. I. Goncharov (✉) · E. O. Osadchenko
Volgograd State University, Volgograd, Russia
e-mail: gimchp@volsu.ru

E. O. Osadchenko
e-mail: gimchp@volsu.ru

A. O. Inshakova
Department of Civil and International Private Law, Volgograd State University, Volgograd, Russia
e-mail: ainshakova@list.ru; gimchp@volsu.ru

Department of Law, Volga Branch of Volgograd State University, Volgograd, Russia

A. O. Inshakova and E. I. Inshakova (eds.), *New Technology for Inclusive and Sustainable Growth*, Smart Innovation, Systems and Technologies 288,
https://doi.org/10.1007/978-981-16-9808-8_33

blocks that store information about transactions. The registry, as in the previous definition, is distributed, and the database is decentralized. A.I. Savelyev gives a similar definition but adds a clarification that the implementation of the technology is based on a cryptographic algorithm [3].

Therefore, it can be formulated that the blockchain is a distributed registry, expressed by blocks collected in a sequential chain, containing certain information and operating on the base of a cryptographic algorithm, integrating a decentralized database.

To study the basis of legal regulation of the technology in question, it is necessary to make certain distinctions. Not only the technology itself falls into the sphere of regulation, but also the relations arising from its use. When regulating the technology itself, terminology, ontology, management, etc., are fixed. When regulating relations, the spheres of use of this technology are regulated, for example, purchase and sale transactions on the Internet. In addition, such areas of regulation may include digital assets; tokenization; international payments; identity information management; electronic voting; secure data exchange, registry maintenance, etc.

The following documents were included in the composition of the regulatory sources studied during this scientific development. Roadmap for the development of "end-to-end" digital technology "Distributed Registry system". ISO 22739:2020 "Blockchain and distributed registry technologies. Dictionary". Part One of the Civil Code of the Russian Federation No.51-FZ of November 30, 1994. Letter of the Federal Tax Service of Russia dated 03.10.2016 No OA-18-17/1027 "On control over the circulation of cryptocurrencies (virtual currencies)". Federal Law No. 86-FZ of 10.07.2002 "On the Central Bank of the Russian Federation (Bank of Russia)". Federal Law No. 173-FZ of 10.12.2003 "On Currency Regulation and Currency Control". Rosfinmonitoring Agreement No. 01–01-14/22440, Federal Tax Service of Russia No. MMV-23-2/77@ dated 10/15/2015 "On Cooperation and Organization of Information Interaction between the Federal Financial Monitoring Service and the Federal Tax Service".

Doctrinal sources represented by scientific works of Russian scientists, including the following scientists: E.G. Bagoyan, A.V. Belitskaya, V.S. Belykh, E.I. Belyaev, O.A. Belyaeva, V.A. Vaypan, O.A. Garashchuk, A.I. Goncharov, M.A. Egorova, A.O. Inshakova, D.E. Matytsin, I.E. Mikheeva, A.I. Savelyev, A.V. Sereda, I.R. Sungatov, E.V. Chaikina, G.O. Shakhnazarov.

33.2 Methodology

The development of the content of this chapter is based on the materialistic worldview and the universal scientific method of historical materialism. General scientific methods of cognition are applied: dialectical, hypothetical-deductive method, generalization, induction and deduction, analysis and synthesis, empirical description. The research also uses private scientific methods: dogmatic, comparative-legal, hermeneutic, structural–functional, etc.

33.3 Results

There is no legal regulation regarding the blockchain technology itself and the relations associated with its application in our jurisdiction. However, some documents on the development of the blockchain application in Russia have been adopted. First of all, we note the program "Digital Economy of the Russian Federation". According to this program, one of the tasks of the Government of the Russian Federation is the development and implementation of the technology in question in various areas of public relations.

When referring the technology in question to objects of civil rights, it is possible to apply the civil legislation of the Russian Federation, first of all, the provisions of Article 128 of the Civil Code of the Russian Federation [4]. It is possible to highlight the Letter of the Federal Tax Service dated 03.10.2016 No. OA-18-17/1027, in which an explanation was given about the turnover of the technology in question and the directions of control over this turnover [5]. Concerning the well-known fiduciary settlement instruments, which many people call "cryptocurrency" in everyday life, it should be pointed out that the turnover of unofficial "monetary" funds and their surrogates on the territory of our state is prohibited by the provisions of Federal Law No. 86-FZ of 10.07.2002 "On the Central Bank of the Russian Federation" [6]. Since the legislator has not settled the issue of what refers to monetary surrogates, what was meant by "cryptocurrency", this leads to contradictions in the practice. That is, as such, there is no direct ban on the use of the technology in question in Russian legislation, but at the same time, there are no legal norms regulating this turnover [7].

The opinion of the Federal Tax Service of the Russian Federation is interesting, in which this service indicates that transactions on the technology in question should be considered currency transactions [8]. At the same time, there are no instructions on the control of the use of decentralized distributed registers technology by the Federal Tax Service of the Russian Federation, the Federal Customs Service of the Russian Federation, the Bank of Russia, and other agents of currency control in the legislation of Russia. In addition to this, it provides for the authority of the Federal Tax Service of the Russian Federation to initiate a request for transactions carried out using blockchain technology within the framework of the agreement between Rosfinmonitoring and the Federal Tax Service of Russia on cooperation and organization of information interaction [2]. In general, it should be borne in mind that there is practically no legal regulation of legal relations using blockchain technology [9].

Further, we are going to consider the regulation of blockchain technology itself. Several standards have been adopted at the international level (Fig. 33.1).

The first standard developed by the International Organization for Standardization was ISO 22739:2020 "Blockchain and Distributed Registry Technologies. Dictionary". It follows from this dictionary that the blockchain is a distributed registry, expressed by blocks collected in a sequential chain and containing certain information and working on the base of a cryptographic algorithm, including a decentralized database [10].

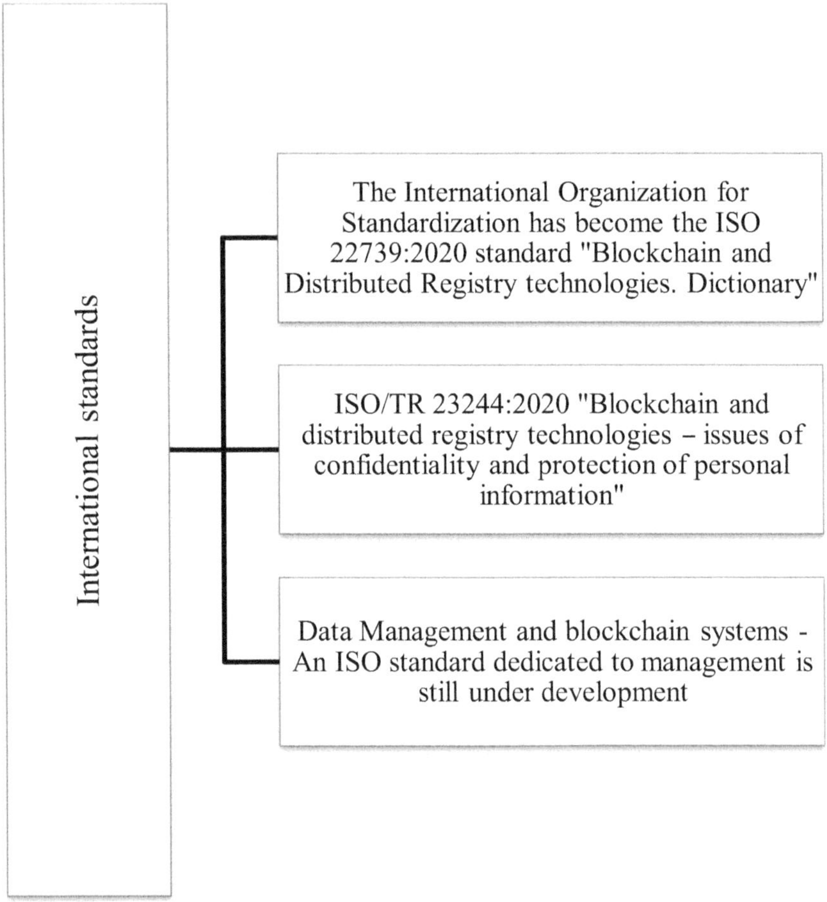

Fig. 33.1 International standards. *Source* compiled by the authors

The legal regulation of blockchain technology in foreign countries is more developed than in our country. Many countries have officially recognized this technology and implemented it in the financial sector. However, there are some states, which have a negative attitude to the technology in question. It is possible to differentiate countries by the criterion of attitude to the blockchain (Fig. 33.2). At the same time, it should be noted that over the past few years, some countries have moved from one group to another. For example, China was initially one of the ardent opponents of the technology in question, any operations and the use of any blockchain technologies in the country were prohibited at the legislative level. Nevertheless, in 2016,

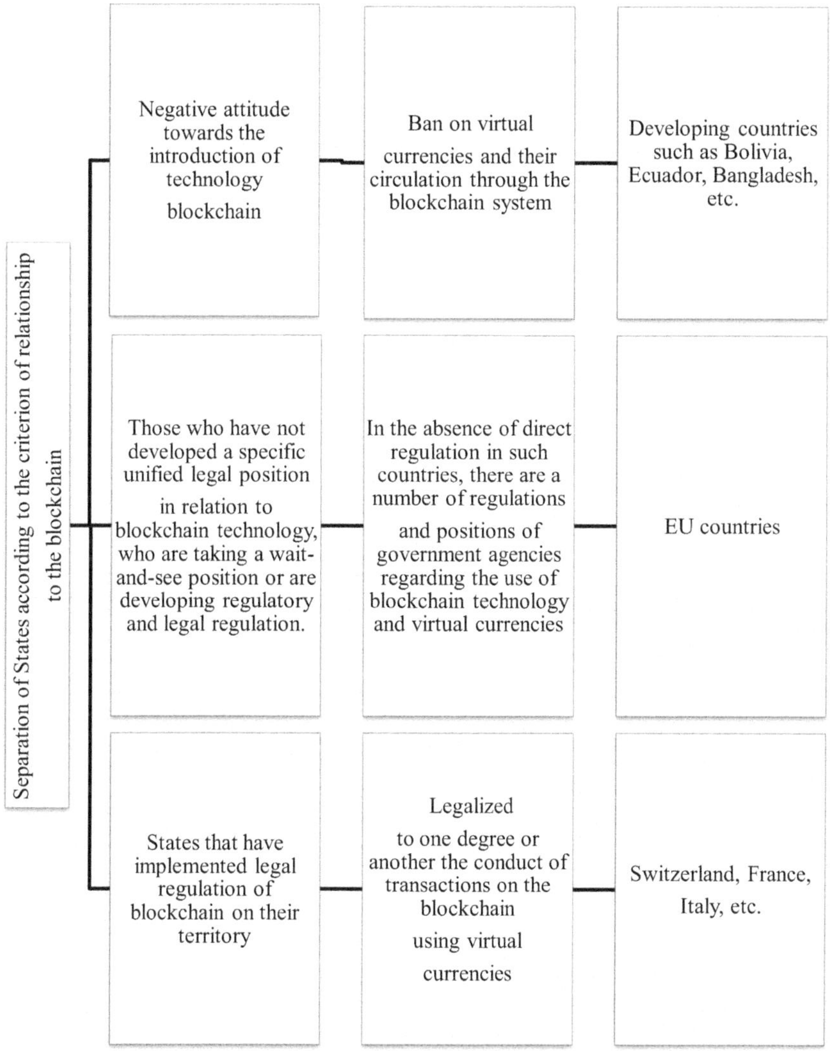

Fig. 33.2 Groups of states according to the criterion of attitude to the blockchain. *Source* compiled by the authors

China completely changed its position and began a policy to stimulate the development of blockchain technology, including this technology found consolidation in the country's thirteenth five-year plan (from 2018 to 2022) [11].

The second group is those states that do not prohibit the use of this technology but do not try to solve the issue of legal regulation in any way, as well as those working toward the legislative consolidation of the blockchain. At the same time, it can be noted that the existing legal framework in these states may partially be aimed at regulating the technology in question, but there is still no direct legal regulation. An example in this context is the legislation of the European Union. In terms of responding to the development of blockchain technology in the world, some changes are made to EU legislation, but there is still no active work.

Switzerland (not a member of the EU) is a country that belongs to the third group. This state has developed a legislative framework to regulate the use of the technology in question. Virtual currency is legalized in this jurisdiction. Another third group of states includes states that have implemented legal regulation of blockchain on their territory and have legalized, to one degree or another, transactions on the blockchain for the circulation of virtual currencies [12]. In particular, in 2017, The Swiss Federal Council has created a “regulatory sandbox”, the purpose of which was to create a favorable environment for startups in the field of financial technology. In 2019, the Swiss government has approved a bill on the adaptation of legal norms to the regulation of the virtual currency and blockchain industry [13]. The third group also includes France. Over the past 5 years, two laws have been adopted in France that have established a special legal regime for the use of the technology in question, the concept of this technology has been fixed, its use for securities turnover, as well as the recognition of blockchain as a technology regulating activities in the field of debt turnover [14]. Italy can currently be assigned to the third group, but back in 2019, this country was conditionally in the second group, then, at the initiative of the Italian government, a law on blockchain and smart contracts was adopted; thereby the blockchain technology was legalized.

Japan might be attributed to one of the most active jurisdictions where blockchain technology is used at the state level. In Japan, there is a special body for regulating this technology—the Digital Assets Commission (active since 2014), and such areas of blockchain as virtual currencies “Bitcoin”, “Ethereum”, etc. are also fully legalized. The US Congress is actively developing a regulatory framework for the development of this technology in the country’s financial sector [15].

In general, the attitude toward blockchain technology in some foreign countries allows us to conclude that the adoption or non-acceptance of the technology in question is directly influenced by the level of development of the financial sector of this country. In the most developed countries with high GDP, there is an active introduction and development of the legal framework regulating the technology in question.

The development of e-commerce in Russia and the world in the last decade has acquired a large-scale character. For a decade, starting from 2011 to 2021, the e-commerce market in Russia showed almost tenfold growth in ruble terms. This is largely facilitated by the development and implementation of advanced digital technologies in all areas of business and administration. The possibilities of blockchain technology are very multifaceted, and, undoubtedly, many more interesting, useful, and profitable applications of this technology will be invented. The peculiarity of the technology under study also lies in the fact that any initiatives can be started by a small group of people or even by one participant for their clients, and then they will be easily distributed to the entire market.

Further, we are going to consider the options for using blockchain (Fig. 33.3).

Separately, should pay attention to the use of blockchain for electronic payments in online commerce. To make a payment when buying via the Internet, you need some kind of payment system (bank). Such a system has certain disadvantages: a commission for conducting operations; strict rules and procedures for handling that do not always suit everyone; technical failures in operation; embezzlement of funds by fraudsters. Blockchain technology is free from these disadvantages and all the disadvantages listed in Fig. 33.4 are absent. With the help of this technology, it is possible to make conditional payments [16].

In recent years, the following main trends in the development of e-commerce in Russia have been most clearly highlighted:

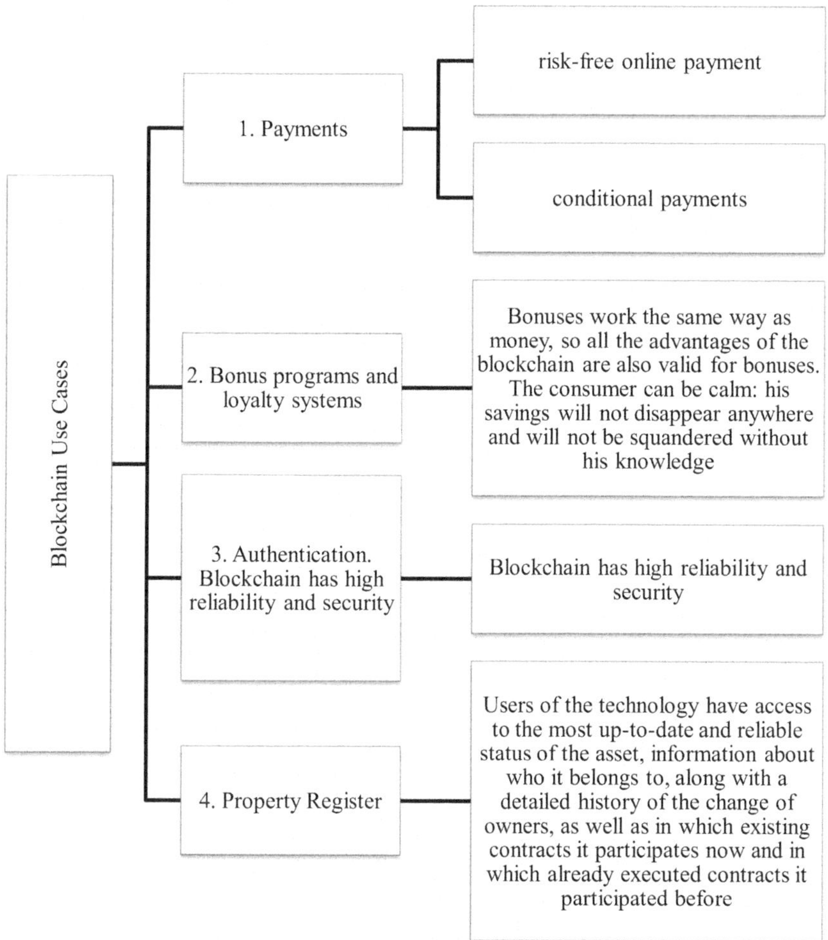

Fig. 33.3 Blockchain use cases. *Source* compiled by the authors

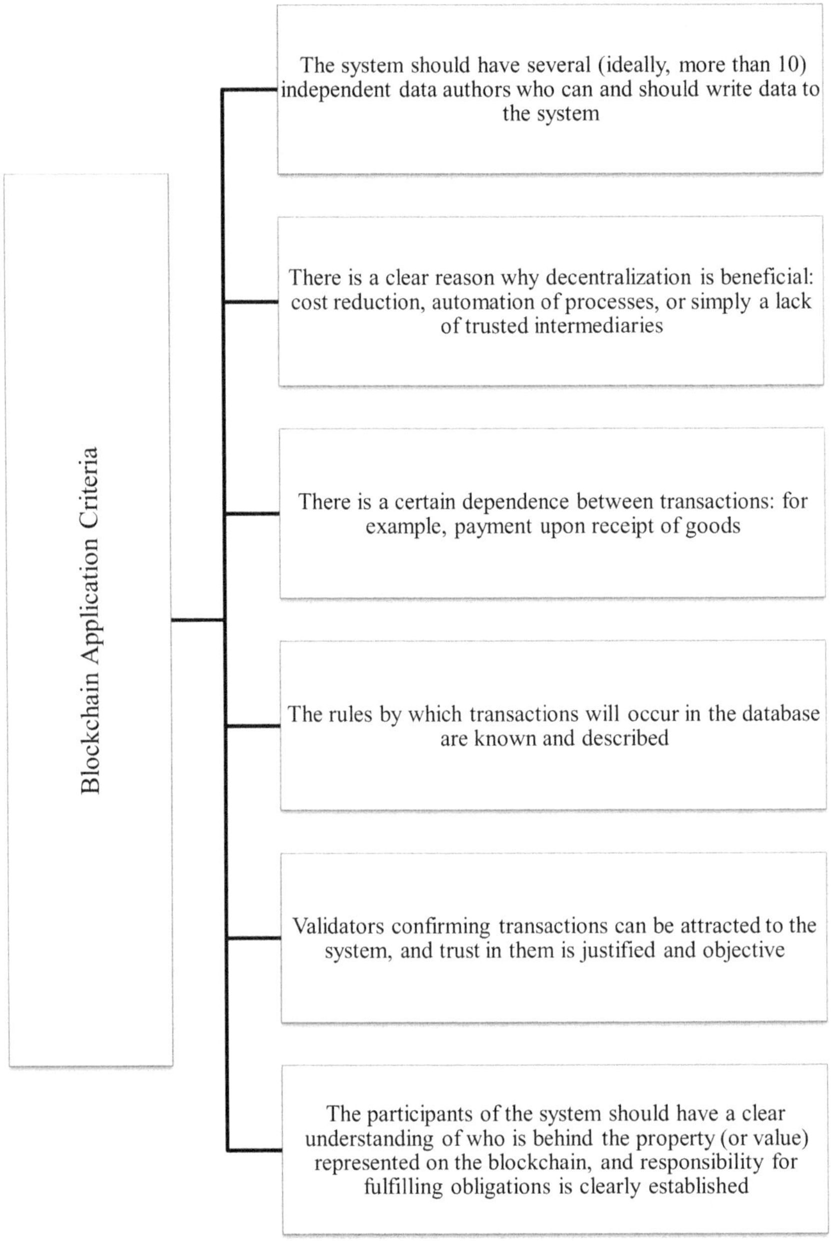

Fig. 33.4 Blockchain application criteria. *Source* compiled by the authors

- the technological improvement of communication and delivery mechanisms, for example, purchases using mobile applications and voice purchases;
- the development of the direction of issuing orders through postamates, for example, by such market participants as Ozon, Goods, Sberbank Logistics, Dixie, etc.
- the trend for the fast delivery of food and ready-made food. So, "Yandex. The shop" promises delivery of basic products and hot food on average in Moscow in 13 min;
- the launch by highly targeted retailers of their own marketplaces, including Detsky Mir and Obuv Rossii companies—development of multifunctional diversified high-tech mega systems providing a wide range of services in the e-commerce market. As an example, we can cite the ecosystem of Sberbank;
- the entry of new players—large Russian networks into the e-commerce market. Dixie and Magnit plan to create their own Internet sites;
- the popularization of B2B sites;
- the focus on the development of online exchange and OTC trading systems;
- the acceleration of the development of various forms of e-commerce in the new business environment, such as distance education services, delivery of online food and grocery orders [17].

Taking into account modern realities, the closure of part of offline trading during the pandemic and the risks of recurrence of these incidents in the future, the growth of these e-commerce segments can accelerate significantly. One of the main features of the development of e-commerce is the creation of multifunctional high-tech mega systems that provide a wide range of services in the e-commerce market by diversifying their activities. An example of such a structure in Russia is the Sberbank ecosystem, which provides various categories of consumers with a wide range of convenient services for everyday life and business, not only in the segment of traditional financial services, but also in various other areas, such as:

- the logistics, represented by the service "Sberbank Logistics";
- the shopping (online marketplace "Beru" and online shopping "Yandex-market");
- the leisure (food delivery service "Sbermarket"; food delivery service DeliveryClub and other services);
- the health;
- the business (service "SBER Marketing", automated trading system "SBERBANK AST", etc.) [18].

The development of digital technologies has made significant changes in the field of exchange trading, contributed to the creation of modern elements of online trading on exchanges: information and trading systems or digital platforms. The Internet and electronic payment and online banking systems simplify the interaction of exchange staff with customers, increase accessibility expand the range of exchange services for a wider range of participants. The expansion of Internet trading opportunities contributes to the emergence of electronic platforms—investment platforms, financial platforms on classical exchanges.

Distributed registry technology allows to digitally recording the main stages of transactions with many commodities traditionally sold on exchanges on a blockchain platform, which ultimately contributes to increased transparency and security of transactions. In this regard, blockchain solutions are of undoubted interest to participants in transactions with classic exchange-traded goods. According to our estimates, in the period 2021–2024, the average annual growth rate of the e-commerce market in Russia will be at least 5%.

In order to use blockchain in online commerce, the technology must meet certain criteria (Fig. 33.4).

By using a distributed registry to track the movement of goods in transport systems in combination with the use of IoT sensors, blockchain systems can provide a relatively easy-to-implement data pipeline that allows all authorized stakeholders to access the same accurate information in real time. This, in turn, contributes to faster and better decision-making by stakeholders throughout the supply chain.

As in other systems, access to information might be controlled using user profiles that specify access rights for each participant to ensure that information about competitors will not be shared with companies that do not have the rights to do so.

33.4 Conclusion

Thus, blockchain technology allows the subjects of trade transactions to use a decentralized distributed registry, to which all participants can access and verify the information contained in the registry at any time, but which no party can control, and also has no ability to change the data recorded and stored in this registry. In practice, smart contracts are actively used in online commerce. In such a smart contract, it may be indicated that now the goods cross a certain line controlled by special equipment, the customs authorities will allow further passage of the goods, the money is automatically transferred from the recipient's bank to the sender's bank—without delay and waiting period. Smart contracts are implemented in the blockchain, while the participants of the transaction see and sign only the part that is relevant to them.

Such opportunities allow us to rethink the entire system of doing business since the fulfillment of many contractual obligations occurs through program code. Due to the existence of the only reliable option in the registry, the costs of checking potential partners are sharply reduced, many disputes will cease to arise, and the circle of participants in trade transactions will significantly expand. When using blockchain, the need to manually follow the "trust, but verify" rule and all associated costs are likely to become outdated. The use of blockchain technology reduces transaction execution costs, reduces risks, promotes the emergence of new business models, and increases the efficiency of transactions. At the same time, the technology makes it possible to significantly expand access to the world market for new participants.

Acknowledgements The work was supported by Russian Science Foundation (project No. 20-18-00314).

References

1. ISO 22739:2020 Blockchain and distributed registry technologies. Dictionary. https://www.iso.org/standard/73771.html (2021)
2. On Cooperation and Organization of Information Interaction between the Federal Financial Monitoring Service and the Federal Tax Service. Rosfinmonitoring Agreement No.01-01-14/22440, Federal Tax Service of Russia No.MMV-23-2/77@ of 15.10.2015. https://www.nalog.gov.ru/rn77/about_fts/fts/ptdsogl/7417409/ (2021)
3. Legal regulation of economic relations in modern conditions of digital economy development: monograph. In: Belitskaya, A.V., Belykh, V.S., Belyaeva, O.A. et al (eds.) Executive editor Vaypan, V.A., Egorova, M.A. Justicinform, Moscow (2019)
4. Part One of the Civil Code of the Russian Federation. Federal Law No. 51-FZ of 30.11.1994. Rossiyskaya Gazeta. 08.12.1994. No.238–239
5. On control over the circulation of cryptocurrencies (virtual currencies). Letter of the Federal Tax Service of Russia of 03.10.2016 No. OA-18-17/1027. http://www.consultant.ru/cons/cgi/online.cgi?req=doc;base=QUEST;n=162766#0 (2021)
6. On the Central Bank of the Russian Federation (Bank of Russia). Federal Law No. 86-FZ of 10.07.2002. Rossiyskaya Gazeta. 13.07.2002. No.127
7. Matytsin, D.E.: Internet-investing as a remote algorithm of the retail investment financing. In: Modern Global Economic System: Evolutional Development Vs. Revolutionary Leap. Springer book series Lecture Notes in Networks and Systems. Springer Nature, Cham (2021)
8. On Currency Regulation and Currency Control. Federal Law No.173-FZ of 10.12.2003. Rossiyskaya Gazeta. 17.12.2003. No.253
9. Inshakova, A.O., Goncharov, A.I., Ershova, I.V.: Digital blockchain registration of legally significant stages of complex good's export-import supplies by business entities of the EAEU and BRICS jurisdictions. In: Inshakova, A., Inshakova, E. (eds.) Competitive Russia: Foresight Model of Economic and Legal Development in the Digital Age. CRFMELD 2019. Lecture Notes in Networks and Systems, vol. 110, pp. 328–336. Springer, Cham (2020). https://doi.org/10.1007/978-3-030-45913-0_37
10. Kalinina, A.E., Inshakova, A.O., Goncharov, A.I.: Polysubject jurisdictional blockchain: electronic registration of facts to reduce economic conflicts. In: Popkova, E.G. (ed.) Ubiquitous Computing and the Internet of Things: Prerequisites for the Development of ICT. Studies in Computational Intelligence, vol. 826, pp. 205–213. Springer Science + Business Media, . Cham (2019)
11. Shakhnazarov, G.O.: Legal regulation of the use of blockchain technology in foreign countries. Colloquium-journal **37**, 144–157 (2019)
12. Sereda, A.V.: On the issue of legal regulation of blockchain technologies: analysis of foreign experience. Probl. Econ. Legal Practice **5**, 141–149 (2019)
13. Mikheeva, I.E.: Review of legal regulation of blockchain technology. Law Econ **5**, 5–10 (2020)
14. Bagoyan, E.G.: Information security and the use of blockchain technology: foreign experience and the need for legal regulation in the Russian Federation. Lawyer **3**, 43–51 (2019)
15. Sungatov, I.R.: Legal regulation of blockchain technology. New Legal Bull. **9**, 24–29 (2020)

16. Savelyev, A.I.: Some risks of tokenization and blockchainization of civil law relations. Law **2**, 15–22 (2018)
17. Garashchuk, O.A.: Legal foundations of blockchain regulation and cryptocurrency circulation in Russia. IT portal. **1**(17). http://itportal.ru/science/economy/pravovye-osnovy-regulirovaniya-blok/ (2018)
18. Belyaev, E.I., Chaikina, E.V.: Innovations and trends in the development of e-commerce. Chronoeconomics **2**, 45–48 (2019)

Chapter 34
Development of Legal Regulation of Key Technologies for the Implementation of the Fourth Industrial Revolution: Digital Tokens as Special Investment Instruments

Agnessa O. Inshakova, Alexander I. Goncharov, and Denis E. Matytsin

Abstract In the modern information space, crypto assets are becoming increasingly popular. For example, tokenized stocks, bonds, other securities and financial instruments have become common in investment turnover. In particular, tokenized shares are securities that are represented by special digital algorithms; they are not quite successfully called that associating them with the word "token". These digital algorithms-codes exist and are accessed only within the framework of blockchain technology. The specified digital code in terms of functionality is a legal substitute for the right to participate in the business of a joint-stock company, or, alternatively, a legal substitute for a monetary claim of a certain person (lender) to another person (borrower). Falsification of a digital token (replacement by some other software algorithm code) is impossible.

A. O. Inshakova (✉)
Head of Department of Civil and International Private Law, ScD in Law, Volgograd State University, Volgograd, Russia
e-mail: ainshakova@list.ru; gimchp@volsu.ru

Department of Law, Volga Branch of Volgograd State University, Volzhsky, Russia

A. I. Goncharov
Volgograd State University, Volgograd, Russia
e-mail: gimchp@volsu.ru

D. E. Matytsin
Insitute of Law, Volgograd State University, Volgograd, Russia
e-mail: Dmatytsin@ya.ru; gimchp@volsu.ru

Department of Civil Law and Procedure, Volga Branch of the International Law Institute, Volzhsky, Russia

A. O. Inshakova and E. I. Inshakova (eds.), *New Technology for Inclusive and Sustainable Growth*, Smart Innovation, Systems and Technologies 288,
https://doi.org/10.1007/978-981-16-9808-8_34

34.1 Introduction

Remote Blockchain investment forms a completely new market for decentralized applications, platforms, and even companies, and there is no doubt that this market needs sources of financing [1]. ICO (Initial Coin Offering) is one of the remote ways to attract investments. Despite some similarities between the terms ICO and IPO (Initial Public Offering), there are actually more differences between these concepts. ICO—this is the process of public offering (sale) of tokens among all those, who are interested in acquiring these tokens to invest in a project using blockchain technology.

In remote banking, a token is traditionally called a compact information storage device ("flash drive"), on which the electronic signature of the bank's client is stored, but still, the term is firmly entrenched in the Internet investment terminology. Unlike remote banking, a token is practically a crypto entry from the registry distributed on the Internet, and conditionally "attached" to each individual investor. The phenomenon of the token is that it can represent almost everything. This means that this cryptographic record turns into a digital asset that can express any rights, obligations, units of value. It all depends on how exactly the token is programmed. As noted above, there are tokens expressing stocks, bonds, various derivatives, instruments that give the right to vote for certain projects, as well as give the right to participate in the management of a business company and confirm the ownership of a certain asset in the investment segment of the Internet. The functionality of the token is limited only by the technical capabilities and imagination of project team specialists. At the same time, it is common for such quasi-ICOs, where entities issue quasi-tokens this does not entail any rights and obligations such a falsified token is neither a payment nor an investment instrument. Legally, this is not an ICO, but fraudulent actions.

The following documents were included in the composition of regulatory sources and judicial acts investigated during this scientific development. The Civil Code of the Russian Federation (Part one). The decision of the Ryazhsky District Court of the Ryazan region of April 26, 2017, in case No. 2-160/2017. The decision of the Oktyabrsky District Court of St. Petersburg dated May 16, 2017, in case No. 2-1993/2017. The decision of the Anapa City Court of Krasnodar Krai dated February 25, 2016, in case No. 2-869/2016. Letter of the Federal Tax Service of Russia dated 03.10.2016 N OA-18-17/1027. Federal Law No. 259-FZ of July 31, 2020 "On Digital Financial Assets, Digital Currency and on Amendments to Certain Legislative Acts of the Russian Federation". Scientific works of Russian scientists, including A.I. Goncharov, A.O. Inshakova, A.E. Kalinina, D.E. Matytsin, etc., represent doctrinal sources.

34.2 Methodology

The development of the content of this chapter is based on the materialistic worldview and the universal scientific method of historical materialism. General scientific methods of cognition are applied: dialectical, hypothetical-deductive method, generalization, induction and deduction, analysis and synthesis, empirical description. The research also uses private scientific methods: dogmatic, comparative-legal, hermeneutic, structural–functional, etc.

34.3 Results

The token is a unique digital technology, not only in terms of its application capabilities, but also of a legal nature, which makes it impossible to agree with the authors who are trying to interpret this concept too one-sidedly. Undoubtedly, the token is a real problem, both for classical legal teaching and for legislative and judicial bodies. An example in this regard can be considered the LOK token (on the Look Rev platform—http://lookrev.com), which simultaneously acts as (1) a means of payment; (2) a service contract for a fixed fee; (3) quasi-capital that entitles the owner to receive dividends. Another example of a multifunctional token is ZrCoin (https://zrcoin.io/), which combines the functions of (1) a stock option; (2) a loan agreement; (3) an option on the underlying asset (zirconium dioxide).

Delving into the content of the term "token", we note the inaccuracy of the word formation ICO (Initial Coin Offering—an initiative coin offer—used to designate this investment instrument. In this part, we can agree with the authors, who also note the inaccuracy of the term ICO. Indeed, firstly, there are no coins and there cannot be, since this is a monopoly of the state. In all likelihood, the word Coin has entered the terminology due to the widespread popularity of one of the crypto assets—Bitcoin—by analogy. In fact, the recipient of the investment offers the investor a certain "cell" through digital computer technologies, into which he needs to enter with his finances in order to receive material benefits based on the results of participation in an entrepreneurial project (profit–benefit). In our opinion, a more correct designation of what we call the abstract word token should be the term IBO: Initial Benefit Offering—an initiative profitable offer.

Secondly, we believe that some consonance of the term ICO with the long-known term IPO is harmful, because it confuses investment terminology, while ICO and IPO have significant differences. This is the main reason why many recipients of investments issue tokens on the Internet, since the issue of classic securities according to IPO rules is a rather lengthy, complex, and expensive procedure [2].

For example, in the United States, the Securities and Exchange Commission closely monitors everyone who intends to conduct an ICO, checks each blockchain startup for compliance with the Howie test. This is how the procedure is usually called in the USA, by which tokens are checked for compliance with the criteria of security. If the token has passed this test and is recognized by the Commission as a security, but the ICO procedure has not been carried out in accordance with the current US securities legislation, the entity issuing such tokens will be brought to civil, administrative, possibly criminal liability in accordance with the Securities Act. In the Russian Federation, there is no similar procedure either in legislation or in law enforcement practice, however, in our country, the procedure for issuing securities is regulated in detail by the Federal Law "On the Securities Market". In addition, administrative (Article 15.17 of the Law on Administrative Offenses of the Russian Federation) and criminal (Article 185 of the Criminal Code of the Russian Federation) liability for violation of the established procedure for issuing securities is provided.

Let us distinguish three enlarged groups of tokens. Firstly, application tokens (Appcoins/protocol tokens/internal tokens) are digital currencies that provide access to distributed network services. For example, the ether token is used in the Ethereum system to purchase "gas", which, in turn, is necessary to launch smart contracts. The Filecoin token allows you to securely store data on the hard drives of thousands of computers through a decentralized system and cryptography, rather than storing them in the cloud of a single provider. In Emercoin (EMC), tokens are required to pay for decentralized network services.

Secondly, privilege tokens. For example, holders of "tokenized shares" in exchange for their investments receive dividends in the form of a percentage of sales or part of transaction fees on the Internet. For example, in the Sia token exchange network, 3.9% of the storage revenue is paid to Siafund owners. At the same time, quasi-stock tokens in Russia and most foreign jurisdictions do not grant corporate rights, but only imitate them.

Third, a token based on the crowd-funding model. This "empty" token does not grant the person investing the money any rights. Such a person does not become an investor, voluntarily donates part of his assets to a certain Blockchain project. It should be borne in mind that donation between legal entities is not allowed under civil law.

The investment turnover of tokens is attractive not only for novice entrepreneurs who are trying to implement their own blockchain project, but also for investors who are promised high returns from token speculation. Let us confirm that a number of successful projects have brought profitability that investors cannot even dream of in the IPO market. For example, Muse GO grew by 2400% only in the period from June to mid-October 2017, Bitquence showed an increase of 1800% over the same period [3], and Spectrocoin rose by 44,770% in a little less than a year since its ICO (from December 2016 to November 2017). (https://icostats.com/roi-since-ico). At the same time, transactions with tokens carry risks that are not comparable to investments in investment objects in the IPO market. Naming successful projects it should be clarified those 9 out of 10 startups that attracted investments through tokenization fail, and only on average 1 out of 10 makes a profit [4]. Examples of the largest unsuccessful projects include DAO [5], KnCMiner [6].

It is especially important that law in almost any country in the world does not protect investors who have invested their funds in tokens. This entails the appearance of a large number of scammers who organize fraudulent token issues, collect money, and do not fulfill the promises they made to investors. Examples of such frauds are: DRC World [7], REcoin [8], OPAIR (https://themerkle.com/top-3-obvious-ico-scams/). Investing in tokens is based only on investors' trust in the persons issuing tokens. It is fundamentally wrong to call them issuers we conditionally call them *tokenants*. Ideally, the recipients of investments issuing tokens should convince investors of the "legal purity" of their project, supported by the guarantee mechanisms of this jurisdiction. In turn, investors should carefully study the investment proposal, the documents provided by the *tokenant* to make sure that the legal norms (a) exist, (b) are observed; (c) if the norms are not observed then there are legal mechanisms to protect the violated rights of investors.

Consider the risks that must necessarily be realized and evaluated by investors when investing in tokens of a particular project.

Firstly, the jurisdiction in which the tokens are issued. This national territory is assessed in the light of the possibility (impossibility) of the investor protecting his rights in case of their violation. The jurisdiction must be explicitly defined in the documents published by the person issuing the tokens. The absence of a provision indicating the law of a particular country, as well as the lack of legal guarantees for the protection of investors' rights, are the first signs that an investor is dealing with a fraudulent issue of tokens. In addition, the jurisdiction should have favorable basic legal conditions: the absence of prohibitions on the turnover of crypto assets and the issuance of tokens; legal regulation of the token market; regulation of identification and marking of the counterparty before the transaction; licensing of organizers of trading crypto assets [9]. Along with this, it is highly desirable that the jurisdiction chosen by the *tokenant* has a developed judicial system and a high level of guarantees of the rights of citizens, as well as foreigners.

Secondly, the registration procedure of the person conducting the issue of tokens—the *tokenant*. The problem here is that the specified person is often a syndicate (association) of individuals, also often remains anonymous. Undoubtedly, the state registration of the *tokenant* as a legal entity in a certain jurisdiction significantly reduces the risks for investors. Because the state authorities control this company and in case of violation, they can hold both the company and the company's management accountable. In addition, the investor clearly sees the figure of a potential defendant, whom he can sue in case of violation of obligations by the *tokenant*. At the same time, investors can check the *tokenant*'s data, including information about the founders, charter, form, organizational structure, share capital, and obtain other legally relevant information necessary for a comprehensive risk assessment. Stand out in this regard: (1) an anonymous person issuing tokens; (2) an individual issuing tokens; (3) an entity not registered as a legal entity, but recognized as such by the law of this jurisdiction; (4) an entity registered and being a legal entity under the law of this jurisdiction [5].

Thirdly, the presence (absence) of legal guarantees in the technical documentation and conditions. In the absence of any requirements for the *tokenant*, bona fide entities take measures to increase the attractiveness of their token in the investment market, in particular, coordinate the procedures for issuing tokens with the legislation in force in this country. Clearly regulated technical documents, as well as the terms of the transaction between the *tokenant* and the investor, including sections on legal guarantees for investors and the responsibility of the recipient of investments with references to the current legislation, indicate that such a project can hardly be qualified as a fraudulent issue of tokens.

The best version of the technical document, the terms of the deal is a civil contract, the subject of which is a digital token. When checking such legal documents, attention should be paid to the following substantive provisions. The first is the correspondence of the legal nature of the agreement and the actual content. Therefore, if the *tokenant* is moving toward the goal of receiving funds from investors, while promising to return these funds with accrued interest, and the contract is described as a loan agreement, and then there is a correspondence between the actual content of the contract and its legal nature. Therefore, it is highly likely that the transaction will be valid in accordance with the legislation of this jurisdiction. The validity of the agreement in the jurisdiction chosen by the *tokenant* is another indicator that the rights of the investor investing in the tokens of this recipient of investment are protected. The second is a detailed regulation of the provisions of the concluded contract, including the details of the parties, the place of conclusion of the contract or the procedure for determining it; the date of conclusion of the contract or the procedure for determining it, etc. The third is the definition in this agreement of the dispute resolution procedure. We mean both the State (jurisdiction) of dispute resolution, and the jurisdiction and competence of a particular state court or arbitration court. The fourth is the procedure for debt repayment in case of project failure. The fifth is the specific scope of rights granted by the token [10]. This is important in the case when the *tokenant* claims that this token grants the right to own shares of a certain corporation but in fact, the token will not be recognized as security in almost any jurisdiction, and therefore, will not grant corporate rights. The United States can be cited as an exception. In connection with the SEC's explanation of The DAO case, a number of tokens may be recognized as SEC—securities [11].

Investing in tokens is still at the stage of development, the same applies to its legal regulation, there are practically no jurisdictions in which the rights of investors are sufficiently protected by law, this is used by individual unscrupulous entities when issuing tokens, hackers, scammers. Unlike the classic issue of securities, an investor, having received a token, does not legally acquire entrepreneurial rights and obligations. However, this does not deprive the investor of the opportunity to initiate a lawsuit in defense of the violated right. Many foreign jurisdictions have similar procedures for the protection of violated rights. Undoubtedly, differences in legal systems, legislation, and law enforcement practice may require contacting a qualified lawyer in the jurisdiction where the tokens are issued.

Suppose, in the Russian Federation, a subject, having issued tokens, disappeared, having received money from investors, let us consider a possible protection procedure. First. In such a situation, it is highly desirable to unite with other victims to file a class action, the chances of winning the case increase significantly. How, for example, did the investors of the Tezos project, which was able to attract more than $ 230 million for ICO [12].

Second. Investors need to identify the defendant. Procedural rules require that the defendant be named in the statement of claim, otherwise, the statement of claim is not considered. If attempts to identify the defendant were unsuccessful, investors can apply for a criminal case. In this case, the investigative authorities will search for the responsible persons. However, it should be understood that the initiation of a criminal case should be considered if there is confidence that the persons who issued the tokens acted intentionally and pursued criminal and illegal goals.

The third. Before filing a claim, it is necessary to determine the jurisdiction and jurisdiction. Further, by filing a statement of claim, investors must make material claims against the defendant. In general, civil law provides sufficient opportunities to protect their violated rights, at the same time, the question remains whether the court will agree with these claims. In the case of fraudulent token issuance, the investor (plaintiff) may file a lawsuit against the defendant for his unjustified enrichment. In this case, the defendant, if the court agrees with the plaintiff's arguments, must return to the investor all the property that constitutes unjustified enrichment (Article 1104 of the Civil Code of the Russian Federation), as well as repay the investor's losses related to the enrichment of the defendant (Article 1105 of the Civil Code of the Russian Federation). The defendant is very likely to stand on the position that he issued tokens using the crowd-funding model (did not assume any obligations), if the court agrees with these arguments, the money will not be returned to investors (Clause 4 of Article 1109 of the Civil Code of the Russian Federation).

An alternative basis for the claim may be the plaintiff's claim for the return by the defendant of what he received because of a transaction concluded under the influence of a material error (Article 178 of the Civil Code of the Russian Federation). The property is subject to return if the court finds that "the error was so significant that this party, reasonably and objectively assessing the situation, would not have made the transaction if it had known about the actual state of affairs". At the same time, if the error was the result of the actions (inaction) of the *tokenant*, the validity of the deal can be challenged in accordance with Article 179 of the Civil Code of the Russian Federation as a transaction made under the influence of deception [13].

Suppose the blockchain project of the *tokenant* was actually provided with certain legal guarantees. All the same, the question remains whether the court will satisfy the investor's claims in case of non-return of investments and non-payment of the promised income. There is no universal procedure for protecting the violated rights of investors, everything depends on the specific type of contractual relationship that has arisen between the token and the investor, the methods of legal protection themselves depend on the structure of the contract in force between the parties. It should be borne in mind that if the name of the contract and its actual content do not coincide, the court does not take into account the name of the contract, but the subject of the contract and the actual content of the rights and obligations of the parties.

The *tokenant* in the Russian jurisdiction will certainly face the fact that the courts do not recognize digital currencies as an object of civil rights, and therefore, believe that the rights of owners (purchasers) of digital currencies and other tokens cannot be protected by legal norms. "Bitcoin does not fall under the objects of civil law listed in Article 128 of the Civil Code of the Russian Federation, since it is not property (goods), money or non-cash money, unregistered securities and property rights. Since there is practically no legal framework in the Russian Federation to regulate payments in "virtual currency". All transactions related to bitcoin transfers are carried out by their owners at their own risk" [14].

Taking into account two circumstances, there are grounds to disagree with such judicial discretion. Firstly, Article 128 of the Civil Code of the Russian Federation includes "other property" as part of the objects, this list is not closed, so it can be reasonably argued that tokens and digital currencies can be attributed to this property and, consequently, to objects of civil law. Secondly, the purchase of Bitcoin tokens is carried out by paying official money, which is traditionally an object of civil law, therefore, the losses incurred by the investor are not virtual at all, but quite real. Given these circumstances, the investor's rights, in this case, should be protected.

You should also pay attention to other judicial interpretations when resolving conflicts in the circulation of tokens (digital currencies). "The absence of cryptocurrencies in the systems or the cancellation of unauthorized transactions, as well as the actual location of the cryptocurrency outside the legal sphere, do not allow the introduction of legal mechanisms to ensure that the parties to the transaction fulfill their obligations" [15]. "Cryptocurrencies, including bitcoin, are money surrogates, contribute to the growth of the shadow economy and can be used by citizens and legal entities in the Russian Federation for illegal purposes" [16–19].

From January 1, 2021, there was a guaranteed opportunity to abandon the not quite correct word token, because Federal Law No. 259-FZ of July 31, 2020 "On Digital Financial Assets, digital Currency and on Amendments to Certain Legislative Acts of the Russian Federation" came into force [20]. Now, absolutely any token is terminologically covered and can be fully replaced by the terms "digital financial assets" and "digital currency" fixed in the legislation. The legislator's guidelines in this new federal law are undoubtedly subject to deep and rapid reflection, which should later be reflected in a separate scientific development.

34.4 Conclusion

Thus, the turnover of tokens—special remote digital investment instruments is characterized by high intensity, a minimum number of cumbersome administrative procedures, low costs compared to classical issues of securities (stocks and bonds). At the same time, a wide range of civil rights objects can be tokenized, they can be quickly involved in circulation, and the fate of any asset can be tracked online even continuously, using a systematic approach of a polysubject jurisdictional blockchain [21]. Legislation to regulate these crypto assets has been difficult and relatively long to form. In 2021, the Russian Federation adopted a full-fledged federal law regulating digital assets and digital currencies. The judicial practice that developed before the adoption of this legislative act was contradictory and fragmented. A thorough analysis of the new federal legislation on digital assets and digital currencies is required in order to improve the procedure for protecting the rights of investors that may be violated in the process of turnover of digital investment instruments in the modern information space.

Acknowledgements The work was supported by Russian Science Foundation (project № 20-18-00314).

References

1. Matytsin, D.E.: Internet-investing as a remote algorithm of the retail investment financing. In: Modern Global Economic System: Evolutional Development vs. Revolutionary Leap. Springer book series "Lecture Notes in Networks and Systems". Springer Nature, Cham (2021)
2. Inshakova, A.O., Goncharov, A.I., Ershova, I.V.: Digital blockchain registration of legally significant stages of complex good's export-import supplies by business entities of the EAEU and BRICS jurisdictions. In: Inshakova, A., Inshakova, E. (eds.) Competitive Russia: Foresight Model of Economic and Legal Development in the Digital Age. CRFMELD 2019. Lecture Notes in Networks and Systems, vol. 110, pp. 328–336. Springer, Cham (2020). https://doi.org/10.1007/978-3-030-45913-0_37
3. The five most profitable ICOs of 2017 (2021). https://golos.id/ico/@altbtc/5-samykh-pribylnykh-ico-2017-goda
4. Jordan Daniell Ethereum's Vitalik Buterin: On "Tokens 1.0". https://www.ethnews.com/ethereums-vitalik-buterin-on-tokens-10 (2021)
5. How to do a successful Initial Coin Offering (and not get sued). https://flagtheory.com/successful-initial-coin-offering/ (2021)
6. Bitcoin Mining Firm KnCMiner Declares Bankruptcy Bitcoin Mining Firm KnCMiner Declares Bankruptcy. https://www.coindesk.com/kncminer-declares-bankruptcy-cites-upcoming-bitcoin-subsidy-halving/ (2021)
7. The organizer of two fraudulent ICOs was arrested in the USA. https://ru.insider.pro/topnews/2017-11-02/organizator-dvuh-moshennicheskih-ico-arestovan-v-ssha/ (2021)
8. In the United States, the first criminal case of ICO fraud was opened. https://www.vedomosti.ru/technology/articles/2017/10/03/736241-ssha-delo-ico (2021)
9. Evgeny Ryabov, Daniil Reitman. Legal risks of ICO and the legal nature of tokens—experts' opinion. https://vc.ru/24993-ico-opinions (2021)
10. Initial Coin Offerings: Know Before You Invest. http://www.finra.org/investors/alerts/initial-coin-offerings-know-before-you-invest(2021)
11. SEC Issues Investigative Report Concluding DAO Tokens, a Digital Asset, Were Securities. https://www.sec.gov/news/press-release/2017-131 (2021)
12. Investors accused the Tezos project of fraud. https://forklog.com/investory-obvinili-proekt-tezos-v-moshennichestve/ (2021)
13. Part One of the Civil Code of the Russian Federation. Federal Law No.51-FZ of 30.11.1994. Rossiyskaya Gazeta. 08.12.1994. No.238-239
14. Decision of the Ryazhsky District Court of the Ryazan region of 26.04.2017 in case No.2-160/2017. http://sudact.ru/regular/doc/qlE6zawJCy6l/#top (2021)
15. Decision of the Oktyabrsky District Court of St. Petersburg of 16.05.2017 in case No.2-1993/2017. http://www.sud-praktika.ru/precedent/234194.html (2021)
16. On control over the circulation of cryptocurrencies (virtual currencies). Letter of the Federal Tax Service of Russia dated 03.10.2016 No.OA-18-17/1027. http://www.consultant.ru/cons/cgi/online.cgi?req=doc;base=QUEST;n=162766#0 (2021).

17. Decision of the Anapa City Court of Krasnodar Krai of 25.022016 in case No. 2–869/2016. (2021). http://sudact.ru/regular/doc/rx4FGJP63uUK/
18. Minister of Finance of Russia: Cryptocurrency is a monetary surrogate. (2021). https://cryptocurrency.tech/ministr-finansov-rossii-kriptovalyuty-eto-denezhnyj-surrogat/
19. Technologies outside the legal field: The Central Bank opposed equating Bitcoin with foreign currencies. (2021). https://russian.rt.com/business/article/430174-nabiullina-centrobank-bitcoin
20. On Digital Financial Assets, Digital Currency and on Amendments to Certain Legislative Acts of the Russian Federation. Federal Law No.259-FZ of 31.07.2020. Rossiyskaya Gazeta. 06.08.2020. No.173
21. Alla E. Kalinina, Agnessa O. Inshakova, Alexander I. Goncharov (2019). Polysubject Jurisdictional Blockchain: Electronic Registration of Facts to Reduce Economic Conflicts. Ubiquitous Computing and the Internet of Things: Prerequisites for the Development of ICT. Studies in Computational Intelligence. Editor Elena G. Popkova. Cham: Springer Science + Business Media, 826, 205–213.

Chapter 35
Cybersecurity System of Business Activity of an Economic Entity: External and Internal Contours

Agnessa O. Inshakova, Alexander I. Goncharov, and Denis A. Appazov

Abstract The authors of this chapter of the monograph substantiate that cybersecurity of entrepreneurial activity of economic entities in the conditions of modern economy, which provides for a high intensity of the use of digital computer technologies, should be built in the system connection of external and internal security circuits. These cybersecurity circuits represent certain procedures for the application of special information technologies that regulate access to the digital environment of companies and minimize the risks of unauthorized interference in their computer databases and management systems. The tools that make up the external cybersecurity circuit of doing business in the interaction of counterparties should combine information technologies: firstly, identification and authentication; secondly, distributed registries; thirdly, smart contracts. Cybersecurity of the intra-corporate business circuit of an economic entity can be provided on the basis of Identity and Access Management information technology (IDM—Identity Management is more common), including SOD control technology—Segregation of Duties and CBA technology—Claim-Based Authentication.

A. O. Inshakova (✉)
Department of Civil and International Private Law, Volgograd State University, Volgograd, Russia
e-mail: ainshakova@list.ru; gimchp@volsu.ru

Department of Law, Volga Branch of Volgograd State University, Volzhsky, Russia

A. I. Goncharov · D. A. Appazov
Volgograd State University, Volgograd, Russia
e-mail: gimchp@volsu.ru

D. A. Appazov
e-mail: gimchp@volsu.ru

A. O. Inshakova and E. I. Inshakova (eds.), *New Technology for Inclusive and Sustainable Growth*, Smart Innovation, Systems and Technologies 288,
https://doi.org/10.1007/978-981-16-9808-8_35

35.1 Introduction

The new economy, which provides for a high intensity of the use of digital computer technologies, has a number of advantages over traditional methods of producing goods, performing works, providing services, since the need for cumbersome transport infrastructure disappears, paper document flow is minimized, the market itself is expanding, becoming global and accessible. Despite all the positive aspects of such an economy, effective ways of its legal regulation are not obvious, since even the participants themselves do not have time to monitor the changes taking place in it [1]. At the same time, it is impossible to leave the digital sphere without legislative regulation, because otherwise not all actions performed in it will receive proper legal protection, compared with similar actions performed in the traditional economy, which will hinder its further development. Consequently, the legislator is faced with the task of identifying the patterns of relations that are forming in the digital sphere of the economy, in order to form a special legal approach for their regulation. In the Russian Federation, the digital development strategy is a high priority and is defined in a special Decree of the President of the Russian Federation until 2030. The Presidium of the Presidential Council for Strategic Development and National Projects has approved the Passport of the national program "Digital Economy of the Russian Federation", which contains a special federal project "Regulatory regulation of the digital environment". One of the threats to the development of the Russian economy, which provides for a high intensity of the use of digital computer technologies, are the problems of ensuring the rights of economic entities in the digital world, including in identifying, preserving user data, creating an environment of trust. In particular, the passport of the federal project "Regulatory Regulation of the digital Environment" states that these problems are being solved, and that results have already been achieved in the formation of a trust environment, but plans to create an information infrastructure and a trust environment have not been fully implemented.

According to the authors, cybersecurity of business activities of economic entities in modern economic conditions should be built in the system connection of external and internal security circuits, which, in turn, represent certain procedures for the use of special information technologies that neutralize the risks of unauthorized interference in their computer databases and management systems.

35.2 Methodology

The development of the content of this chapter is based on the materialistic worldview and the universal scientific method of historical materialism. General scientific methods of cognition are applied: dialectical, hypothetical-deductive method, generalization, induction and deduction, analysis and synthesis, empirical description. The research also uses private scientific methods: dogmatic, comparative-legal, hermeneutic, structural–functional, etc.

35.3 Results

35.3.1 The External Cybersecurity Contour of Doing Business

The problem of identification and authentication of business entities is also complicated by the fact that there is no unified approach to its solution. On the one hand, it is proposed to create a universal digital identifier, for example, on the model of a qualified enhanced electronic signature, which is actively used in relations with government organs. On the other hand, it is impossible to restrict business in using different methods and levels of identification and authentication, as this may affect the security of remote digital deals between business entities. Entrepreneurs, in turn, need a simplified identification procedure for their customers, remote collection, and updating of data about them.

The Government of the Russian Federation has adopted the Concept of introducing an identity card of a citizen of the Russian Federation in the Russian Federation using a plastic card with an electronic data carrier [2]. Nevertheless, later, the release of universal electronic cards in the Russian Federation was stopped, with the help of which they wanted to organize the provision of state and municipal services in electronic form, provide interdepartmental electronic interaction, and create basic information resources [3]. In addition, by a decree of the Government of the Russian Federation, the Ministry of Internal Affairs of Russia was ordered to postpone measures to introduce an identity card in the territory of the Russian Federation. Issued in the form of a plastic card with an electronic data carrier, as the main identity document of a citizen of the Russian Federation, therefore, in the near future it is not worth waiting for the replacement of a paper passport with an electronic one [4].

There is no unified system of identification and authentication of subjects on the Internet yet. This is explained by the fact that the digital economy is developing dynamically, including thanks to e-commerce, and the establishment of strict legal and technical frameworks will only hinder its development [5]. The founding agreement of ICANN ("The Corporation for the Management of Domain Names and IP Addresses" is an independent organization that takes an active role in creating a secure Internet space") states that the "private sector" is at the head of the development of the Internet, and the main goal is to create a simple, predictable, consistent legal environment for e-commerce [6]. Now, there are only framework acts at the international level regulating the interaction of entrepreneurs and other participants in the digital sphere. The Model Law on Electronic Commerce of the United Nations Commission on International Trade Law (UNCITRAL) in paragraph 57 of Article 7 establishes a "flexible approach" to the method of identification and its reliability through an agreement between the "originator and addressee of the data message". The chosen method should not contradict the requirements established by law and the corresponding purpose for which it was necessary. In paragraph 58 of the same article, technical and commercial factors are fixed, which must be taken into account when choosing a method (method) for identifying the parties. Among them, it is worth highlighting the following criteria: the type and volume of the transaction, the frequency of transactions between the parties, the "degree of acceptance or rejection" of the identification method, the availability of alternative identification methods, and the costs of using them. From this, it can be concluded that the parties within the framework of this law are free to choose the means of interaction, and there is no need for a universal identifier. This is confirmed by paragraph 59 of the same article, which specifies the purpose of establishing only guidelines, which in turn can be disclosed in the form of mandatory norms of national law or remain at the discretion of the parties. It is worth noting that despite the freedom granted to the parties by this document, it notes that the agreement on the electronic technical means itself will not give legal force to the messages, that the last word still remains with the "applicable law outside the framework of the Model Law" ("Model Law on Electronic Commerce") [7]. Does this mean that there is no point in creating a universal system of requirements for authentication and identification methods, that the very idea of creating a single identifier is impossible? Now, it can be unequivocally stated that there is no universal solution and the legislation provides for the possibility of providing identification in any available way, but attempts to create a unified system are being made [8].

In the Russian Federation, such a system is being formed—the "Unified Identification and Authentication System". "Unified Identification and Authentication System" was developed by the Ministry of Digital Development, Communications and Mass Communications of Russia as part of the formation of the e-government infrastructure in order to streamline and centralize the processes of registration, identification, authentication, and authorization of users for the provision of public services. The definition of "Unified Identification and Authentication System" is given in Clause 19 of Article 2 of the Federal Law "On Information, Information Technologies, and Information Protection". According to this definition, "identification and authentication system is a federal state information system, the order of use of which is established by the Government of the Russian Federation, and which provides, in cases provided for by the legislation of the Russian Federation, authorized access to information contained in information systems" [9]. The user is offered to create a single account with various authentication methods (electronic digital signature, two-factor authentication, etc.). This record will be entered into one of the registers (individuals, legal entities, etc.)—at the same time, for information security purposes, data is not accumulated, but only the synchronization of credentials that are maintained by various departments takes place. This ensures trust in the identity of the user without the face-to-face appearance of the person or his representative and the presentation of the necessary analog document. It turns out that in the economy created in the Russian Federation by digitalization, data about a person, about a legal entity is digitized, followed by linking them to an account on the public services portal through the "Unified Identification and Authentication System"—a "digital double" is created [10].

There are prospects for such a system for business entities: user data obtained by remote identification with subsequent authentication through the "Unified Identification and Authentication System", acting as an end-to-end technology, are reliable at a sufficiently high level. In 2017, it was planned to open access to personal data of citizens to commercial organizations. But for foreign agents, access to the domestic system remains closed, therefore, it will not be possible to use their digital profile at the international level. The idea of such a system is promising in itself, but in real practice, it faces difficulties in execution. Until now, the information systems of many departments have not been ready for digital profiles "Unified Identification and Authentication System" "2.0" (for example, in terms of the requirement of the Ministry of Digital Development, Communications and Mass Communications of Russia to update information about citizens during the one day). Thus, the creation of a unified system of identification and authentication of participants in the new economy is possible, but so far it either requires huge costs for technical equipment and appropriate regulatory regulation for information security purposes, or another approach that has not yet been developed. At the same time, it must be remembered that at the international and national levels, the principle of freedom of choice of means of identification and authentication by the participants of legal relations themselves prevails.

The entry into the channel of remote interaction and the exit from it of economic entities with the help of the above technologies can be carried out quite efficiently and safely, stably repeating as necessary and depending on the intensity of their entrepreneurial activity. Here, the task of preserving the digital data that is formed precisely as a result of this remote interaction of counterparties is highlighted. This task should be solved using blockchain technology [11].

This well-known distributed registry technology is a computer algorithm for a continuous sequential chain of blocks containing information. New blocks, being added to the end of the chain, strengthen the remaining links. In these blocks, information about transactions is recorded, which is protected by cryptographic means from hacking. The main advantage of using the blockchain system in business is that a specialized intermediary is excluded, since it allows you to record reliable data about the ownership of an asset existing in digital form to a certain person and transfer it to another person. This is achieved due to the fact that all transactions are interconnected due to sequential encryption (the transaction receives a cryptographic identifier (hash), which is added to the header of the record of the next transaction). If you try to change the data about a single transaction, the entire chain will be "compromised" and will be rejected by the participants of the system. So it is almost impossible to make changes since it is necessary to obtain the consent of all participants in the system. In particular, such a system allows you to protect counterparties from fraud. With the help of blockchain, for example, end-to-end traceability of goods is achieved, its movement to the destination, the persons through whom it moves during the delivery process.

For example, De Beers Corporation has been using distributed registry technology since 2018 to track chipped mini containers with diamonds from their mining location to a sales representative, thus protecting itself from fraud. At the same time, blockchain is a derivative technology, it is a way of fixing and exchanging data in digital form. In our opinion, distributed registry technology should be used in combination with smart contracts. Smart contracts are executed in the form of a special program code that is executed automatically when certain conditions occur [12].

Thus, the tools that make up the external cybersecurity circuit of doing business in the interaction of counterparties should combine information technologies: firstly, identification and authentication; secondly, distributed registries; thirdly, smart contracts.

35.3.2 The Internal Cybersecurity Loop of Doing Business

In the practice of entrepreneurial activity, identification is necessary not only for companies as proper subjects of contractual relations, but also important for their employees within the corporation. Identity and Access Management (IAM) technology is a set of technologies and software products that meet the tasks of managing the lifecycle of accounts and managing access to various Internet services in the company. But the more common designation is Identity Management (IDM), which means managing user accounts or electronic images to provide them with access to certain company systems. The main threat to information security is the risks of theft or distortion of information due to the presence of "ownerless" accounts or records with excessive privileges in digital systems.

So, there is a whole layer of judicial practice: an employee copied customer databases for subsequent resale, a fired employee deleted the entire database, etc. Despite the fact that in such situations, the employer can go to court to recover damages, most often he does not receive compensation. For example, according to the materials of the case of Poster One LLC and Razumeyko O.P. on the recovery of funds, the dismissed employee destroyed the client database, which is the intellectual property of Poster One LLC, and did not transfer his working documents to a new employee. Despite all the evidence of illegal actions of the employee, nevertheless, the court sided with the employee, denying the plaintiff satisfaction of the claim [13]. In addition to the disappointing business practice of collecting damages, the question arises about the harm caused by such actions, which sometimes cannot be restored in monetary terms. With the introduction of the PDM system, these problems are solved, since account management is automated in accordance with the company's rules. Installing such a system is quite expensive, since its use implies the presence of a full-scale corporate role model of users. All information resources of the enterprise are taken into account, as well as the business functions of the staff and the order of access for each of them to each resource are described. But the installation of such a system is justified for large companies. For example, IDM can automatically select an e-mail server based on the user's region when creating a mailbox, generate account names according to the specified transliteration rules, fill in organizational attributes for the cost accounting system, etc. The IDM system ensures that no unidentified accounts remain, and access to parts of the digital infrastructure corresponds to the position of the employee, the tasks of his specialty. The system reacts to changes in personnel events: hiring, promotion, vacation, dismissal, etc., thereby, it can respond promptly to any process, thereby avoiding downtime of new employees who cannot access the company's information environment. Also, an additional automation mechanism in the IDM system is the self-service interface—employees can request access rights to systems, change passwords in their accounts, without waiting for a response from an IT specialist. At the same time, all actions that are performed in the digital infrastructure of the company are recorded in the archival resources of the system. This makes it possible, if necessary, relatively inexpensive to detect and remove digital traces during the investigation of incidents, including their subsequent qualification as offenses. Along with this, in the current period, the IDM system integrates the control of SOD conflicts (Segregation of Duties). These are special powers of employees, which cannot be given to two persons at the same time. In addition, IDM integrates risk assessment, namely, the modern IDM system has learned to calculate risks for each role, right, and each information resource.

The solution to the complex task of monitoring the actions of employees in the IDM system consists in regulating the identification, authentication, and authorization procedures in the network, which ensures the necessary internal cybersecurity of the business entity. So, to get started, an employee needs to go through the identification procedure. Then, each time he logs in, he authenticates. In modern Russia, the Claim-Based Authentication model is mainly used in adaptive systems (CBA). Such a model consists of three components: the trusting party, the identification center, and the user. The trusting party is a web service that requests an "approval" from the user (a unit of identification information). The identification Center, in turn, issues electronic identifiers with a set of "statements" in which information about the user is encrypted—all this is certified by the digital signature of the center. The Identification Center replaces user authentication with trust in the "statements" provided by them. All further responsibility for user authorization falls on the Identification Center, for this purpose authentication methods such as logins/passwords, information cards, which are actually electronic passports, can be used.

The main advantage of the IDM system over the manual management of the company's information and digital environment is that users have an extended profile. Such a profile can be used to compare the actions and behavior of users in real time with a historical baseline indicator, as a result of which significant deviations from "normal" behavior are revealed, indicating the presence of problems and high risks of internal cybersecurity.

From the point of view of legal regulation, such a system does not contradict the provisions of Russian legislation. The Labor Code of the Russian Federation regulates issues related to the protection of the personal data of employees. The system specifies data related to the employee's position, as well as his surname, first name, patronymic, working postal address, telephone data, so that the provisions of Article 10 of the Federal Law "On Personal Data" are not violated [14]. Within the meaning of Article 87 of the Labor Code of the Russian Federation, the employer has the right to establish "the procedure for storing and using personal data of employees", and he is also obliged to ensure the protection of such data. The IDM system is aimed precisely at ensuring general information security, including protecting the company from the leakage of information about the personal data of its employees [15].

Thus, cybersecurity within the corporate business contour of an economic entity can be provided on the basis of Identity and Access Management information technology (IDM—Identity Management is more common), including SOD—Segregation of Duties control technology and CBA technology—Claim-Based Authentication.

35.4 Conclusion

In the digital economy, there are various approaches to the identification of business entities, such as the creation of a single identification system for all or the recognition of alternative identification methods with the establishment of legal regulation for them when using. The use of blockchain technology allows business entities to perform proper interaction of the parties, remotely conclude smart contracts that are minimally exposed to the illegal influence of the human factor. Despite the fact that all processes in distributed ledger technology are regulated by mathematical algorithms, they certainly need legal regulation, at least in order to recognize or not recognize the legal protection of participants in transactions concluded through these information technologies. In particular, it is necessary to regulate identification in remote Internet transactions in more detail and control the dissemination of data about the subjects in these deals. Along with this, using Identity and Access Management technology (IDM—Identity Management is more common), business entities can ensure the security of the company's internal digital environment, which is mandatory in conditions when the main exchange of documentation between departments of large organizations takes place in electronic format.

Acknowledgements The work was supported by Russian Science Foundation (project № 20-18-00314).

References

1. Matytsin, D.E.: Internet-investing as a remote algorithm of the retail investment financing. In: Modern Global Economic System: Evolutional Development vs. Revolutionary Leap. Springer book series. Lecture Notes in Networks and Systems. Springer Nature, Cham (2021)
2. On approval of the Concept of Introduction in the Russian Federation of an identity card of a citizen of the Russian Federation, issued in the form of a plastic card with an electronic data carrier, and an action plan for the implementation of the Concept (together with the "Concept of introduction in the Russian Federation of an identity card of a citizen of the Russian Federation, issued in the form of a plastic card with an electronic data carrier, as the main document, identity document of a citizen of the Russian Federation on the territory of the Russian Federation").

Decree of the Government of the Russian Federation dated 09/19/2013 No. 1699-r. Official Internet portal of legal information. http://www.pravo.gov.ru (2021)
3. On Amendments to Certain Legislative Acts of the Russian Federation and Invalidation of Certain Provisions of Legislative Acts of the Russian Federation. Federal Law No. 471-FZ of 28.12.2016. Rossiyskaya Gazeta. 30.12.2016. No.298
4. On measures to assist Election commissions in the exercise of their Powers in the Preparation and Conduct of Elections of Deputies of the State Duma of the Federal Assembly of the Russian Federation of the Seventh Convocation": Resolution of the Government of the Russian Federation No.315 of 16.04.2016. Rossiyskaya Gazeta. 22.04.2016. No.87
5. Matytsin, D.E., Rusakova, E.P.: Strategy of quality management in industry 4.0 and formation of cognitive economy based on industrial and manufacturing engineering in the Russian Federation and countries of the EU. Int. J. Qual. Res. **15**, 4 (2020). https://doi.org/10.24874/IJQR15.04-03
6. The "Founding Agreement of the Internet Corporation for Assigned Names and Numbers" was approved by the ICANN Board on 09.08.2016 and submitted to the Secretary of State of California on 03.10.2016. https://www.icann.org/resources/pages/articles-of-incorporation-2020-03-04-ru
7. "Model Law on Electronic Commerce adopted by the United Nations Commission on International Trade Law" (UNCITRAL) (2021). UN General Assembly Resolution No.A/51/628 of 16.12.1996. Guide to Enactment. https://base.garant.ru/5426911/
8. Agnessa O. Inshakova and Alexander I. Goncharov (2020). Negotiation: Features of Regulation and Prospects for Law Enforcement. In Inshakova, A.O., Bogoviz, A.V. (Ed.) Alternative Methods of Judging Economic Conflicts in the National Positive and Soft Law. A volume in the series Popkova, E.G. (Ed.) Advances in Research on Russian Business and Management. Charlotte, NC, USA, Information Age Publishing. https://www.infoagepub.com/ products/Alternative-Methods-of-Judging-Economic-Conflicts-in-the-National-Positive-and-Soft-Law.
9. On Information, Information Technologies and Information Protection (with amendments and additions coming into force on 01.10.2021). Federal Law No. 149-FZ of 27.07.2006. Rossiyskaya Gazeta. 29.07.2006. No.165
10. Inshakova, A.O., Goncharov, A.I., Ershova, I.V.: Digital blockchain registration of legally significant stages of complex good's export-import supplies by business entities of the EAEU and BRICS jurisdictions. Lecture Notes in Networks and Systems **110**, 328–336 (2020)
11. Kalinina, A.E., Inshakova, A.O., Goncharov, A.I.: Polysubject jurisdictional blockchain: electronic registration of facts to reduce economic conflicts. In: Popkova, E.G. (ed.) Ubiquitous Computing and the Internet of Things: Prerequisites for the Development of ICT. Studies in Computational Intelligence, vol. 826, pp. 205–213. Cham: Springer Science + Business Media (2019)
12. Inshakova, A.O., Goncharov, A.I., Salikov, D.A.: Electronic-digital smart contracts: modernization of legal tools for foreign economic activity. Lecture Notes in Networks and Systems **91**, 3–13 (2020)
13. Decision No. 2-674/12 of April 25, 2012 Timiryazevsky district Court. Archive of Timiryazevsky District Court. The city of Moscow
14. On Personal data. Federal Law No. 152-FZ of 27.07.2006 (ed. of 02.07.2021). Rossiyskaya Gazeta. 29.07.2006. No.165
15. The Labor Code of the Russian Federation of 30.12.2001 No.197-FZ. Rossiyskaya Gazeta. 31.12.2001. No.256

The manufacturer's authorised representative in the EU is Springer Nature Customer Service Centre GmbH, Europaplatz 3, 69115 Heidelberg, Germany. If you have any concerns regarding our products, please contact ProductSafety@springernature.com

Printed and bound by CPI Group (UK) Ltd, Croydon, CR0 4YY
30/09/2026
02225248-0010